Transport Processes in Multicomponent Plasma

Transport Processes in Multicomponent Plasma

V.M. Zhdanov

CRC Press
Taylor & Francis Group
Boca Raton London New York

CRC Press is an imprint of the
Taylor & Francis Group, an **informa** business

To my wife Galina

First published 2002 by Taylor & Francis
11 New Fetter Lane, London EC4P 4EE

Simultaneously published in the USA and Canada
by Taylor & Francis Inc.
29 West 35th Street, New York, NY 10001

Taylor & Francis is an imprint of the Taylor & Francis Group

Typeset by the author

Padstow, Cornwall

This book has been printed from camera-ready copy supplied by the author.

British Library Cataloguing in Publication Data
A catalogue record for this book is available from the British Library

Library of Congress Cataloguing in Publication Data
A catalogue record for this book has been requested

ISBN 0–415–27920–8

Contents

Preface

The book offered to the reader deals with the classical kinetic theory of transport phenomena in a multicomponent plasma. The term "transport phenomena" is known to describe the processes of matter, momentum and energy transfer in gases and plasma under the influence of spatial inhomogeneity (density, velocity and temperature gradients) or under the influence of external electric and magnetic fields. The classical description of these phenomena focuses on pair interactions of plasma particles, which determines the significant role of the collision integral in the kinetic equation irrespective on whether it is written down in a Boltzmann form as for neutral gases or in the Landau form as in the case of Coulomb interactions of charged particles. The results obtained within the framework of the classical theory mainly concern nonequilibrium states of a system, when the macroscopic parameters of the plasma weakly change at distances and times of the order of the mean free path and mean time between collisions of particles. As a rule, the plasma is considered to be in homogeneous and rectilinear magnetic and electric fields. At the same time, the presence of an inhomogeneous and curved magnetic field such as the field used in toroidal confinement systems makes possible a nontrivial generalization of the results of the classical theory to describe the specific transport of particles and energy in such systems ("the neoclassical transport theory"). Certainly, various forms of anomalous transport in plasma, determined by collective effects (such as instability and turbulence) are not considered by the classical theory.

The book is an extended and updated edition of the author's "Transport Phenomena in a Multicomponent Plasma" first published in Russian by Energoizdat Publishing House in 1982. Earlier books and reviews on the theory of transport phenomena in plasma dealt mainly with fully ionized electron–ion plasma or with partially ionized three-component plasma of electrons, ions and neutral particles. However, plasma used in experimental and industrial devices as well as plasma of natural origin, such as ionospheric one, turn out practically to be multicomponent. Even in a comparatively "pure" hydrogen plasma of thermonuclear fusion devices impurities originating from the walls of a discharge chamber play a dominant role and strongly influence radiation, diffusion, electrical conductivity and other parameters of the plasma. The complex composition of the mixture of charged and neutral particles must be taken into consideration when dealing with plasma of combustion products in the channels of magnetohydrodynamic generators of energy, when analyzing phenomena in gas discharges and many other cases.

To the author's knowledge, no new monographs on multicomponent plasma have been recently published and this makes a new edition on transport processes in a multicomponent plasma timely and justifiable considering the latest developments in the field. The contents of the book in general remain as previously in agreement with the earlier established and used methods of the classical kinetic theory of gases and plasma though. In the new edition, the author presents some topics at greater length hoping to make it helpful in graduate and postgraduate studies of certain subjects. A chapter on the transport phenomena in a multicomponent mixture of neutral particles has been added. Also, some new sections on the nonequilibrium thermodynamics of multicomponent plasma have appeared, and examples of applications of theoretical results have been described separately, which contributed to a greater volume as compared with the Russian edition of 1982.

Other books on the kinetic theory of plasma make a stress on the treatment of transport phenomena using, as a rule, the Chapman–Enskog method and the method of expansion of the distribution function over the tensorial spherical harmonic. In contrast, this work, similar to the Russian edition, systematically applies the moment method suggested by Grad in 1949. The generalization of this method for a multicomponent gas mixture and a nonisothermal multispecies plasma was made in a number of works in 1960–70s. Very early in his scientific career, the author became an adherent of the application of the moment method in the kinetic theory of gases and plasma and used it in investigations which culminated with the first edition of the present book in 1982. The efficiency of Grad's method (Hermitian moment method) was then demonstrated independently in a known review by Hirshman and Sigmar (Nuclear Fusion, 1981) on the neoclassical theory of transport processes in a multicomponent plasma, and also in a two-volume monograph by R. Balescu (Transport Processes in Plasmas, 1988) on the classical and neoclassical theories of transport processes in plasma. The book by R. Balescu, however, considered only the two-component fully ionized plasma. Yet, the advantages of the moment method are revealed the most when it is used with respect to multicomponent systems and polyatomic gases as it has been shown already in early works by the author.

In addition to the analysis of electron transport properties exhaustively considered in many works on plasma physics, the book also concentrates on transport phenomena related to heavy particles. Besides partially ionized gas mixtures whose properties are important in the analysis of processes in magnetohydrodynamic and gas discharge devices and the description of phenomena in the ionospheric plasma, investigation of transport in a fully ionized plasma with several ion species can be of interest as it is relevant to the diffusion of impurities across the magnetic field in various schemes of magnetic confinement of plasma. An important recent application of there results is their use in the problem of edge plasma.

I would like to express my gratitude to Yu. Klimontovich, A. Rukhadze, W. Ebeling and P.P.J.M. Schram for their useful advice and comments. I would also like to pay tribute to the late E.A. Mason whose fruitful discussion of the kinetic theory of polyatomic gases I appreciate.

Many thanks are due to M.Ya. Alievsky and P.N. Yushmanov for their help in preparation of this book.

V. Zhdanov

Introduction

The term "multicomponent plasma" implies a partially or fully ionized gas mixture of an arbitrary number of species of neutral and charged particles satisfying the condition of quasineutrality. The peculiar collective Coulomb interaction of charged particles as well as the presence of electric and magnetic fields result in a greater variety of phenomena in the ionized gases as compared with a common gas mixture, yet general kinetic methods of analysis are equally applicable in the case of plasma. This refers primarily to the phenomena of electrical conductivity, diffusion, viscosity and thermal conductivity known under the common term of transport phenomena.

The classical Boltzmann equation is as a rule the starting point in analyzing the transport phenomena. Based on this equation, a comparatively finite theory of transport processes was developed for dilute monoatomic gases and gas mixtures, the hydrodynamic equations were substantiated within its framework and expressions for all kinetic coefficients, i.e. transport coefficients, were derived. The major results in this field are familiar [1–4]. It is known that the problems in theory, caused by the presence of charged particles in the plasma, are due to the divergence of some integrals used to express the effective collision cross sections for the Coulomb potential. In practice, such difficulties are overcome by the use of the screened Coulomb potential or by the formal cutoff of the impact parameter at a length of the order of the Debye radius. The rigorous kinetic theory of plasma developed in the 1960–70s and based on the derivation of kinetic equations from the Liouville equation permitted a more consistent account of the effect of many particles on the collision of charged particles in plasma [5–8]. This approach defines the limits of applicability of the collision integrals of Boltzmann and Landau, whose use is essentially related to screening the Coulomb field of the charges, and also makes it possible to consider more general kinetic equations allowing for the dynamic polarization of the medium which is essential in considering strongly nonequilibrium states of the plasma. The new effects are also exhibited in considering the contribution to the collision integral in large scale fluctuations [8] which, in particular, affects the kinetic coefficients of the plasma in a turbulent state.

Such effects do not play an important part in the phenomena discussed as the author assumes that plasma deviations from the local thermodynamic equilibrium are small, as a rule. Therefore, conventional methods based on the solution of the Boltzmann kinetic equation with the help of the concept of the Debye screening of charged particles are quite justified in describing them. In this approach the kinetic theory of transport phenomena in the plasma proves to be a generalization of the theory developed for a gas mixture in which electrons and ions along with atoms or molecules are taken as independent species of the mixture whereas the specific behaviour of the charged particles in the external electric and magnetic fields are also taken into consideration [3,9–11].

When statistically treating plasma in terms of a single-particle approximation, its state can be fully defined by specifying the distribution function $f_\alpha(\mathbf{v}, \mathbf{r}, t)$, which implies the density of the particles of species α in the six-dimensional phase space with the $\mathbf{v}$ and $\mathbf{r}$ coordinates, where $\mathbf{v}$ is the particle velocity, $\mathbf{r}$ is its coordinate and t is time. Further on, the macroscopic behaviour of the plasma is of primary interest. The macroscopic parameters of the plasma species will be defined as velocity moments with respect to the distribution function $f_\alpha(\mathbf{v}, \mathbf{r}, t)$. The hierarchy of equations for the moments results directly from the kinetic equation, which f_α satisfies, when it is multiplied by a sequence of orthogonal tensorial polynomials with respect to the particle velocity and then is integrated over the velocity space. The distribution function can be expanded over the same infinite sequence of polynomials if the distribution correspondent to a specified state of the gas is taken as the zero approximation, i.e. the weight function. In the conventional Grad's method [12–14] the local Maxwell distribution is taken as the zero approximation, while the tensorial Hermite polynomials of the dimensionless particle velocities are used as polynomials. Because of the orthogonality of the polynomials, the expansion coefficients appear to be expressed in terms of the moments of the distribution function. Limiting ourselves to a finite number of terms in the expansion, we get a chance to express the higher order moments via the set of moments left in the expansion and thus to make the system of moment equations closed. That is basically the main idea of the moment method whose generalization for the multicomponent plasma and molecular gas will be used further in the discussion.

The moment equations or transport equations are greatly simplified if we use the assumption of plasma parameters slowly changing in space and time of the order of mean free path and mean time between collisions of the particles. Then the linear transport relationships linking the values of the nonequilibrium macroscopic plasma parameters and the gradients of the basic hydrodynamic variables as well as external forces affecting the particles can be derived. The proportionality factors in these relations correspond to the so called kinetic coefficients or transport coefficients. Calculation of these values is one of the main purposes of this book.

The $13N$-moment approximation is one of the frequently used approximations of the Grad's method [12–16]. In the expansion of the distribution function only those moments which have an explicit physical meaning are kept. Based on this approach the kinetic theory of transport processes in a partially ionized gas was considered [17–21]. When applied to slightly nonequilibrium and slowly changing states of the plasma, this approximation gives all the required linear transport relations, yet it does not ensure the adequate accuracy in the calculations of kinetic coefficients. Most likely that is the reason why the Grad's method is not so widely used in the kinetic theory of plasma compared with the Chapman–Enskog method [1–3] or the expansion over spherical harmonics [9–11]. In this book the author shows that the use of the linearized equations for the moments of higher order does permit one, within the framework of the Grad's method, to obtain kinetic coefficients with the same accuracy ensured by other methods while preserving all the advantages of the moment method.

About the contents of the book.

The first two chapters are introductory. Chapter 1 derives the Boltzmann equation for a multicomponent gas mixture and considers the effective cross sections of elastic scattering of particles, the problems of elimination of divergence in cross sections of Coulomb interactions and collision integral of charged particles. Chapter 2 deals with the general transport equation for a multicomponent plasma, conservation equations and validation of multitemperature and multifluid plasma models. A section of this chapter concentrates on nonequilibrium thermodynamics of plasma.

A number of significant peculiarities of the macroscopic behaviour of plasma in external fields can be made clear in terms of the so called quasihydrodynamic approximation considered in Chapter 3. This approximation is used to derive the generalized Ohm's law for an arbitrary multicomponent plasma and to analyze diffusion in a weakly ionized plasma and transverse transport of particles in a magnetized plasma.

Chapter 4 is a systematic discussion of the moment method and its application with the aim to derive macroscopic transport equations in nonisothermal multicomponent plasma and linear transport relationships following from it. One- and two-temperature plasma approximations are also considered in detail. We demonstrate a possibility to use higher approximations of the moment method in calculations of the kinetic coefficients of the plasma as well as the relation of the results to the well-known data obtained by the Chapman–Enskog method.

Chapter 5 considers the transport phenomena in an isothermal multicomponent gas mixture without a magnetic field. The relationships given and, in particular, those referring to the diffusion, barodiffusion and thermal diffusion in a multicomponent mixture have, in our view, some advantages over those obtained with the help of the Chapman–Enskog method.

Chapters 6–8 are devoted to the transport processes in nonisothermal multispecies plasma in a magnetic field. Along with general expressions of the viscous-stress tensor, heat flux and diffusion fluxes of species, Chapter 6 substantiates a technique for separation of the transport equations of electrons from the equation for heavy particles which permits one to calculate the electron transport properties irrespective of the other properties in any approximation. Chapter 7 gives a detailed analysis of the transport processes in a partially ionized two-temperature gas. Chapter 8 discusses the transport phenomena in a fully ionized multispecies plasma of electrons and several species of ions, each in different charge states. The procedure used in the latter case applies an approximation to the distribution function of the plasma particles which in terms of accuracy corresponds to the well-known results by Braginskii for a simple plasma. A special section dealt with the application of the results to the analysis of diffusion and heat transport in toroidal systems for the so called Pfirsch–Schlüter regime.

Chapter 9 considers some practical applications of the results obtained in the previous chapters. Their choice is somewhat arbitrary and is intended to demonstrate various aspects of using the results of the kinetic theory of multicomponent plasma in specific problems.

Finally, Chapter 10 treats transport phenomena in a polyatomic gas mixture accounting for internal degrees of freedom of molecules. Attention is paid to transport equations and expressions of the kinetic coefficients due to rotational and vibrational excitations of heavy particles. The results obtained can be applied to a limiting case of a weakly ionized polyatomic gas mixture in which processes of inelastic interactions resulting in electronic excitation, ionization, recombination, etc. can still be neglected.

This book uses the International System of Units (SI). Temperature is measured in degrees Kelvin.

Chapter 1

Kinetic equation

1.1 Distribution function and macroscopic parameters of plasma

The state of a plasma can be described by a single-particle distribution function $f_\alpha(\mathbf{v}, \mathbf{r}, t)$ defined so that $f_\alpha(\mathbf{v}, \mathbf{r}, t)\, d\mathbf{v} d\mathbf{r}$ is an average number of particles of species α at time t in the vicinity of point $\mathbf{r}$ of the volume element $d\mathbf{r} = dx dy dz$, the velocity of which near $\mathbf{v}$ is in the interval $d\mathbf{v} = dv_x dv_y dv_z$.

The basic macroscopic parameters of plasma species such as the local number density n_α, mass density ρ_α, mean velocity $\mathbf{u}_\alpha$ and temperature T_α are introduced by the relations

$$
\begin{aligned}
n_\alpha &= \int f_\alpha d\mathbf{v}, \qquad \rho_\alpha = m_\alpha n_\alpha = \int m_\alpha f_\alpha d\mathbf{v}, \\
n_\alpha \mathbf{u}_\alpha &= \int \mathbf{v} f_\alpha d\mathbf{v}, \quad \frac{3}{2} n_\alpha k T_\alpha = \frac{m_\alpha}{2} \int (\mathbf{v} - \mathbf{u})^2 f_\alpha d\mathbf{v},
\end{aligned}
\tag{1.1.1}
$$

where m_α is the mass of a particle of species α and k is the Boltzmann constant.

The parameters of plasma as a whole, i.e. the number density n, mass density ρ, mass-average velocity (or centre-of-mass velocity) $\mathbf{u}$ and temperature T can be obtained by summation over α, so that

$$
\begin{aligned}
n &= \sum_\alpha n_\alpha, \qquad \rho = \sum_\alpha \rho_\alpha, \\
\rho \mathbf{u} &= \sum_\alpha \rho_\alpha \mathbf{u}_\alpha, \quad nT = \sum_\alpha n_\alpha T_\alpha.
\end{aligned}
\tag{1.1.2}
$$

Let us introduce a new variable, the peculiar velocity of a particle $\mathbf{c} = \mathbf{v} - \mathbf{u}$. Having defined the diffusion velocity $\mathbf{w}_\alpha = \mathbf{u}_\alpha - \mathbf{u}$, we get

$$\mathbf{J}_\alpha = \rho_\alpha \mathbf{w}_\alpha = m_\alpha \int \mathbf{c} f_\alpha d\mathbf{v}, \tag{1.1.3}$$

where $\mathbf{J}_\alpha$ is the mass diffusion flux of species α. From the definition of the mass-average velocity $\mathbf{u}$ it follows that

$$\sum_\alpha \rho_\alpha \mathbf{w}_\alpha = 0. \tag{1.1.4}$$

It would be appropriate now to introduce the definitions of partial and total pressure tensors, $P_{\alpha rs}$ and P_{rs}, respectively

$$P_{\alpha rs} = m_\alpha \int c_r c_s f_\alpha d\mathbf{v}, \quad P_{rs} = \sum_\alpha P_{\alpha rs}. \tag{1.1.5}$$

In terms of physics, $P_{\alpha rs}$ is the momentum flux density of species α. Separating the diagonal part of the tensor, we can represent it as

$$P_{\alpha rs} = p_\alpha \delta_{rs} + \pi_{\alpha rs}, \tag{1.1.6}$$

where

$$\begin{aligned} p_\alpha &= n_\alpha k T_\alpha = \frac{m_\alpha}{3} \int c^2 f_\alpha d\mathbf{v}, \\ \pi_{\alpha rs} &= m_\alpha \int \left(c_r c_s - \frac{1}{3} \delta_{rs} c^2 \right) f_\alpha d\mathbf{v}. \end{aligned} \tag{1.1.7}$$

The value p_α denotes the partial pressure of particles of species α in the plasma and $\pi_{\alpha rs}$ corresponds to the partial viscous-stress tensor (or dissipative pressure tensor).

For plasma as a whole

$$\begin{aligned} P_{rs} &= p\delta_{rs} + \pi_{rs}, \\ p = \sum_\alpha p_\alpha &= nkT, \quad \pi_{rs} = \sum_\alpha \pi_{\alpha rs}. \end{aligned} \tag{1.1.8}$$

Now we shall introduce the vector of the kinetic energy flux density of species α, commonly called the heat flux vector

$$\mathbf{q}_\alpha = \int \frac{m_\alpha c^2}{2} \mathbf{c} f_\alpha d\mathbf{v}. \tag{1.1.9}$$

For the total heat flux in plasma, we have

$$\mathbf{q} = \sum_\alpha \mathbf{q}_\alpha = \sum_\alpha \int \frac{m_\alpha c^2}{2} \mathbf{c} f_\alpha d\mathbf{v}. \tag{1.1.10}$$

The macroscopic parameters introduced thus far are in fact a few first moments of the distribution function. The general description of the moment of order n is

$$M^{(n)}_{\alpha r_1 \ldots r_n} = m_\alpha \int c_{r_1 \ldots r_n} f_\alpha d\mathbf{v}, \tag{1.1.11}$$

where the mass density ρ_α is the zero order moment, the mass diffusion flux $\mathbf{J}_\alpha$ is the first order moment, the partial pressure tensor $P_{\alpha rs}$ represents the second order moment and the doubled heat flux $2q_{\alpha r} = M^{(3)}_{\alpha rss}$ is the reduced third order moment.

All the plasma constituents were earlier assumed to be structureless particles. However, a multicomponent plasma may actually comprise excited atoms, molecules and molecular ions which together provide various internal motions corresponding to electronic, rotational and vibrational degrees of freedom and whose mixture should be properly accounted for in the form of the distribution function using a generalized index i. In turn, the latter subdivides into indices i_1, i_2..., where the population $\{i_\nu\}$ meets the requirements of a full set of quantum numbers for the νth internal modes.

In this semi-classical approach, when the translational degrees of freedom are treated classically and the internal degrees are considered in quantum mechanical terms, the plasma state can be described with the help of the distribution function

$$f_{\alpha i} = f_\alpha (\mathbf{v}, \mathbf{r}, i, t). \tag{1.1.12}$$

In determining the macroscopic values, summation with respect to index i must be included. For example,

$$n_\alpha = \sum_i \int f_{\alpha i} d\mathbf{v}, \qquad n_\alpha \mathbf{u}_\alpha = \sum_i \int \mathbf{v} f_{\alpha i} d\mathbf{v}. \tag{1.1.13}$$

By analogy, the partial pressure tensor is found as

$$P_{\alpha rs} = \sum_i \int m_\alpha c_r c_s f_{\alpha i} d\mathbf{v}. \tag{1.1.14}$$

Here the diagonal part of the tensor $P_{\alpha rs}$ is no longer consistent with the conventional form of hydrostatic partial pressure. As previously, such a tensor can be split into two parts

$$P_{\alpha rs} = P_\alpha \delta_{rs} + \pi_{\alpha rs}, \tag{1.1.15}$$

where

$$P_\alpha = \frac{m_\alpha}{3} \sum_i \int c^2 f_{\alpha i} d\mathbf{v},$$
$$\pi_{\alpha rs} = m_\alpha \sum_i \int \left(c_r c_s - \frac{1}{3} \delta_{rs} c^2 \right) f_{\alpha i} d\mathbf{v}. \tag{1.1.16}$$

Some alterations arise in defining the plasma temperature. To be more accurate, instead of T_α we shall introduce the concept of the mean energy of species α per plasma unit volume. It will be given by the sum of mean energies corresponding to the translational and internal degrees of freedom of particles of species α

$$n_\alpha E_\alpha = n_\alpha E_\alpha^{tr} + n_\alpha E_\alpha^{int}, \tag{1.1.17}$$

where

$$n_\alpha E_\alpha^{tr} = \sum_i \int \frac{m_\alpha c^2}{2} f_{\alpha i} d\mathbf{v}, \quad n_\alpha E_\alpha^{int} = \sum_i \int E_{\alpha i} f_{\alpha i} d\mathbf{v}. \tag{1.1.18}$$

Here $E_{\alpha i}$ is the internal energy of a particle of species α in an ith quantum state.

One may easily see that

$$P_\alpha = \frac{2}{3} n_\alpha E_\alpha^{tr}. \tag{1.1.19}$$

The total mean energy per plasma unit volume is then defined by

$$nE = \sum_\alpha n_\alpha E_\alpha = \sum_\alpha \sum_i \int \left(\frac{m_\alpha c^2}{2} + E_{\alpha i} \right) f_{\alpha i} d\mathbf{v}. \tag{1.1.20}$$

It is this particular value which determines the temperature of the molecular plasma in slightly nonequilibrium states, as we shall later see.

Finally, when defining the heat flux vector of species α in the plasma, we should take into account the densities of the energy fluxes referring to the transport of both translational and internal energies of the particles, in which case

$$\mathbf{q}_\alpha = \mathbf{q}_\alpha^{tr} + \mathbf{q}_\alpha^{int}, \tag{1.1.21}$$

where

$$\mathbf{q}_\alpha^{tr} = \sum_i \int \frac{1}{2} m_\alpha c^2 \mathbf{c} f_{\alpha i} d\mathbf{v},$$
$$\mathbf{q}_\alpha^{int} = \sum_i \int E_{\alpha i} \mathbf{c} f_{\alpha i} d\mathbf{v} \tag{1.1.22}$$

and for the total heat flux we have

$$\mathbf{q} = \sum_\alpha \mathbf{q}_\alpha = \sum_\alpha \sum_i \int \left(\frac{1}{2} m_\alpha c^2 + E_{\alpha i} \right) \mathbf{c} f_{\alpha i} d\mathbf{v}. \tag{1.1.23}$$

1.2 Boltzmann's kinetic equation

The evolution of the single-particle distribution function $f_\alpha(\mathbf{v}, \mathbf{r}, t)$ in space and time is described by the so called kinetic equation. If plasma is taken as a multicomponent gas mixture of neutral and charged particles in the external force field, the nonequilibrium phenomena in it can be analyzed in terms of the Boltzmann's kinetic equation [1,2] written as

$$\frac{\partial f_\alpha}{\partial t} + \mathbf{v} \cdot \nabla f_\alpha + \frac{1}{m_\alpha} \mathbf{F}_\alpha \cdot \nabla_v f_\alpha = J_\alpha, \tag{1.2.1}$$

where ∇ and ∇_v denote the divergence operators in Cartesian coordinates expressed in terms of space and velocity variables and $\mathbf{F}_\alpha$ is the external force acting on a particle of species α.

The left-hand side of Eq. (1.2.1) is in fact a substantial time derivative Df_α/Dt when differentiating along the particle trajectory in the six-dimensional phase space. In the non-collisional case, following the Liouville theorem, $Df_\alpha/Dt = 0$. Collisions bring about the violation of the latter condition since, as a result of every collision treated as a pointlike event, the particle abruptly changes its velocity, i.e. disappears in one point of the phase space to re-appear in another. This fact is taken into consideration on the right-hand side of Eq. (1.2.1) by introducing a collision integral J_α.

The Boltzmann collision integral accounts only for binary collisions or, to put it differently, its derivation absolutely neglects possible simultaneous interactions of three particles and more complicated complexes. It would be appropriate here to recall the basic considerations used to derive the expression for J_α .

First it is to be noted that in considering various pairs of interacting particles in a multicomponent plasma

$$J_\alpha = \sum_\beta J_{\alpha\beta}, \tag{1.2.2}$$

where $J_{\alpha\beta}$ is the collision integral of particles of species α and β.

For the present we shall limit ourselves to elastic collisions of particles when they are treated as classical pointlike force centres. In addition, any external forces affecting the particles are taken to be so small as compared with the forces of particle interaction that their effect on the collision dynamics can be neglected.

The conservation laws of linear momentum and energy for the collision of two particles of m_α and m_β masses can be written as

$$\begin{gathered} m_\alpha \mathbf{v}_\alpha + m_\beta \mathbf{v}_{1\beta} = m_\alpha \mathbf{v}'_\alpha + m_\beta \mathbf{v}'_{1\beta}, \\ \frac{1}{2} m_\alpha \mathbf{v}_\alpha^2 + \frac{1}{2} m_\beta \mathbf{v}_{1\beta}^2 = \frac{1}{2} m_\alpha \mathbf{v}_\alpha'^2 + \frac{1}{2} m_\beta \mathbf{v}_{1\beta}'^2, \end{gathered} \tag{1.2.3}$$

where the primes serve to mark the after-collision velocities and the subscript index 1 is introduced to differentiate the colliding particles at $\alpha = \beta$. The notion of "before" and "after" collisions corresponds to the initial and final stages of the free motion of the particles when the influence of the interaction upon their movement becomes inessential.

It might be convenient to introduce the velocity of the centre-of-mass of a pair of particles

$$\mathbf{G} = \frac{m_\alpha \mathbf{v}_\alpha + m_\beta \mathbf{v}_{1\beta}}{m_\alpha + m_\beta} = \frac{m_\alpha \mathbf{v}'_\alpha + m_\beta \mathbf{v}'_{1\beta}}{m_\alpha + m_\beta}. \tag{1.2.4}$$

It follows from the first relation (1.2.3) that this velocity does not change during collisions. Let us also introduce the relative velocities of the particles before and after the collision

$$\mathbf{g} = \mathbf{v}_{1\beta} - \mathbf{v}_\alpha, \qquad \mathbf{g}' = \mathbf{v}'_{1\beta} - \mathbf{v}'_\alpha. \tag{1.2.5}$$

In terms of these variables

$$\begin{gathered} m_\alpha \mathbf{v}_\alpha = m_\alpha \mathbf{G} - \mu_{\alpha\beta} \mathbf{g}, \quad m_\alpha \mathbf{v}'_\alpha = m_\alpha \mathbf{G} - \mu_{\alpha\beta} \mathbf{g}', \\ m_\beta \mathbf{v}_{1\beta} = m_\beta \mathbf{G} + \mu_{\alpha\beta} \mathbf{g}, \quad m_\beta \mathbf{v}'_{1\beta} = m_\beta \mathbf{G} + \mu_{\alpha\beta} \mathbf{g}', \end{gathered} \tag{1.2.6}$$

where

$$\mu_{\alpha\beta} = \frac{m_\alpha m_\beta}{m_\alpha + m_\beta}$$

is the reduced mass of the particles taking part in a collision.

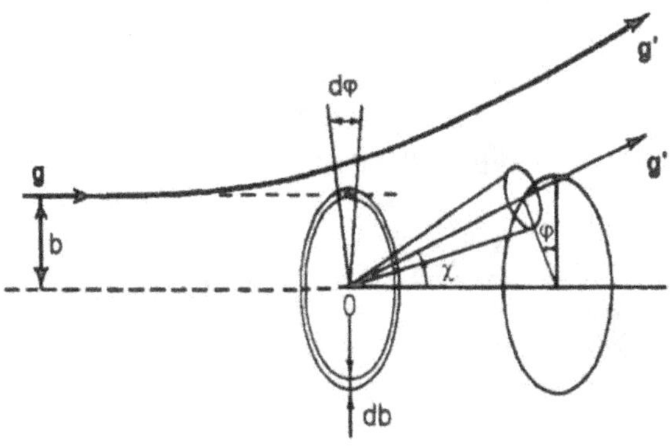

Figure 1.1. Diagram of binary collision

Substituting Eq. (1.2.6) into the second Eq. (1.2.3) and taking account of (1.2.4) leads to the condition

$$g = g^{'}, \tag{1.2.7}$$

which implies that the resulting vector of the relative velocity $\mathbf{g}$ transforms into $\mathbf{g}^{'}$ without changing its value $g = |\mathbf{g}|$. The change of direction of $\mathbf{g}^{'}$ with respect to $\mathbf{g}$ is governed by the polar and azimuthal angles χ and φ, respectively (Fig. 1.1). Then

$$\mathbf{g}^{'} = g\mathbf{i}\cos\chi + g\mathbf{j}\sin\chi\cos\varphi + g\mathbf{k}\sin\chi\sin\varphi, \tag{1.2.8}$$

where $\mathbf{i}$, $\mathbf{j}$, $\mathbf{k}$ are perpendicular unit vectors with $\mathbf{g}$ directed along $\mathbf{i}$.

It follows from Eqs. (1.2.4)–(1.2.6) that the inversions of velocities are obtained by a mere interchange of the primed and unprimed values. This property as well as the linearity of equations ensures equal Jacobians of direct and inverse transformations. One can easily show that the relationships

$$d\mathbf{v}_{\alpha} d\mathbf{v}_{1\beta} = d\mathbf{G} d\mathbf{g}, \qquad d\mathbf{v}_{\alpha}^{'} d\mathbf{v}_{1\beta}^{'} = d\mathbf{G} d\mathbf{g}^{'}, \tag{1.2.9}$$

and

$$d\mathbf{v}_{\alpha} d\mathbf{v}_{1\beta} = d\mathbf{v}_{\alpha}^{'} d\mathbf{v}_{1\beta}^{'}, \tag{1.2.10}$$

prove to be valid [1].

For further discussion, it would be helpful to introduce the concept of differential scattering cross section. Consider a monoenergetic beam of particles incident on the force centre at the initial velocity $\mathbf{g}$ with all possible values of the impact parameter b (Fig. 1.1). The number of particles scattered per unit time into the solid angle element

$d\Omega = \sin\chi d\chi d\varphi$ is proportional to the beam density I_0 and the value $d\Omega$. Then the differential scattering cross section $\sigma(g, \Omega)$ is defined in such a way that $I_0\sigma d\Omega$ should be equal to the number of particles scattered per unit time into the solid angle element $d\Omega$. Figure 1.1 shows that the same number of particles pass through the ring element $bdbd\varphi$, therefore,

$$I_0\sigma(g, \Omega)\, d\Omega = I_0 bdbd\varphi,$$

from which it follows that

$$\sigma(g, \chi, \varphi) = \frac{b}{\sin\chi}\left|\frac{db}{d\chi}\right|. \qquad (1.2.11)$$

It is be noted that for particles treated as pointlike force centres due to the spherical symmetry of their interaction potential, the scattering cross section proves to be the function of the module of the relative particle velocity g and the scattering angle χ alone and it does not depend on φ.

Now an explicit expression of the collision integral $J_{\alpha\beta}$ can be obtained.

Let $K^+_{\alpha\beta}d\mathbf{v}d\mathbf{r}dt$ be an average number of collisions for a time interval dt, as a result of which the final state of a single particle from a colliding pair leads to the element of the phase volume $d\mathbf{v}d\mathbf{r}$ in the vicinity of point $(\mathbf{v}, \mathbf{r})$. Correspondingly, $K^-_{\alpha\beta}d\mathbf{v}d\mathbf{r}dt$ is the average number of collisions during the same time interval when the original state of one of the colliding particles is within the same volume. As the full change of the number of particles per volume element $d\mathbf{v}d\mathbf{r}$ per dt is equal to $(Df_\alpha/dt)\, d\mathbf{v}d\mathbf{r}dt$ in the absence of collisions, in the case of collisions it is equalled to the expression $\sum\limits_\beta \left(K^+_{\alpha\beta} - K^-_{\alpha\beta}\right) d\mathbf{v}d\mathbf{r}dt$ from which it follows that

$$J_{\alpha\beta} = K^+_{\alpha\beta} - K^-_{\alpha\beta}. \qquad (1.2.12)$$

To calculate the average number of collisions, we shall assume the lack of correlation between the position and velocity of two colliding particles (cf. the "persistent chaos" assumption). The time interval dt is taken to be small compared with the characteristic time scale of the changes of macroscopic values and big enough compared with the duration of collision.

Now consider a collision between particles of species β and species α whose centre is shifted to the origin of the coordinate system (Fig. 1.1). Let the impact parameter corresponding to that collision be within the range b and $b+db$ and the azimutal angle within the range φ and $\varphi+d\varphi$.

For this collision to occur during the time interval dt, the centre of the particle of species β proceeding at velocity g at the beginning of the time interval t should be inside the cylinder with the base square $bdbd\varphi$ and the length equal to gdt. The total average number of particles of species β with velocities within the interval $d\mathbf{v}_{1\beta}$ in the "collision cylinder" is

$$f_\beta\left(\mathbf{v}_{1\beta}, \mathbf{r}, t\right) gbdbd\varphi d\mathbf{v}_{1\beta} dt.$$

Assuming the positions and velocities of the particles of both species α and β to be statistically independent, each of the cylinders can be related to every particle of species α whose average number in the interval $d\mathbf{v}_\alpha$ and volume $d\mathbf{r}$ is $f_\alpha\left(\mathbf{v}_\alpha, \mathbf{r}, t\right) d\mathbf{v}_\alpha d\mathbf{r}$. The resulting total average number of collisions between the species α and β during dt in the volume $d\mathbf{r}$ is defined as

$$f_\alpha\left(\mathbf{v}_\alpha, \mathbf{r}, t\right) f_\beta\left(\mathbf{v}_{1\beta}, \mathbf{r}, t\right) gbdbd\varphi d\mathbf{v}_\alpha d\mathbf{v}_{1\beta} d\mathbf{r} dt. \tag{1.2.13}$$

The above expression is valid when calculating $K_{\alpha\beta}^-$ as every collision of the particle of species α and species β brings about the change in the particle velocity and hence the loss of one particle of a given set. The total velocity $K_{\alpha\beta}^- d\mathbf{v}_\alpha d\mathbf{r} dt$, corresponding to the loss of particles of species α resulting from collisions with particles of species β, is obtained by the integration of expression (1.2.13) with respect to all the velocities $\mathbf{v}_{1\beta}$, impact parameter b and angle φ. Ultimately,

$$K_{\alpha\beta}^- = \iiint f_\alpha f_{1\beta} gbdbd\varphi d\mathbf{v}_{1\beta},$$

or, in terms of Eq. (1.2.11),

$$K_{\alpha\beta}^- = \iint f_\alpha f_{1\beta} g \sigma_{\alpha\beta}\left(g, \chi\right) d\Omega d\mathbf{v}_{1\beta}, \tag{1.2.14}$$

where the reductions $f_\alpha = f_\alpha\left(\mathbf{v}_\alpha, \mathbf{r}, t\right)$ and $f_{1\beta} = f_\beta(\mathbf{v}_{1\beta}, \mathbf{r}, t)$ are made.

By analogy, $K_{\alpha\beta}^+$ can be calculated considering the collisions in which the particles with the initial velocities outside the interval $d\mathbf{v}_\alpha$ finally get inside it. As a result,

$$K_{\alpha\beta}^+ = \iint f_\alpha' f_{1\beta}' g' \sigma'\left(g', \chi\right) d\Omega d\mathbf{v}_{1\beta}', \tag{1.2.15}$$

where $f_\alpha' = f_\alpha(\mathbf{v}_\alpha', \mathbf{r}, t)$ and $f_{1\beta}' = f_\beta(\mathbf{v}_{1\beta}', \mathbf{r}, t)$.

To reduce the collisional integral to the final form we shall use the reciprocity relation for direct and inverse collisions which follows from the symmetry of the elastic scattering process [1]:

$$g\sigma_{\alpha\beta}(g,\chi)\,d\Omega d\mathbf{v}_\alpha d\mathbf{v}_{1\beta} = g'\sigma'_{\alpha\beta}(g',\chi)d\Omega d\mathbf{v}'_\alpha d\mathbf{v}'_{1\beta}. \tag{1.2.16}$$

Combining Eqs. (1.2.14) and (1.2.15) and taking into account Eq. (1.2.16), we obtain

$$J_{\alpha\beta} = \iint (f'_\alpha f'_{1\beta} - f_\alpha f_{1\beta}) g\sigma_{\alpha\beta}(g,\chi)\,d\Omega d\mathbf{v}_{1\beta}. \tag{1.2.17}$$

Thus, Boltzmann's kinetic equation is ultimately written as

$$\frac{\partial f_\alpha}{\partial t} + \mathbf{v}_\alpha \cdot \nabla f_\alpha + \frac{1}{m_\alpha}\mathbf{F}_\alpha \cdot \nabla_v f_\alpha$$
$$= \sum_\beta \iint \left(f'_\alpha f'_{1\beta} - f_\alpha f_{1\beta}\right) g\sigma_{\alpha\beta}(g,\chi)\,d\Omega d\mathbf{v}_{1\beta}. \tag{1.2.18}$$

Here the force F_α acting on a particle of species α is generally described as

$$\mathbf{F}_\alpha = e_\alpha(\mathbf{E} + \mathbf{v}_\alpha \times \mathbf{B}) + \mathbf{X}_\alpha, \tag{1.2.19}$$

where $\mathbf{E}$ and $\mathbf{B}$ are electric and magnetic fields respectively, e_α is the particle charge and $\mathbf{X}_\alpha$ is a force of non-electromagnetic nature which for simplicity are considered independently of $\mathbf{v}$.

The generalized kinetic equation of the Boltzmann type can also be written for the case when particles of a gas mixture (or plasma) have internal degrees of freedom. Such an equation is called the Wang-Chang and Uhlenbeck kinetic equation [3,4]. Its derivation is in fact not different from the procedure discussed above. It is essential, though, that in collisions the particles go from one velocity range to another and the changes of their internal states $i,j \rightleftharpoons k,l$ accompany the process. The relationship of symmetry between the direct and inverse collisions is mainly used in the final derivation of the collision integral [4]

$$g\sigma_{\alpha\beta}(ij \mid kl, g,\chi)\,d\mathbf{v}_\alpha d\mathbf{v}_{1\beta} d\Omega$$
$$= g'\sigma'_{\alpha\beta}(kl \mid ij, g',\chi) d\mathbf{v}'_\alpha d\mathbf{v}'_{1\beta} d\Omega, \tag{1.2.20}$$

where $\sigma_{\alpha\beta}(ij \mid kl, g, \chi)$ is the cross section of the direct collision corresponding to the collision processes $\mathbf{g} \to \mathbf{g}'$ and $ij \to kl$ and $g'\sigma'_{\alpha\beta}(kl \mid ij, g', \chi)$ is the cross section of the inverse collision at $\mathbf{g}' \to \mathbf{g}$ and $kl \to ij$. Then from the conservation laws of momentum and energy for inelastic scattering it follows that

$$g'^2 = g^2 - \left(\frac{2}{\mu_{\alpha\beta}}\right)\Delta E_{\alpha\beta}, \tag{1.2.21}$$

where

$$\Delta E_{\alpha\beta} = E_{\alpha k} + E_{1\beta l} - E_{\alpha i} - E_{1\beta j}. \tag{1.2.22}$$

As previously, the relationship between the vectors $\mathbf{g}$ and $\mathbf{g}'$ is determined by Eq. (1.2.8) and the substitution of g for g' on the right-hand side.

It is to be pointed out that relation (1.2.20) is strictly valid only for the nondegenerate internal states of the particles, i.e. it may turn out to be incorrect for molecules with rotational degrees of freedom. In that case a more justifiable quantum mechanical approach is used to derive the Waldmann–Snider kinetic equation [5,6]. However, as was shown by Waldmann [7], this equation is transformed into the Wang-Chang and Uhlenbeck equation on the assumption of isotropic orientation of the angular momentum of the molecule in space and further summation along these orientations. One should not fail to change the normalization of the distribution function and take into account the statistical weights corresponding to the rotational states of the molecules and to introduce the effective averaged cross section over the degenerate states [5,7].

To sum up, the kinetic equation for a gas mixture (plasma) with internal degrees of freedom of particles can be written as

$$\begin{aligned} &\frac{\partial f_{\alpha i}}{\partial t} + \mathbf{v}_\alpha \cdot \nabla f_{\alpha i} + \frac{1}{m_\alpha}\mathbf{F}_\alpha \cdot \nabla_v f_{\alpha i} \\ = &\sum_\beta \sum_{jkl} \iint (f'_{\alpha k} f'_{1\beta l} - f_{\alpha i} f_{1\beta j}) g \sigma_{\alpha\beta}(ij \mid kl, g, \chi)\, d\Omega d\mathbf{v}_{1\beta}. \end{aligned} \tag{1.2.23}$$

1.3 Effective cross sections

The problem of defining differential cross sections of elastic and inelastic interactions of particles is a specific task which must be solved for the general case in terms of quantum mechanics. However, calculations of kinetic coefficients (transport coefficients) generally demand effective cross sections integrated over scattering angles. Such is the case of, say, the diffusion cross section or momentum transfer cross section

$$\begin{aligned} Q^{(1)}_{\alpha\beta}(g) &= \int_\Omega \sigma_{\alpha\beta}(g,\chi)(1-\cos\chi)\, d\Omega \\ &= 2\pi \int_0^\pi \sigma_{\alpha\beta}(g,\chi)(1-\cos\chi)\sin\chi d\chi. \end{aligned} \tag{1.3.1}$$

In the general case, the expressions to be considered below will include the transport cross sections of an arbitrary lth order with positive integer l

$$Q^{(l)}_{\alpha\beta}(g) = \int_{\Omega} \sigma_{\alpha\beta}(g,\chi)\left(1-\cos^l\chi\right) d\Omega$$
$$= 2\pi \int_0^{\pi} \sigma_{\alpha\beta}(g,\chi)\left(1-\cos^l\chi\right)\sin\chi d\chi. \qquad (1.3.2)$$

For the elastic collisions of particles the dependence of the scattering angle χ upon the impact parameter b at arbitrary values of g can be determined within the framework of the classical approach to the problem of the scattering of a particle with the reduced mass $\mu_{\alpha\beta}$ in the stationary central force field with the potential $U_{\alpha\beta}(r)$. In this case [8]

$$\chi = |\pi - 2\Theta_0|, \qquad (1.3.3)$$

where Θ_0 is the angle corresponding to the point of the closest approach of the particles r_0 and defined by

$$\Theta_0 = b\int_{r_0}^{\infty} \frac{dr}{r^2}\left[1-\frac{b^2}{r^2}-\frac{2U(r)}{\mu g^2}\right]^{-1/2}. \qquad (1.3.4)$$

The value of r_0 is found from the condition

$$r_0^2 - b^2 - 2r_0^2\frac{2U(r_0)}{\mu g^2} = 0. \qquad (1.3.5)$$

The dependence $\sigma(g,\chi)$ can be sought for using the calculated $\chi(b)$ with the help of Eq. (1.2.11).

In the simplest case of interaction such as the collision of two hard elastic spheres of diameters d_α and d_β the solution to be obtained from Eqs. (1.3.3)–(1.3.5) is [1]

$$\cos\frac{\chi}{2} = \frac{b}{d_{\alpha\beta}}, \qquad b \le d_{\alpha\beta}, \qquad (1.3.6)$$

which, by the way, can be generally obtained from elementary geometrical considerations. Here $d_{\alpha\beta} = (d_\alpha + d_\beta)/2$ and

$$\sigma_{\alpha\beta} = \frac{1}{4}d^2_{\alpha\beta}, \qquad Q^{(1)}_{\alpha\beta} = \pi d^2_{\alpha\beta},$$
$$Q^{(l)}_{\alpha\beta} = \pi d^2_{\alpha\beta}\left[1-\frac{1}{2}\frac{1+(-1)^l}{1+l}\right]. \qquad (1.3.7)$$

As a model potential for the elastic interaction of particles usable in calculations, the inverse power-law potential is frequently applied

$$U_{\alpha\beta}(r) = \left(\frac{\varkappa_{\alpha\beta}}{r}\right)^{\nu}, \tag{1.3.8}$$

where ν is called the index of repulsion (at $\nu \to \infty$ this model transforms into the hard sphere model). In this case, the expression of the effective cross section $Q^{(l)}_{\alpha\beta}$ has the form [1,9]

$$Q^{(l)}_{\alpha\beta} = 2\pi\varkappa^2_{\alpha\beta}\left(\frac{\mu_{\alpha\beta}g^2}{2\nu}\right)^{-2/\nu}\int_0^{\infty}\left[1-\cos^l\chi(z,\nu)\right]zdz.$$

The value $\chi(z,\nu)$ is defined from the relation

$$\chi = \pi - 2\int_0^{y_0}\left[1 - y^2 - \nu^{-1}\left(\frac{y}{z}\right)^{\nu}\right]^{-1/2}dy$$

when y_0 satisfies the equation

$$1 - y_0^2 - \nu^{-1}\left(\frac{y_0}{z}\right)^{\nu} = 0,$$

and, therefore,

$$Q^{(l)}_{\alpha\beta} = 2\pi\varkappa^2_{\alpha\beta}\left(\frac{\mu_{\alpha\beta}g^2}{2\nu}\right)^{-2/\nu}A_l(\nu), \tag{1.3.9}$$

where $A_l(\nu)$ is the integral whose value depends on ν alone. Some values of $A_l(\nu)$ for $l = 1,2$ are given in Table 1.1. It is noted from Eq. (1.3.9) that

$$Q^{(l)}_{\alpha\beta} \sim g^{-4/\nu}.$$

It should be pointed out that the dependence holds true for the potentials describing the attraction between the particles. One of the interactions between the charged particles (electrons, ions) and neutral atoms (molecules) corresponds to the so called polarization interaction. It is related to the electric dipole moment of atoms arising from the Coulomb field of the incident charged particle. The potential energy of the interaction of the induced dipole moment of the atom and the electron is equal to

$$U(r) = -\frac{2\alpha_d e^2}{4\pi\varepsilon_0 r^4}, \tag{1.3.10}$$

Table 1.1. Values of $A_l(\nu)$ for the pointlike repulsion centres

ν	$A_1(\nu)$	$A_2(\nu)$
4	0.298	0.308
6	0.306	0.283
8	0.321	0.279
10	0.333	0.278
12	0.346	0.279
14	0.356	0.280
20	–	0.286
24	–	0.289
∞	0.500	0.333

where α_d is the atom polarizability and ε_0 is the permittivity in free space. The latter corresponds to the case of $Q_{\alpha\beta} \sim g^{-1}$.

This kind of dependence agrees with the experimental data of elastic cross sections of electron scattering at high energies for some atoms and molecules (H, He, H_2) [10,11]. The effect of exchange interaction between electrons becomes essential for light atoms at low energies and can be calculated only by quantum mechanical techniques. Some peculiar patterns in the cross section behavior appear in electron scattering on atoms of heavy inert gases. This is manifested as an abrupt drop in the scattering cross section at low energies of electrons and a dip minimum (the Ramsauer effect) and the following growth of the cross section at mid energies [12,13]. In practice, one often has to approximate the theoretically calculated or experimental dependences of effective cross sections on the velocity by simple power relationships to take into the quantitative account the contribution of electron–atom interactions in the transport coefficients. It is evident that for different atoms and molecules and for various energy ranges such approximations could differ greatly. The reader can find much useful information on the effective elastic cross section of electron scattering on atoms and molecules in [14,15].

The law of polarization interaction (1.3.10) proves to be valid when describing the elastic collisions of ions in atoms and molecules and it works for ion scattering on other gas atoms. However, with the ions scattered in their own gas the effect of resonant charge transfer becomes of primary importance and leads to a marked growth of cross sections. The theoretical analysis shows that in this case the diffusion cross section of scattering $Q^{(1)}(g)$ is approximately equal to the doubled cross section of charge transfer [16]

$$Q^{(1)}(g) \approx 2Q_{res}(g) \, . \tag{1.3.11}$$

The latter can be determined from experiments on ion beam scattering. In a wide range of energies, the dependence $Q_{res}(g)$ is quite well approximated by [17]

$$Q_{res}^{1/2}(E) = a_1 - a_2 \ln E, \qquad E = \frac{\mu g^2}{2}. \tag{1.3.12}$$

Finally, a few words about the atom–atom and atom–molecule collisions between neutral particles. The model potential of interaction frequently used in this case, since it allows one to calculate the effective cross sections $Q_{\alpha\beta}^{(l)}(g)$, is that of Lennard–Jones [18]

$$U_{\alpha\beta}(r) = 4\varepsilon_{\alpha\beta}\left[\left(\frac{\sigma_{\alpha\beta}}{r}\right)^{12} - \left(\frac{\sigma_{\alpha\beta}}{r}\right)^{6}\right], \tag{1.3.13}$$

where $\sigma_{\alpha\beta}$ is the distance at which the potential function reverses its sign and $\varepsilon_{\alpha\beta}$ is the minimal potential value. It should be pointed out that if σ and ε are experimentally defined from the temperature dependences of transport coefficients for pure gases, the following combination rules are applied for the interaction of different pairs of particles

$$\sigma_{\alpha\beta} = \frac{1}{2}(\sigma_\alpha + \sigma_\beta), \qquad \varepsilon_{\alpha\beta} = (\varepsilon_\alpha \varepsilon_\beta)^{1/2}. \tag{1.3.14}$$

Sometimes, to describe the repulsion more realistically, if the Lennard–Jones potential does not grow steeply enough, one should introduce the exponential dependence of $U(r)$ at small distances. For example, the modified Buckingham $(6 - \exp)$ model of the potential has this feature though such a potential is more difficult to calculate [19].

The models $(6 - 12)$ and $(6 - \exp)$ may be valid in the limited ranges of temperatures. Thus, the use of $(6 - 12)$ potential brings the error of 30 and 50% into the average transport section if the gas temperature increases over 2000 K or 3000 K, respectively. The use of the exponential repulsion potential at $T > 2000$ K is more appropriate. The simple two-parameter Born–Mayer potential [20]

$$U(r) = U_0 \exp(-ar)$$

is usually used in this case.

The values of parameters U_0 and a are determined from experiments. The combination rules for the interaction of different pairs of particles are written as

$$a(\alpha,\beta) = \frac{a(\alpha,\alpha) + a(\beta,\beta)}{2},$$

$$U_0(\alpha,\beta) = \sqrt{U_0(\alpha,\alpha)\, U_0(\beta,\beta)}.$$

We are not concerned with the effective cross sections of inelastic collisions of particles in a plasma referring the keen reader elsewhere [21]. A few models of inelastic interaction of molecules with rotational degrees of freedom are presented in Chapter 10.

Now we'll turn to the interaction of charged particles. For the Coulomb potential

$$U_{\alpha\beta} = \frac{e_\alpha e_\beta}{4\pi\varepsilon_0 r} \tag{1.3.15}$$

calculations based on Eqs. (1.3.3)–(1.3.5) give

$$\tan\frac{\chi}{2} = \frac{b}{b_0}, \qquad b_0 = \frac{|e_\alpha e_\beta|}{4\pi\varepsilon_0 \mu_{\alpha\beta} g^2}. \tag{1.3.16}$$

For the differential scattering cross section we come in this case to the well-known Rutherford formula

$$\sigma(g,\chi) = \frac{(b_0/2)^2}{\sin^4(\chi/2)} = \left(\frac{e_\alpha e_\beta}{8\pi\varepsilon_0\mu_{\alpha\beta}g^2}\right)^2 \frac{1}{\sin^4(\chi/2)}. \tag{1.3.17}$$

Surprisingly, the same result is obtained from the quantum mechanical approach.

Note that the substitution of Eq. (1.3.17) into the expression for the diffusion cross section (1.3.1) results in the fact that at small values of χ the subintegral expression would change as χ^{-1} and, hence, the integral would diverge at the lower limit. The divergence can be due to the long range pattern of the Coulomb interaction. To formally eliminate the divergence, we shall resort to the fact that any charged particle in the plasma generates a redistributed space charge around it as if compensating for the electric field produced by the particle. In fact, at a certain distance called the Debye radius r_D the electrical field of a separate charged particle ceases to act on other charged particles, in which case it is appropriate to speak of the shielded Coulomb potential of charged particles

$$U_{\alpha\beta} = \frac{e_\alpha e_\beta}{4\pi\varepsilon_0 r} e^{-r/r_D}. \tag{1.3.18}$$

The solution of Poisson's equation in this case shows that in a multicomponent plasma [22]

$$r_D^{-2} = \sum_\alpha \frac{n_\alpha e_\alpha^2}{\varepsilon_0 k T_\alpha}. \tag{1.3.19}$$

To calculate the differential scattering cross section $\sigma_{\alpha\beta}(g,\chi)$ for the potential of the kind in (1.3.18) is difficult but basically possible. When defining the effective cross sections (1.3.2), the same result is obtained if we use a Coulomb potential cut off at the Debye length instead of the potential in the form (1.3.18)

$$U_{\alpha\beta}(r) = \frac{e_\alpha e_\beta}{4\pi\varepsilon_0 r}, \quad r \le r_D,$$
$$U_{\alpha\beta}(r) = 0, \qquad r > r_D, \tag{1.3.20}$$

which means that at $b \le b_{\max} = r_D$ the cross section is described by the formula (1.3.15) whereas for collisions with $b > r_D$ there is no scattering. The minimum scattering angle $\chi_{\min}$ corresponding to the maximum value of the impact parameter $b_{\max} = r_D$ is defined by

$$\tan\frac{\chi_{\min}}{2} = \frac{b_0}{r_D}. \tag{1.3.21}$$

For the potential (1.3.15) the diffusion cross section is calculated as

$$\begin{aligned} Q_{\alpha\beta}^{(1)}(g) &= 2\pi \int_{\chi_{\min}}^{\pi} (1-\cos\chi) \frac{\left(\frac{b_0}{2}\right)^2}{\sin^4\left(\frac{\chi}{2}\right)} \sin\chi d\chi \\ &= -4\pi b_0^2 \ln \frac{1}{\sin\left(\frac{\chi_{\min}}{2}\right)}, \end{aligned}$$

or, taking into account Eq. (1.3.21),

$$Q_{\alpha\beta}^{(1)}(g) = 2\pi b_0^2 \ln\left[1 + \left(\frac{r_D}{b_0}\right)^2\right].$$

The similar calculation gives the result

$$Q_{\alpha\beta}^{(2)}(g) = 4\pi b_0^2 \left\{ \ln\left[1 + \left(\frac{r_D}{b_0}\right)^2\right] - \frac{\left(\frac{r_D}{b_0}\right)^2}{1 + \left(\frac{r_D}{b_0}\right)^2} \right\}.$$

We stress that the approximation considered is valid only if $\chi_{\min} \ll 1$ or $b_0/r_D \ll 1$. In this case, the approximation looks like

$$Q_{\alpha\beta}^{(1)}(g) \approx 4\pi b_0^2 \ln\left(\frac{r_D}{b_0}\right). \tag{1.3.22}$$

In practice, averaging is preferable because of a weaker dependence upon b_0 or g under the logarithm in Eq. (1.3.22)

$$\overline{Q}_{\alpha\beta}^{(1)}(g) = 4\pi b_0^2 \ln \Lambda, \tag{1.3.23}$$

where $\ln \Lambda$ is the so called Coulomb logarithm and

$$\Lambda = r_D \left(\frac{1}{b_0}\right) = \frac{4\pi\varepsilon_0}{|e_\alpha e_\beta|}\mu_{\alpha\beta} r_D \overline{g_{\alpha\beta}^2}. \tag{1.3.24}$$

If $\ln \Lambda \gg 1$ we can neglect the term proportional to $x^2/\left(1+x^2\right)$ where $x = r_D/b_0$ in the expression for $\overline{Q}_{\alpha\beta}^{(2)}(g)$, then

$$\overline{Q}_{\alpha\beta}^{(2)} = 8\pi b_0^2 \ln \Lambda. \tag{1.3.25}$$

Interestingly, the values of the effective cross section $Q^{(l)}(g)$ for $l = 1, 2$ which were obtained with the shielded Coulomb potential (1.3.18) may be written in the form [23]

$$\begin{aligned} \overline{Q}_{\alpha\beta}^{(1)}(g) &= 4\pi b_0^2 \left(\ln \Lambda + \ln 2 - \frac{1}{2} - \gamma\right), \\ \overline{Q}_{\alpha\beta}^{(2)}(g) &= 8\pi b_0^2 \left(\ln \Lambda + \ln 2 - 1 - \gamma\right), \end{aligned} \tag{1.3.26}$$

where $\gamma = 0.5772$ is the Euler constant.

For $\ln \Lambda \gg 1$ we can neglect the constant terms in (1.3.26) and these expressions are equivalent to the $Q_{\alpha\beta}^{(l)}(g)$ integrals obtained by the cutting off procedure.

To calculate the kinetic coefficients in plasma for a general case, in what follows, we shall use averaging of effective cross sections and the Maxwell distribution over velocities defined at the temperature of an appropriate plasma species:

$$f_\alpha^{(0)} = n_\alpha \left(\frac{\gamma_\alpha}{2\pi}\right)^{3/2} \exp\left(-\frac{\gamma_\alpha v_\alpha^2}{2}\right), \tag{1.3.27}$$

where

$$\gamma_\alpha = \frac{m_\alpha}{kT_\alpha}. \tag{1.3.28}$$

By analogy, we can calculate the mean values of the arbitrary function of the module of the relative velocity. In particular,

$$n_\alpha n_\beta \overline{F}(g) = \iint F\left(|\mathbf{v}_\alpha - \mathbf{v}_\beta|\right) f_\alpha^{(0)}\left(v_\alpha^2\right) f_\beta^{(0)}\left(v_\beta^2\right) d\mathbf{v}_\alpha d\mathbf{v}_\beta. \tag{1.3.29}$$

Passing over to the variables in the centre-of-mass, we have

$$\begin{aligned} f_\alpha^{(0)}\left(v_\alpha^2\right) f_\beta^{(0)}\left(v_\beta^2\right) &= n_\alpha n_\beta \frac{\left(\gamma_\alpha \gamma_\beta\right)^{3/2}}{(2\pi)^3} \exp\left(-\frac{\gamma_\alpha}{2} v_\alpha^2 - \frac{\gamma_\alpha}{2} v_\alpha^2\right) \\ &= n_\alpha n_\beta \left(\frac{\gamma_\alpha + \gamma_\beta}{2\pi}\right)^{3/2} \left(\frac{\gamma_{\alpha\beta}}{2\pi}\right)^{3/2} \\ &\quad \times \exp\left[-\frac{(\gamma_\alpha + \gamma_\beta)}{2} G^2 - \frac{\gamma_{\alpha\beta}}{2} g^2\right], \end{aligned}$$

where

$$\gamma_{\alpha\beta} = \frac{\gamma_\alpha \gamma_\beta}{\gamma_\alpha + \gamma_\beta}.$$

Using the relationship (1.2.9) and integrating over $\mathbf{G}$, we find

$$\overline{F}(g) = 4\pi \left(\frac{\gamma_{\alpha\beta}}{2\pi}\right)^{3/2} \int F(g)\, g^2 \exp\left(-\frac{\gamma_{\alpha\beta}}{2} g^2\right) dg \tag{1.3.30}$$

and, specifically,

$$\overline{g}_{\alpha\beta} = \left(\frac{8}{\pi \gamma_{\alpha\beta}}\right)^{1/2}, \qquad \overline{g_{\alpha\beta}^2} = \frac{3}{\gamma_{\alpha\beta}}. \tag{1.3.31}$$

Substituting $\overline{g_{\alpha\beta}^2}$ (1.3.31) into (1.3.24), we obtain

$$\Lambda_{\alpha\beta} = \frac{12\pi\varepsilon_0}{|e_\alpha e_\beta|} \frac{\mu_{\alpha\beta}}{\gamma_{\alpha\beta}} r_D. \tag{1.3.32}$$

If $T_\alpha = T_\beta = T$, the result is

$$g_{\alpha\beta} = \left(8kT/\pi\mu_{\alpha\beta}\right)^{1/2}, \qquad \overline{g_{\alpha\beta}^2} = 3kT/\mu_{\alpha\beta}. \tag{1.3.31$'$}$$

$$\Lambda = \frac{12kT\pi\varepsilon_0}{|e_\alpha e_\beta|} r_D. \tag{1.3.32$'$}$$

Let us give the formulae for the calculation of Λ at $Z = 1$ expressed in SI units

$$\Lambda = 1.24 \cdot 10^7 \left(\frac{T^3}{n_e}\right)^{1/2}.$$

The values of $\ln \Lambda$ at $Z = 1$ are given in Table 1.2. The corrections due to the quantum effects are introduced for the temperatures exceeding $T = 4.2 \cdot 10^5$K[24].

Table 1.2. The values of $\ln \Lambda$

T, K	Electron number density n_e, m^{-3}							
	10^{12}	10^{15}	10^{18}	10^{19}	10^{20}	10^{22}	10^{24}	10^{27}
10^2	9.43	5.97	2.52	1.37	-	-	-	-
10^3	12.8	9.43	5.97	4.82	3.67	1.37	-	-
$2 \cdot 10^3$	13.9	10.5	7.01	5.86	4.71	2.40	-	-
$3 \cdot 10^3$	14.5	11.1	7.62	6.47	5.32	3.02	-	-
10^4	16.3	12.8	9.43	8.28	7.12	4.82	2.52	-
10^5	19.7	16.3	12.8	11.7	10.6	8.28	5.97	2.52
10^6	22.8	19.3	15.9	14.8	13.6	11.3	8.96	5.54
10^7	25.1	21.6	18.1	17.0	15.9	13.6	11.2	7.85
10^8	27.4	24.0	20.5	19.4	18.2	15.9	13.6	10.1

1.4 The collision integral for charged particles

A specific feature of the Coulomb interaction between charged particles is a slow decrease of the interaction force with the increasing distance between the particles. In other words, such plasma is a system with long range interaction. This implies that in fact the motion of the particles is strongly influenced by far collective interactions. Relatively big values of the Coulomb logarithm, meeting the requirement of $b_0/r_D \ll 1$, prove that the main contribution to the transport cross section is made by collisions with large impact parameters or small scattering angles ($\chi \ll \pi/2$). It is easily shown that the condition $b_0 \ll r_D$ corresponds to the requirement that the number of charged particles in the Debye sphere around the separated charge be great enough. This means that each charged particle interacts simultaneously with many electrons and ions. A small deviation in every far collision allows one to treat it as an additive and simultaneous small perturbation. The net effect of these deviations becomes more prominent than individual close collisions accompanied by scattering at greater angles. Although only close collisions are of common binary type, their resulting influence is negligible. Nevertheless, since the final result of scattering at far collisions can be represented by a sum of deviations after the interaction of a charged particle and each of the other particles, the whole process can be reduced to a kind of "a single interaction per some time interval", i.e. to a single kind of binary collision.

When deriving a collision integral, it is quite common to use a purely probabilistic approach and an assumption of the Markovian stochastic process of scattering. Thus, we come to the familiar equation of Fokker–Planck describing the change of the distribution function of charged particles resulting from continuous and overlapping weak interactions of these particles. The process as a whole can by represented as a diffusion in the space of particle velocities. The derivation of the appropriate equation is given in a number of works on plasma physics [25,26].

The general expression of the collision integral $J_{\alpha\beta}$ is written in the form

$$J_{\alpha\beta} = \frac{\partial}{\partial v_{\alpha r}} \left(D^{\beta}_{\alpha rs} \frac{\partial f_\alpha}{\partial v_{\alpha s}} \right) - \frac{\partial}{\partial v_{\alpha r}} \left(A^{\beta}_{\alpha r} f_\alpha \right). \tag{1.4.1}$$

The coefficient $D^{\beta}_{\alpha rs}$ is called the diffusion coefficient in the space of particle velocities and the coefficient $A^{\beta}_{\alpha r}$ is the force coefficient. The explicit form of the coefficients can be found on the basis of another commonly used derivation of $J_{\alpha\beta}$ with the underlying Boltzmann's form of the collision operator (1.2.17) [27,15].

As scattering at small angles gives rise to a slight difference in velocities (momenta) of the interacting particles, we can expand the values $f'_\alpha(\mathbf{v}'_\alpha)$ and $f'_\beta(\mathbf{v}'_{1\beta})$ of the integrand in (1.2.17) in a Taylor series in the vicinity of $f_\alpha(\mathbf{v}_\alpha)$ and $f_\beta(\mathbf{v}_{1\beta})$ so that

$$\begin{aligned} f'_\alpha\left(\mathbf{v}'_\alpha\right) &= f_\alpha(\mathbf{v}_\alpha) + \frac{\partial f_\alpha}{\partial v_{\alpha r}} \left(v'_{\alpha r} - v_{\alpha r} \right) \\ &+ \frac{1}{2} \frac{\partial^2 f_\alpha}{\partial v_{\alpha r} \partial v_{\alpha s}} \left(v'_{\alpha r} - v_{\alpha r} \right) \left(v'_{\alpha s} - v_{\alpha s} \right) + \ldots . \end{aligned} \tag{1.4.2}$$

Expansion is similarly performed for $f'_{1\beta}\left(\mathbf{v}'_{1\beta}\right)$.

Keeping the terms up to the second order in the expanded expression and integrating it over the scattering angles, we get the differential operator of the second order formally consistent with the Fokker–Planck operator (1.4.1). The divergence of the integrals derived at the large impact parameter or the small scattering angle can be corrected by means of the earlier described procedure of cutting off at the Debye radius, i.e. the value $\chi_{\min}$ determined from Eq. (1.3.21) is used as a lower limit at small scattering angles. The divergence can also arise at the upper limit because of the assumption made concerning the infinitesimal momentum transferred in collisions. This assumption is violated at $b \sim b_0$. Then, according to Eq. (1.3.16), we can take $\chi_{\max} = \pi$ without introducing large errors, since the contribution to the effective cross section of collisions with deviations at great angles is negligible if $r_D/b_0 \gg 1$. As a result, for coefficients $D^{\beta}_{\alpha rs}$ and $A^{\beta}_{\alpha r}$ we can write

$$D^{\beta}_{\alpha rs} = \frac{2\pi}{m^2_{\alpha}} \left(\frac{e_{\alpha}e_{\beta}}{4\pi\varepsilon_0}\right)^2 \ln\Lambda \int U^{\beta}_{\alpha rs} f_{\beta}(\mathbf{v}_{\beta})\, d\mathbf{v}_{\beta},$$

$$A^{\beta}_{\alpha r} = \frac{2\pi}{m^2_{\beta}} \left(\frac{e_{\alpha}e_{\beta}}{4\pi\varepsilon_0}\right)^2 \ln\Lambda \int U^{\beta}_{\alpha rs} \frac{\partial f_{\beta}(\mathbf{v}_{\beta})}{\partial v_{\beta s}} d\mathbf{v}_{\beta}, \tag{1.4.3}$$

where

$$U^{\beta}_{\alpha rs} = \frac{g^2_{\alpha\beta}\delta_{rs} - g_{\alpha\beta r}g_{\alpha\beta s}}{g^3_{\alpha\beta}} = \frac{\partial^2 |g_{\alpha\beta}|}{\partial g_{\alpha\beta r}\partial g_{\alpha\beta s}}.$$

Another form of the collision integral for charged particles corresponds to the kind first obtained by Landau for a simple electron–ion plasma [28], in which case, using the expressions for $D^{\beta}_{\alpha rs}$ and $A^{\beta}_{\alpha r}$ (1.4.3), one has

$$J_{\alpha\beta} = \frac{2\pi}{m_{\alpha}} \left(\frac{e_{\alpha}e_{\beta}}{4\pi\varepsilon_0}\right)^2 \ln\Lambda \frac{\partial}{\partial v_{\alpha r}} \int U^{\beta}_{\alpha rs} \times \left[\frac{f_{1\beta}(\mathbf{v}_{1\beta})}{m_{\alpha}} \frac{\partial f_{\alpha}(\mathbf{v}_{\alpha})}{\partial v_{\alpha s}} - \frac{f_{\alpha}(\mathbf{v}_{\alpha})}{m_{\beta}} \frac{\partial f_{1\beta}(\mathbf{v}_{1\beta})}{\partial v_{1\beta s}}\right] d\mathbf{v}_{\beta}. \tag{1.4.4}$$

In such a form the collision integral $J_{\alpha\beta}$ is well known as the Landau collision integral.

The range of application of the collision integral as given by Landau could be established only within the framework of the general theory which could consistently account for effects of many particles on collisions of charged particles in plasma. Such a theory based on the approximated techniques of solving the Liouville equation (e.g. the hierarchy of BBKI [27,29–31], the diagram method of Prigogine and Balescu [32]) uses the fact of a small parameter available: the ratio of the mean energy of the Coulomb interaction and the mean kinetic energy of the relative particle motion. An account of plasma dynamic polarization leads to a collision integral whose form is close in structure to the Landau integral but which accounts for the dielectric permeability of the medium explicitly in the integrand as well as its dependence on the wave number (the Balescu–Lenard collision integral [32]).

The analysis of the subintegral function for small wave numbers or large impact parameters proves that for the states that do not differ greatly from the thermodynamic equilibrium, the collision integral appears converging, with the effective cutoff parameter determined by the Debye radius. It can also be shown that for a strongly nonequilibrium plasma the application of the Balescu–Lenard collision integral results in qualitatively new effects due to the fact that it describes the phenomenon of interaction of plasma particles and waves which could propagate in plasma. These effects first shown by Silin [33] can substantially arise in nonisothermal plasma in which the electron temperature greatly exceeds the ion temperature ($T_e \gg T_i$). The ion–sound waves propagate in this sort of plasma and the interaction of waves and particles contributes notably to the collision integral and influences the value of electron kinetic coefficients [27,34]. The modified collision integrals also occur in consideration of rapidly changing processes in plasma and in the case of strong fields affecting the very act of particle collisions [27]. In addition, new effects are displayed when considering the contribution to the collision integral of large-scale fluctuations which in turn affects the values of kinetic coefficients of plasma in a turbulent state [30].

Here we do not intend to study all these effects, the interaction of charged particles in the kinetic equations is well described by means of the collision integral in the Landau form (1.4.4). Moreover, by calculating the average momentum and energy transfer in Coulomb collisions it can be shown that the direct calculation in terms of the Boltzmann collision integral and formal cut-off of the impact parameter at the Debye length bring about the same result which is obtained on the basis of the Landau integral. Our conclusion remains valid with respect to all other calculations of the kinetic coefficients in plasma; therefore, for practical purposes we shall further use the Boltzmann representation of the collision integral.

Chapter 2

Transport equations

2.1 General transport equation

Equations of evolution of the macroscopic plasma parameters (i.e. transport equations) can be directly obtained from the kinetic equation (1.2.1) by multiplying it by a dynamic variable $\psi_\alpha(\mathbf{v}_\alpha, \mathbf{r}, t)$ and integrating it with respect to the range of particle velocities

$$\int \psi_\alpha \left(\frac{\partial f_\alpha}{\partial t} + \mathbf{v}_\alpha \cdot \nabla f_\alpha + \frac{1}{m_\alpha} \mathbf{F}_\alpha \cdot \nabla_v f_\alpha \right) d\mathbf{v} \; = \; \sum_\beta \int \psi_\alpha J_{\alpha\beta} d\mathbf{v}. \tag{2.1.1}$$

Let us define the averaging procedure of an arbitrary value $\varphi_\alpha(\mathbf{v}_\alpha, \mathbf{r}, t)$ with the help of the distribution function $f_\alpha(\mathbf{v}_\alpha, \mathbf{r}, t)$

$$n_\alpha \langle \varphi_\alpha \rangle = \int \varphi_\alpha f_\alpha d\mathbf{v}. \tag{2.1.2}$$

The left-hand side of (2.1.1) can be transformed using the relations

$$\begin{aligned} \int \psi_\alpha \frac{\partial f_\alpha}{\partial t} d\mathbf{v} &= \frac{\partial}{\partial t} \int \psi_\alpha f_\alpha d\mathbf{v} - \int \frac{\partial \psi_\alpha}{\partial t} f_\alpha d\mathbf{v} \\ &= \frac{\partial}{\partial t} n_\alpha \langle \psi_\alpha \rangle - n_\alpha \left\langle \frac{\partial \psi_\alpha}{\partial t} \right\rangle, \end{aligned} \tag{2.1.3}$$

$$\begin{aligned} \int \psi_\alpha (\mathbf{v}_\alpha \cdot \nabla) f_\alpha d\mathbf{v} &= \int \nabla \cdot (\psi_\alpha \mathbf{v}_\alpha f_\alpha) d\mathbf{v} - \int (\mathbf{v}_\alpha \cdot \nabla \psi_\alpha) f_\alpha d\mathbf{v} \\ &= \nabla \cdot (n_\alpha \langle \psi_\alpha \mathbf{v}_\alpha \rangle) - n_\alpha \langle \mathbf{v}_\alpha \cdot \nabla \psi_\alpha \rangle, \end{aligned} \tag{2.1.4}$$

$$\int \psi_\alpha \frac{\partial f_\alpha}{\partial v_x} d\mathbf{v} = \iint [\psi_\alpha f_\alpha]_{v_x \to -\infty}^{v_x \to \infty} dv_y dv_z - \int \frac{\partial \psi_\alpha}{\partial v_x} f_\alpha d\mathbf{v}$$

$$= -n_\alpha \left\langle \frac{\partial \psi_\alpha}{\partial v_x} \right\rangle. \tag{2.1.5}$$

The latter expression is written only for one of the Cartesian components. In (2.1.5) the first term resulting from the integration by part vanishes as it is assumed that $\psi_\alpha f_\alpha \to 0$ at $v_x \to \infty$ and $v_x \to -\infty$. Besides, it is implied that for the external force of (1.2.19) the condition $\partial F_{\alpha l}/\partial c_{\alpha l} = 0$ is valid. For the constituent of $\mathbf{F}_\alpha$ which is velocity-dependent (the Lorentz force) this requirement is met because the Cartesian l-component of the vector product does not include the lth component of the particle velocity.

As a result, we arrive at the transport equation of the kind

$$\frac{\partial}{\partial t} n_\alpha \langle \psi_\alpha \rangle + \nabla \cdot (n_\alpha \langle \psi_\alpha \mathbf{v}_\alpha \rangle) - n_\alpha \left[\left\langle \frac{\partial \psi_\alpha}{\partial t} \right\rangle + \langle \mathbf{v}_\alpha \cdot \nabla \psi_\alpha \rangle \right.$$

$$\left. + \frac{1}{m_\alpha} \langle \mathbf{F}_\alpha \cdot \nabla_v \psi_\alpha \rangle \right] = \sum_\beta \int \psi_\alpha J_{\alpha\beta} d\mathbf{v}. \tag{2.1.6}$$

If ψ_α appears to be the function of the particle velocity $\mathbf{v}_\alpha$ alone and does not depend on time or coordinates, the first two terms in the square brackets vanish.

It is noted that some moments of the distribution function of interest (1.1.11) have been defined with the variable $\mathbf{c} = \mathbf{v} - \mathbf{u}$ rather than $\mathbf{v}$. This makes it appropriate to demonstrate another means of writing the transport equation used when ψ_α is a function of the peculiar velocity of a particle, i.e. $\psi_\alpha = \psi_\alpha(\mathbf{c}_\alpha, \mathbf{r}, t)$ and the original kinetic equation is put down in terms of $\mathbf{c}_\alpha$ which is valid in the frame of reference when the mass-average velocity of plasma is taken to be zero. Then the operator ∇_v is replaced by ∇_c, and instead of $\partial f_\alpha/\partial t$ we have

$$\frac{\partial f_\alpha}{\partial t} + \frac{\partial \mathbf{c}_\alpha}{\partial t} \cdot \nabla_c f_\alpha = \frac{\partial f_\alpha}{\partial t} - \frac{\partial \mathbf{u}}{\partial t} \cdot \nabla_c f_\alpha,$$

and instead of $\mathbf{v}_\alpha \cdot \nabla f_\alpha$ the substitute is made

$$\mathbf{v}_\alpha \cdot \nabla f_\alpha + \frac{\partial f_\alpha}{\partial c_{\alpha r}} v_{\alpha s} \frac{\partial c_{\alpha r}}{\partial x_s}$$

$$= (\mathbf{c}_\alpha + \mathbf{u}) \cdot \nabla f_\alpha - \frac{\partial f_\alpha}{\partial c_{\alpha r}} (c_{\alpha s} + u_s) \frac{\partial u_r}{\partial x_s},$$

where r and s are indices corresponding to the components along the axes of the Cartesian coordinates and the repeated subscripts r, s indicate that these expressions are to be summed over the values $r, s = 1, 2, 3$. As a result, the kinetic equation in the variable of $\mathbf{c}_\alpha = \mathbf{v}_\alpha - \mathbf{u}$ acquires the form

$$\frac{df_\alpha}{dt} + \mathbf{c}_\alpha \cdot \nabla f_\alpha + \frac{1}{m_\alpha}\mathbf{F}^*_\alpha \cdot \nabla_c f_\alpha - c_{\alpha s}\frac{\partial f_\alpha}{\partial c_{\alpha r}}\frac{\partial u_r}{\partial x_s} = \sum_\beta J_{\alpha\beta}\,, \quad (2.1.7)$$

where $d/dt = \partial/\partial t + \mathbf{u}\cdot\nabla$ and $\mathbf{F}^*_\alpha$ is determined by the relationship

$$n_\alpha \mathbf{F}^*_\alpha = n_\alpha \mathbf{F}_\alpha - \rho_\alpha \frac{d\mathbf{u}}{dt}. \quad (2.1.8)$$

After multiplying (2.1.7) on the left- and right-hand sides by $\psi_\alpha(\mathbf{c}_\alpha, \mathbf{r}, t)$, integrating with respect to velocities and using the relationships analogous to (2.1.3)–(2.1.5), we obtain

$$\frac{dn_\alpha\langle\psi_\alpha\rangle}{dt} + n_\alpha\psi_\alpha\nabla\cdot\mathbf{u} + \nabla\cdot(n_\alpha\langle\psi_\alpha\mathbf{c}_\alpha\rangle) - n_\alpha\left[\left\langle\frac{d\psi_\alpha}{dt}\right\rangle\right.$$

$$\left. + \langle\mathbf{c}_\alpha\cdot\nabla\psi_\alpha\rangle + \frac{1}{m_\alpha}\langle\mathbf{F}^*_\alpha\cdot\nabla_c\psi_\alpha\rangle - \left\langle c_{\alpha s}\frac{\partial\psi_\alpha}{\partial c_{\alpha r}}\right\rangle\frac{\partial u_r}{\partial x_s}\right] = R_\alpha, \quad (2.1.9)$$

where averaging is performed with the help of the distribution function $f_\alpha(\mathbf{c}_\alpha, \mathbf{r}, t)$ and the right-hand side is defined as

$$R_\alpha = \sum_\beta \int \psi_\alpha J_{\alpha\beta} d\mathbf{c}_\alpha. \quad (2.1.10)$$

If ψ_α depends only on the variable $\mathbf{c}_\alpha$, then the first two terms in the square brackets (2.1.9) turn to zero.

The transport equation (2.1.9) is essential for our further consideration.

The right-hand side of the equation can be interpreted as the rate of changes in the average physical value $\psi_\alpha(\mathbf{c}_\alpha, \mathbf{r}, t)$ resulting from the collisions of the particles. Consider a way of putting down the right-hand side arising when the reciprocity relation (1.2.16) for direct and inverse collisions is used. By definition, we have

$$R_\alpha = \sum_\beta \iiint \psi_\alpha(f'_\alpha f'_{1\beta} - f_\alpha f_{1\beta}) g\sigma_{\alpha\beta} d\Omega d\mathbf{c}_\alpha d\mathbf{c}_{1\beta}. \quad (2.1.11)$$

Let us investigate the part of the expression under the summation sign

$$\iiint \psi_\alpha f'_\alpha f'_{1\beta} g \sigma_{\alpha\beta} d\Omega d\mathbf{c}_\alpha d\mathbf{c}_{1\beta}.$$

Replacing the primed and unprimed values and taking (1.2.16), we can easily see that the expression is equivalent to

$$\iiint \psi'_\alpha f_\alpha f_{1\beta} g \sigma_{\alpha\beta} d\Omega d\mathbf{c}_\alpha d\mathbf{c}_{1\beta},$$

and, thus,

$$R_\alpha = \sum_\beta \iiint (\psi'_\alpha - \psi_\alpha) f_\alpha f_{1\beta} g \sigma_{\alpha\beta} d\Omega d\mathbf{c}_\alpha d\mathbf{c}_{1\beta}. \qquad (2.1.12)$$

In what follows the transport equations will be considered for the plasma as a whole, they will be found by summation over α. Let us first deduce two useful expressions for the right-hand side of the general transport equation in our case. To this end, we shall make summation with respect to α

$$\sum_\alpha R_\alpha = \sum_\alpha \sum_\beta \iiint (\psi'_\alpha - \psi_\alpha) f_\alpha f_{1\beta} g \sigma_{\alpha\beta} d\Omega d\mathbf{c}_\alpha d\mathbf{c}_{1\beta}. \qquad (2.1.13)$$

Replacing α and β and exchanging $\mathbf{c}_\alpha$ and $\mathbf{c}_{1\beta}$ as well as $\mathbf{c}'_\alpha$ and $\mathbf{c}'_{1\beta}$ would not change the value. Therefore, along with (2.1.13) we have

$$\sum_\alpha R_\alpha = \sum_\alpha \sum_\beta \iiint (\psi'_{1\beta} - \psi_{1\beta}) f_\alpha f_{1\beta} g \sigma_{\alpha\beta} d\Omega d\mathbf{c}_\alpha d\mathbf{c}_{1\beta}. \qquad (2.1.13')$$

Summation of the two expressions brings about the result

$$\sum_\alpha R_\alpha = \frac{1}{2} \sum_\alpha \sum_\beta \iiint (\psi'_\alpha + \psi'_{1\beta} - \psi_\alpha - \psi_{1\beta}) \times f_\alpha f_{1\beta} g \sigma_{\alpha\beta} d\Omega d\mathbf{c}_\alpha d\mathbf{c}_{1\beta}. \qquad (2.1.14)$$

Finally, by replacing the primed and unprimed values and using the reciprocity relation, we find

$$\sum_\alpha R_\alpha = -\frac{1}{2} \sum_\alpha \sum_\beta \iiint (\psi'_\alpha + \psi'_{1\beta} - \psi_\alpha - \psi_{1\beta}) \times f'_\alpha f'_{1\beta} g \sigma_{\alpha\beta} d\Omega d\mathbf{c}_\alpha d\mathbf{c}_{1\beta}. \qquad (2.1.15)$$

The half-sum of (2.1.14) and (2.1.15) gives

$$\sum_\alpha R_\alpha = \frac{1}{4}\sum_\alpha\sum_\beta \iiint (\psi'_\alpha + \psi'_{1\beta} - \psi_\alpha - \psi_{1\beta})$$

$$\times (f_\alpha f_{1\beta} - f'_\alpha f'_{1\beta}) g\sigma_{\alpha\beta} d\Omega d\mathbf{c}_\alpha d\mathbf{c}_{1\beta}. \qquad (2.1.16)$$

2.2 Conservation equations

An evident consequence of the general transport equations (2.1.6) or (2.1.9) are conservation equations of mass, momentum and energy for the species α in a mixture. Assuming $\psi_\alpha = m_\alpha$ in Eq. (2.1.6), we find

$$\frac{\partial \rho_\alpha}{\partial t} + \nabla \cdot (\rho_\alpha \mathbf{u}_\alpha) = 0. \qquad (2.2.1)$$

This equation is called the continuity equation for species α of a mixture. Its right-hand side is identically equal to zero due to the condition (2.1.12) because of the mass conservation in the elastic pair collision of particles of both species α and β.

The continuity equation can also be written in the form of the conservation equation for the number of particles of species α

$$\frac{\partial n_\alpha}{\partial t} + \nabla \cdot n_\alpha (\mathbf{u} + \mathbf{w}_\alpha) = \dot{n}_\alpha, \qquad (2.2.2)$$

where $\dot{n}_\alpha$ is the reaction rate of species α production per unit volume. In the case of electrons and ions, the right-hand side of (2.2.2) is different from zero when accounting for ionization and recombination in the plasma; for neutral atoms and molecules and also molecular ions, a possible source of formation and disappearance of particles of a particular kind can be due to various chemical reactions. For collision not accompanied by any reaction $\dot{n}_\alpha = 0$ and, in fact, Eq. (2.2.2) is coincident with Eq. (2.2.1) which differs from the previous one only by the factor m_α.

Now let us take $\psi_\alpha = m_\alpha \mathbf{c}$ and $m_\alpha c_\alpha^2/2$ in Eq. (2.1.9). As a result,

$$\frac{d}{dt}\rho_\alpha w_{\alpha r} + \rho_\alpha w_{\alpha r} \nabla \cdot \mathbf{u} + \rho_\alpha w_{\alpha s}\frac{\partial u_r}{\partial x_s} + \frac{\partial P_{\alpha rs}}{\partial x_s} - n_\alpha \langle F^*_{\alpha r}\rangle = R_{\alpha r}, \qquad (2.2.3)$$

$$\frac{3}{2}\left(\frac{dp_\alpha}{dt} + p_\alpha \nabla \cdot \mathbf{u}\right) + P_{\alpha rs}\frac{\partial u_r}{\partial x_s} + \nabla \cdot \mathbf{q}_\alpha - n_\alpha \langle \mathbf{c}_\alpha \cdot \mathbf{F}^*_\alpha\rangle = Q_\alpha, \qquad (2.2.4)$$

where

$$R_{\alpha r} = \sum_{\beta} \iiint m_\alpha (c'_{\alpha r} - c_{\alpha r}) f_\alpha f_{1\beta} g \sigma_{\alpha\beta} d\Omega d\mathbf{c}_\alpha d\mathbf{c}_{1\beta}, \qquad (2.2.5)$$

$$Q_\alpha = \sum_{\beta} \iiint \frac{m_\alpha}{2} (c'^2_\alpha - c^2_\alpha) f_\alpha f_{1\beta} g \sigma_{\alpha\beta} d\Omega d\mathbf{c}_\alpha d\mathbf{c}_{1\beta}, \qquad (2.2.6)$$

and

$$n_\alpha \langle \mathbf{F}^*_\alpha \rangle = n_\alpha \langle \mathbf{F}_\alpha \rangle - \rho_\alpha \frac{d\mathbf{u}}{dt}. \qquad (2.2.7)$$

The averaged external force acting on the particles of species α is described by

$$\langle \mathbf{F}_\alpha \rangle = e_\alpha (\mathbf{E} + \mathbf{u}_\alpha \times \mathbf{B}) + \mathbf{X}_\alpha. \qquad (2.2.8)$$

To derive the momentum and energy conservation equations for the plasma as a whole we should take into account the fact that the values $m_\alpha \mathbf{c}_\alpha$ and $m_\alpha c^2_\alpha/2$ are summation invariants of elastic collisions of the particles which follow from the conservation laws (1.2.3). Then, owing to (2.1.14) or (2.1.16), we get

$$\sum_\alpha R_{\alpha r} = 0, \qquad \sum_\alpha Q_\alpha = 0. \qquad (2.2.9)$$

Finally, after summation of (2.2.1) and (2.2.3)–(2.2.4) over α the hydrodynamics equations of a mixture are obtained, i.e. the continuity equation, motion and energy equations of the plasma as a whole:

$$\frac{d\rho}{dt} + \rho \nabla \cdot \mathbf{u} = 0, \qquad (2.2.10)$$

$$\rho \frac{du_r}{dt} + \frac{\partial P_{rs}}{\partial x_s} - \sum_\alpha n_\alpha \langle F_{\alpha r} \rangle = 0, \qquad (2.2.11)$$

$$\frac{3}{2}\left(\frac{dp}{dt} + p\nabla \cdot \mathbf{u}\right) + P_{rs}\frac{\partial u_r}{\partial x_s} + \nabla \cdot \mathbf{q} - \sum_\alpha n_\alpha \langle \mathbf{c}_\alpha \cdot \mathbf{F}_\alpha \rangle = 0, \quad (2.2.12)$$

where the condition (1.1.4) is essentially used when writing the equations.

It should be noted that the continuity equation for the total number of particles in the plasma on condition that $\sum \dot{n}_\alpha = 0$ can be written as

$$\frac{\partial n}{\partial t} + \nabla \cdot (n\mathbf{u}) + \nabla \cdot \sum_\alpha n_\alpha \mathbf{w}_\alpha = 0. \tag{2.2.13}$$

Let us introduce a definition of the electric current density in the plasma

$$\mathbf{j} = \sum_\alpha n_\alpha e_\alpha \mathbf{w}_\alpha. \tag{2.2.14}$$

Then the terms dependent on the external forces $\mathbf{F}_\alpha$ can be represented as

$$\sum_\alpha n_\alpha \langle \mathbf{F}_\alpha \rangle = \mathbf{j} \times \mathbf{B} + \rho^e \mathbf{E}' + \sum_\alpha n_\alpha \mathbf{X}_\alpha, \tag{2.2.15}$$

$$\sum_\alpha n_\alpha \langle \mathbf{c}_\alpha \cdot \mathbf{F}_\alpha \rangle = \mathbf{j} \cdot \mathbf{E}' + \sum_\alpha n_\alpha \mathbf{w}_\alpha \cdot \mathbf{X}_\alpha, \tag{2.2.16}$$

where $\mathbf{E}' = \mathbf{E} + \mathbf{u} \times \mathbf{B}$ is the electric field in the reference frame moving at velocity $\mathbf{u}$ and $\rho^e = \sum_\alpha n_\alpha e_\alpha$ is the charge density in the plasma. In what follows the plasma is assumed to be quasineutral and meets the requirement $\rho^e = 0$.

Taking into account Eqs. (1.1.6) and (2.2.15), we can reduce the motion equation of plasma as a whole to the familiar form

$$\rho \frac{d\mathbf{u}}{dt} + \nabla p + \nabla \hat{\pi} - \sum_\alpha n_\alpha \mathbf{X}_\alpha - \mathbf{j} \times \mathbf{B} = 0, \tag{2.2.17}$$

where $\nabla \hat{\pi} = \partial \pi_{rs} / \partial x_s$ corresponds to the "force of viscous friction" acting on the plasma per unit volume.

As regard the energy equation, various forms of its representation are possible. Taking into account the relation $p = nkT$ and excluding the derivatives $\partial n/\partial t$ by means of (2.2.13), we can write (2.2.12) as

$$\begin{aligned} &nc_v \frac{dT}{dt} + nkT \nabla \cdot \mathbf{u} + \pi_{rs} \frac{\partial u_r}{\partial x_s} + \nabla \cdot \mathbf{q} \\ &-\frac{3}{2} kT \nabla \cdot \sum_\alpha n_\alpha \mathbf{w}_\alpha - \sum_\alpha n_\alpha \mathbf{w}_\alpha \cdot \mathbf{X}_\alpha - \mathbf{j} \cdot \mathbf{E}' = 0, \end{aligned} \tag{2.2.18}$$

where $c_v = (3/2)\, k$ is the heat capacity per particle.

In some cases, it is more convenient to use the balance equation of the total energy of the plasma obtained if (2.2.17) is multiplied by $\mathbf{u}$ and added to the energy equation in the form of (2.2.12). The result will be

$$\frac{\partial}{\partial t}\left(\frac{1}{2}\rho u^2+\frac{3}{2}p\right)+\frac{\partial}{\partial x_r}\left[\left(\frac{1}{2}\rho u^2+\frac{5}{2}p\right)u_r+\pi_{rs}u_s+q_r\right]$$
$$=\sum_\alpha n_\alpha \mathbf{w}_\alpha\cdot\mathbf{X}_\alpha+\mathbf{j}\cdot\mathbf{E}. \tag{2.2.19}$$

Let us recall that the transport equations under consideration are valid, generally speaking, for the particles having no internal degrees of freedom so that $(3/2)\,p=nE^{tr}$ represents the mean thermal energy of the plasma particles per unit volume due to the kinetic energy of their translational motion and $E^{tr}=(3/2)\,kT$. For the case of the molecular plasma whose molecules and molecular ions possess internal degrees of freedom, the values $(3/2)\,p$ and $(5/2)\,p$ should be replaced by nE and $nE+p$, where $E=E^{tr}+E^{int}$, with $nE^{int}=\sum_\alpha n_\alpha E_\alpha^{int}$ being the total mean energy per unit volume related to the internal energy state of the particles. The problem of a more comprehensive study of internal degrees of freedom and inelastic collisions of particles in the molecular plasma will be considered at greater length in Chapter 10.

We intend to look into other ways of deriving the motion and energy equations for a particular species of the plasma. Through some simple transformations (using the continuity equation (2.2.1)) the equation (2.2.3) can be reduced to the form structurally analogous to the motion equation (2.2.17) but in the form adequate for the species α of the plasma

$$\rho_\alpha\frac{d_\alpha u_{\alpha r}}{dt}+\frac{\partial P^*_{\alpha rs}}{\partial x_s}-n_\alpha\langle F_{\alpha r}\rangle=R_{\alpha r}. \tag{2.2.20}$$

Here $P^*_{\alpha rs}=P_{\alpha rs}-\rho_\alpha w_{\alpha r}w_{\alpha s}$ and the reduction $d_\alpha/dt=\partial/\partial t+u_{\alpha r}\partial/\partial x_r$ is used.

In turn, the energy equation (2.2.4) with the exclusion of the derivative dn_α/dt by means of (2.2.2) at $\dot{n}_\alpha=0$, can be written as

$$n_\alpha c_v\left(\frac{dT_\alpha}{dt}+\frac{2}{3}T_\alpha\nabla\cdot\mathbf{u}\right)+\nabla\cdot\mathbf{q}_\alpha+\pi_{\alpha rs}\frac{\partial u_r}{\partial x_s}$$
$$-\frac{3}{2}kT_\alpha\nabla\cdot(n_\alpha\mathbf{w}_\alpha)-n_\alpha\mathbf{w}_\alpha\cdot\mathbf{F}^*_\alpha=Q_\alpha. \tag{2.2.21}$$

The values of $\mathbf{R}_\alpha$ and Q_α are determined by expressions (2.2.5) and (2.2.6) and represent, respectively, the rate of change of the momentum and energy per unit volume for species α as a result of collisions between species α and all species β.

The conservation equations (2.2.1)–(2.2.4) as well as the full conservation equations (2.2.10)–(2.2.12) do not form a closed system, since in addition to conventional hydrodynamic variables ρ_α, $\mathbf{u}_\alpha$ and p_α(or T_α)

there are moments of higher orders, viz. the viscous-stress tensor $\pi_{\alpha rs}$ and the heat flux vector $\mathbf{q}_\alpha$.

Two moments of still higher order may appear in the transport equations which may be written for such variables. These moments can be reduced to the variables for which the transport equations have already been derived and the right-hand sides of these equations (the so called moments with respect to collision integrals) can be calculated only by applying certain approximations of the distribution function. One of the simplest approaches corresponding to the so called quasihydrodynamic approach in plasma will be treated in the next chapter. A more detailed description of the transport processes in the multicomponent plasma is based on expanding the distribution function of the plasma particles in series over irreducible tensorial Hermite polynomials with the weight function which corresponds to the local Maxwell distribution (the Grad's moment method). Further chapters will treat possible applications of this technique. However, before we pass to it, let us consider some general relationships ensuing from the kinetic equation and describing the evolution of such an important statistical and kinetic value as gas entropy in time and space.

2.3 Entropy balance equation

In kinetic theory, the total entropy density, i.e. entropy per unit volume of a mixture, is introduced by the expression [1,2]

$$ns = -k \sum_\alpha \int f_\alpha (\ln f_\alpha - 1)\, d\mathbf{v} = -k \sum_\alpha n_\alpha \langle \ln f_\alpha - 1 \rangle . \quad (2.3.1)$$

The entropy balance equation can be directly derived from the transport equation (2.1.6) if the value $\psi_\alpha = \ln f_\alpha - 1$ is used as a dynamic variable. In this case,

$$\begin{aligned} &\frac{\partial}{\partial t} n_\alpha \langle \ln f_\alpha - 1 \rangle + \nabla \cdot [n_\alpha \langle \mathbf{v}_\alpha (\ln f_\alpha - 1) \rangle] \\ &- \int \left[\frac{\partial f_\alpha}{\partial t} + \mathbf{v}_\alpha \cdot \nabla f_\alpha + \frac{1}{m_\alpha} \mathbf{F}_\alpha \cdot \nabla_v f_\alpha \right] d\mathbf{v}_\alpha \\ = &\sum_\beta \int (\ln f_\alpha - 1)\, J_{\alpha\beta} d\mathbf{v}_\alpha . \end{aligned} \quad (2.3.2)$$

The subintegral expression in the square brackets in (2.3.2) corresponds to the left-hand side of the kinetic equation (1.2.1). Integrating with respect to the velocity range brings about the left-hand side of the continuity equation (2.2.2) which is identically equal to zero when there is no reaction. Multiplying the terms left in (2.3.2) by $-k$ and summing over α results in the equation

$$\frac{\partial ns}{\partial t} + \nabla\cdot(ns\mathbf{u} + \mathbf{J}_s) = \sigma \tag{2.3.3}$$

if (2.3.1) is adequately taken into consideration. Here

$$\mathbf{J}_s = -k\sum_{\alpha}\int \mathbf{c}_\alpha f_\alpha\,(\ln f_\alpha - 1)\,d\mathbf{v}_\alpha \tag{2.3.4}$$

is the non-convective entropy flux and

$$\sigma = -k\sum_{\alpha}\sum_{\beta}\int \ln f_\alpha J_{\alpha\beta} d\mathbf{v}_\alpha \tag{2.3.5}$$

is local entropy production.

The latter, on the basis of relation (2.1.16), is easily transformed into

$$\sigma = \frac{k}{4}\sum_{\alpha}\sum_{\beta}\iiint (f_\alpha' f_\beta' - f_\alpha f_\beta)\ln\frac{f_\alpha' f_\beta'}{f_\alpha f_\beta} g\sigma_{\alpha\beta} d\Omega d\mathbf{v}_\alpha d\mathbf{v}_\beta. \tag{2.3.6}$$

Note that the subintegral expression in (2.3.6) comprises the function of $(x-y)\ln(x/y)$ which may either be positive or turn to zero at $x = y$. Thanks to this,

$$\sigma \geq 0. \tag{2.3.7}$$

The entropy balance equation (2.3.3), along with the condition (2.3.7), corresponds to the local formulation of the second law of thermodynamics [2].

Let us integrate Eq. (2.3.3) with respect to a certain volume V filled by plasma

$$\frac{dS}{dt} + \int_{\Omega} (ns\mathbf{u} + \mathbf{J}_s)\,d\Omega = \int_V \sigma d\mathbf{r},$$

where $S = \int ns d\mathbf{r}$ is the total entropy of the system under consideration and Ω is the surface limiting the volume concerned. For an isolated system, i.e. without exchanging the thermal energy and matter with the environment, the integral with respect to the surface on the left-hand side turns to zero and, as a result,

$$\frac{dS}{dt} = \int_V \sigma d\mathbf{r} \geq 0. \tag{2.3.8}$$

This inequality is well known in the kinetic gas theory as Boltzmann's H-theorem and it corresponds to the integral representation of the second law of thermodynamics [1,2].

In accordance with (2.3.8) the entropy of an isolated system can only increase in time and at $t \to \infty$ it approaches a certain limit. In this limit $dS/dt = 0$ which is possible only if the integrals in the expression of σ (2.3.6) turn to zero. The latter occurs only if

$$f_\alpha f_{1\beta} = f'_\alpha f'_{1\beta} \tag{2.3.9}$$

or, to put it another way, when

$$\ln f_\alpha + \ln f_{1\beta} = \ln f'_\alpha + \ln f'_{1\beta}. \tag{2.3.9$'$}$$

It should be pointed out that when using the reciprocity relation (1.2.16), the condition (2.3.9) can be re-written as

$$f_\alpha f_{1\beta} g \sigma_{\alpha\beta} d\Omega d\mathbf{v}_\alpha d\mathbf{v}_{1\beta} = f'_\alpha f'_{1\beta} g' \sigma'_{\alpha\beta} d\Omega d\mathbf{v}'_\alpha d\mathbf{v}'_{1\beta}, \tag{2.3.10}$$

which describes the principle of the detailed balancing following which the number of collisions during the transition $\mathbf{g} \to \mathbf{g}'$ in the equilibrium state of the gas exactly equals the number of collisions with transition $\mathbf{g}' \to \mathbf{g}$.

Since $\ln f_\alpha$ is an additive invariant, it turns out to be represented as a linear combination of the additive collision invariants

$$\ln f_\alpha = a_\alpha^{(1)} m_\alpha + \mathbf{a}^{(2)} m_\alpha \mathbf{c}_\alpha + a^{(3)} \frac{m_\alpha c_\alpha^2}{2}. \tag{2.3.11}$$

The constants $a_\alpha^{(1)}$ may vary for different species of particles as m_α itself is retained in every collision. On the contrary, the constants $\mathbf{a}^{(2)}$ and $a^{(3)}$ must be equal for all species because during their collisions only their total momentum and energy of a particle pair are conserved. The values of these constants can be expressed in terms of n_α, $\mathbf{u}$ and T (while the temperatures of all the plasma species in equilibrium are assumed to be equal), using the definitions of macroscopic parameters of plasma of (1.1.1) and (1.1.2).

Ultimately, the equilibrium Maxwell distribution of species α of the gas mixture moving with mass-average velocity $\mathbf{u}$ follows from (2.3.11)

$$f_\alpha^{(0)} = n_\alpha \left(\frac{m_\alpha}{2\pi kT}\right)^{3/2} \exp\left[-\frac{m_\alpha (\mathbf{v}_\alpha - \mathbf{u})^2}{2kT}\right]. \tag{2.3.12}$$

Now it is appropriate to consider a case when the particles of species possess internal degrees of freedom. When deriving the entropy balance equation, the generalized kinetic equation in the form of (1.2.23) can be taken as a basis. When defining the entropy density and flux entropy by analogy with (2.3.1) and (2.3.4) and considering the summation over the internal states, we come to the entropy balance equation which fully agrees with the form of (2.3.3). In the case of equilibrium, the entropy production is cancelled which corresponds to the requirement fulfilled in

$$f_{\alpha i} f_{1\beta j} = f'_{\alpha k} f'_{1\beta l}. \tag{2.3.13}$$

As in the case considered earlier, $\ln f_{\alpha i}$ proves to be an additive invariant, i.e. it can be represented as

$$\ln f_{\alpha i} = a_\alpha^{(1)} m_\alpha + \mathbf{a}^{(2)} m_\alpha \mathbf{c}_\alpha + a^{(3)} \left(\frac{m_\alpha c_\alpha^2}{2} + E_{\alpha i} \right). \tag{2.3.14}$$

The condition (2.3.14) holds for all i, j, k, l, therefore, the constants $a_\alpha^{(1)}, \mathbf{a}^{(2)}$ and $a^{(3)}$ do not depend on i and can be expressed on the basis of macroscopic parameters of the mixture by means of (1.1.13), (1.1.19) and definition of (1.1.2) after summation over i. As a result, the equilibrium Maxwell–Boltzmann distribution function is obtained

$$f_{\alpha i}^{(0)} = n_\alpha \left(\frac{m_\alpha}{2\pi kT} \right)^{3/2} Q_\alpha^{-1} \exp\left(-\frac{m_\alpha c_\alpha^2}{2kT} - \varepsilon_{\alpha i} \right), \tag{2.3.15}$$

where

$$\varepsilon_{\alpha i} = E_{\alpha i}/kT, \qquad Q_\alpha = \sum_i \exp\left(-\varepsilon_{\alpha i}\right). \tag{2.3.16}$$

The expression (2.3.15) is derived under the assumption of $a^{(3)} = (kT)^{-1}$ which makes it agree with the conventional thermodynamic definition of temperature, as the total mean energy per unit plasma volume is taken in this case to be

$$nE = \sum_\alpha \left(n_\alpha E_\alpha^{tr} + n_\alpha E_\alpha^{int} \right) = \frac{3}{2} nkT + \sum_\alpha n_\alpha \langle \varepsilon_\alpha \rangle kT, \tag{2.3.17}$$

where

$$\langle \varepsilon_\alpha \rangle = Q_\alpha^{-1} \sum_i \varepsilon_{\alpha i} \exp\left(-\varepsilon_{\alpha i}\right). \tag{2.3.18}$$

Here the function of Q_α plays the part of the statistical sum over the internal states of particles of species α [3]. The requirement (2.3.17) becomes in fact the definition of the temperature for the plasma with excited internal degrees of freedom of the particles.

Usually the Boltzmann's H-theorem is proved for homogeneous states of gas mixtures in which case the parameters of Maxwell or Maxwell–Boltzmann distributions are independent of both time and coordinates which fully corresponds to the state of absolute thermodynamic equilibrium. Following the local formulation of the second law of thermodynamics for the local entropy production to turn into zero, the conditions (2.3.9) or (2.3.13) must be fulfilled. The distributions of the kind (2.3.12) or (2.3.15) also follow from these conditions except for parameters n_α, $\mathbf{u}$ and T, which depend on time and coordinates in the general case. This kind of distributions can be classified as locally equilibrium.

To realize locally equilibrium states, it is necessary to have at least two characteristic time scales, viz. the time of establishing the equilibrium in the system which is dependent on its volume and, besides, a notably less relaxation time which characterizes the equilibrium set in a macroscopically small volume and, hence, is independent of the whole volume of the system. A dilute gas mixture can meet the above requirements since any characteristic time between collisions in it greatly exceeds the time of collisions of particles but is considerably less than the characteristic time interval during which an essential change of macroscopic parameters of the gas occurs. In the case of spatially inhomogenious states, a small parameter becomes in essence the ratio of the characteristic mean free path and the characteristic scales of spatial inhomogeneity of the macroscopic parameters or the so called Knudsen number ($Kn \ll 1$).

A possibility to introduce locally equilibrium distributions of Maxwell-Boltzmann type in the zero approximation and to apply the perturbation technique in terms of the Knudsen number is known to be one of the main peculiar features of the Chapman–Enskog method which is widely used in the kinetic theory of non-uniform gases [4]. An application of locally equilibrium distributions as a weight function when expanding over the irreducible tensorial Hermite polynomials also proves to be very convenient when applying the Grad's moment method [5], since a greater freedom of choosing the distribution parameters is allowed. For example, in contrast to a conventional gas mixture, the plasma with a lighter mass constituent (electrons) lets one use the Maxwell distributions with differing temperatures of species and particle velocities shifted relative to their own macroscopic velocities of the species. The so called multitemperature and multifluid plasmas which we shall consider in the next section are examples of such models of plasma [6,7].

2.4 Multitemperature and multifluid models of plasma

A specific feature of a plasma as compared with a gas mixture is that there are electrons available in it whose masses are much smaller than those of ions and neutral plasma particles. The transfer of energy in collisions of light and heavy particles turns out to be proportional to the small ratio of m_e/m_β $(\beta \neq e)$ (see next). A slow energy exchange between the subsystems of electrons and heavy particles in the plasma results in the fact that the equilibrium in each of the subsystems sets up earlier that the complete equilibrium is achieved. This means that during a rather long period of time which can be comparable to the macroscopic time in the problem under consideration, there arises a state of quasi-equilibrium in the plasma which is characterised by different temperature of the species $(T_e \neq T_\beta)$. The process of relaxation into a complete thermodynamical equilibrium is described then by the relaxation equation for T_α to be considered in the following chapter.

In the situation under study $\ln f_\alpha$ can be represented as

$$\ln f_\alpha = a_\alpha^{(1)} m_\alpha + \mathbf{a}^{(2)} m_\alpha \mathbf{c}_\alpha + a_\alpha^{(3)} \frac{m_\alpha c_\alpha^2}{2}, \tag{2.4.1}$$

where the constant $a_\alpha^{(3)}$ is determined from the temperature of each species T_α, rather than the common temperature of the mixture as was done previously. The resulting local Maxwellian acquires the form

$$f_\alpha^{(0)} = n_\alpha \left(\frac{m_\alpha}{2\pi k T_\alpha}\right)^{3/2} \exp\left[-\frac{m_\alpha (\mathbf{v}_\alpha - \mathbf{u})^2}{2kT_\alpha}\right], \tag{2.4.2}$$

which agrees with the so called multitemperature model of plasma. If in the subsystem of heavy species the particles of considerably smaller mass that those of the rest can be separated, then to introduce a specific concept of temperature for them becomes as natural as in the case of the electron component. In practice, the temperatures can be assumed to be approximately the same for these heavy plasma species in which the energy exchange occurs rather intensively due to a small difference between the masses of the species particles. However, for convenience the temperatures of all the species can be taken formally different. Some differences in them will be revealed later in the analysis of the relevant equations of relaxation.

Though the temperatures of plasma species are taken to be different in the general case, the application of Maxwell distributions shifted with respect to the mass-average velocity of the mixture $\mathbf{u}$ including the variable $\mathbf{c}_\alpha = \mathbf{v}_\alpha - \mathbf{u}$ when defining the basic macroscopic plasma parameters (e.g. moments of the distribution function) corresponds to the so called one-fluid model of the plasma. The so called multifluid models are also used in the theory of plasma.

The treatment of a possible two-fluid model for a fully ionized plasma containing a single kind of ions is presented in particular in [6]. An essential feature of such an approach is that because of the small relation of m_e/m_β one can simplify the cross integral of collisions $J_{e\beta}$ and $J_{\beta e}$ in the kinetic equations for both electrons and ions and to combine their terms so that the effect of small terms on the choice of the zero approximation is excluded. The resulting kinetic equations for electrons are formally separated in form from the equation for ions. A similar situation in the multicomponent plasma is encountered when the mass of particles of one heavy species happens to be noticeably less than the mass of the other atoms and ions. In all these cases the H-theorem may be approximately used for each of the separated equations. Thus, $\ln f_\alpha$ becomes an additive invariant of the form

$$\ln f_\alpha = a_\alpha^{(1)} m_\alpha + a_\alpha^{(2)} m_\alpha \mathbf{c}_\alpha^* + a_\alpha^{(3)} \frac{m_\alpha c_\alpha^{*2}}{2}, \tag{2.4.3}$$

where $\mathbf{c}_\alpha^* = \mathbf{v}_\alpha - \mathbf{u}_\alpha$, and all coefficients $a_\alpha^{(1)}, a_\alpha^{(2)}$ and $a_\alpha^{(3)}$ depend on α and are approximately defined in terms of macroscopic parameters $n_\alpha, \mathbf{u}_\alpha$ and T_α^* referred to each of the species. Then

$$\frac{3}{2} n_\alpha k T_\alpha^* = \frac{m_\alpha}{2} \int c_\alpha^{*2} f_\alpha d\mathbf{v}. \tag{2.4.4}$$

As a result, the local equilibrium Maxwellian acquires the form

$$f_\alpha^{(0)} = n_\alpha \left(\frac{m_\alpha}{2\pi k T_\alpha^*} \right)^{3/2} \exp\left[-\frac{m_\alpha (\mathbf{v}_\alpha - \mathbf{u}_\alpha)^2}{2kT_\alpha^*} \right]. \tag{2.4.5}$$

It is a typical feature of the multifluid plasma model which helps make the use of the variables $n_\alpha, \mathbf{u}_\alpha$ and T_α^* as the original values instead of the commonly taken variables $n_\alpha, \mathbf{u}$ and T_α and variables $\pi_{\alpha rs}^*$ and $\mathbf{q}_\alpha^*$ instead of $\pi_{\alpha rs}$ and $\mathbf{q}_\alpha$. When doing so, the macroscopic parameters of the species $p_\alpha^*, \pi_{\alpha rs}^*$ and $\mathbf{q}_\alpha^*$ are defined as

$$p_\alpha^* = n_\alpha k T_\alpha^*,$$

$$\pi_{\alpha rs}^* = m_\alpha \int \left(c_{\alpha r}^* c_{\alpha s}^* - \frac{1}{3}\delta_{rs} c_\alpha^{*2} \right) f_\alpha d\mathbf{v}, \qquad (2.4.6)$$

$$\mathbf{q}_\alpha^* = \frac{m_\alpha}{2} \int \mathbf{c}_\alpha^* c_\alpha^{*2} f_\alpha d\mathbf{v}.$$

By analogy, all the other moments of the distribution function can be represented with the help of the variable $\mathbf{c}_\alpha^* = \mathbf{v}_\alpha - \mathbf{u}_\alpha$ instead of $\mathbf{c}_\alpha = \mathbf{v}_\alpha - \mathbf{u}$.

It is easy to show that the new variables $p_\alpha^*, \pi_{\alpha rs}^*$ and $\mathbf{q}_\alpha^*$ are related to the formerly used variables as [6]

$$p_\alpha^* = p_\alpha - \frac{1}{3}\rho_\alpha w_\alpha^2, \quad \pi_{\alpha rs}^* = \pi_{\alpha rs} - \rho_\alpha \left(w_{\alpha r} w_{\alpha s} - \frac{1}{3}\delta_{rs} w_\alpha^2 \right),$$

$$q_{\alpha r}^* = q_{\alpha r} - \frac{5}{2} p_\alpha w_{\alpha r} - \pi_{\alpha rl} w_{\alpha l} + \rho_\alpha w_{\alpha r} w_\alpha^2. \qquad (2.4.7)$$

The balance equations of momentum and energy for species α are written in the new variable as

$$\rho_\alpha \frac{d_\alpha u_{\alpha r}}{dt} + \frac{\partial P_{\alpha rs}^*}{\partial x_s} - n_\alpha \langle F_{\alpha r} \rangle = R_{\alpha r},$$
$$\frac{3}{2} k n_\alpha \frac{d_\alpha T_\alpha^*}{dt} + p_\alpha^* \nabla \cdot \mathbf{u}_\alpha + \pi_{\alpha ls}^* \frac{\partial u_{\alpha s}}{\partial x_l} + \nabla \cdot \mathbf{q}_\alpha^* = Q_\alpha^*, \qquad (2.4.8)$$

where

$$Q_\alpha^* = Q_\alpha - \mathbf{w}_\alpha \cdot \mathbf{R}_\alpha.$$

In what follows the states of plasma will be considered in which the terms squared with respect to the moments $\rho_\alpha \mathbf{w}_\alpha, \pi_{\alpha rs}$ and $\mathbf{q}_\alpha$ will be neglected as compared with the linear terms. It is easy to see that in such an approximation the use of new and earlier used transport equations brings about quite identical results. It is evident that this requires deriving the right-hand sides of Eqs. (2.2.20) and (2.2.21). Below they are calculated from the simplest approximation to the distribution function (see Chapter 3). Further calculation will be performed on the basis of the Grad's moment method in the chapters that follow.

2.5 Nonequilibrium thermodynamics of a multicomponent plasma

Continuity and energy equations obtained in Section 2.2 may be used to derive explicit expressions of the entropy flux density and local entropy production in the equation of entropy balance (2.3.3). Thus, we are led to a number of relationships used in nonequilibrium thermodynamics whose results prove a helpful addition to the conclusions of the kinetic theory [2,4]. The original assumption of the nonequilibrium thermodynamics is known to be the statement that the local entropy (entropy per unit volume) in the case of a slight deviation of the system from equilibrium is the same function of the basic thermodynamic parameters, i.e. the energy density, specific volume and particle concentration as in the complete thermodynamic equilibrium of a system, which means that the conventional thermodynamic relations of the equilibrium thermodynamics written for the local variables remain valid even in the slight nonequilibrium state of the system [2].

To obtain an important original relationship of the entropy density change against the basic thermodynamic variables, we can use the kinetic expression of entropy density (2.3.1) with a substitution of the local Maxwellian of the kind (2.3.12). It is seen that in the case of the complete statistic equilibrium, to which the Maxwellian with time and coordinate independent thermodynamic parameters corresponds, the statistic and kinetic expressions of the entropy density must be closely related. The use of the local Maxwellian must result in the similar relationship for the entropy density ns except for the local variables which change from point to point as well as in time. Such an expression for ns can be shown to be valid with an accuracy up to the linear terms with respect to small derivations of the distribution function from local equilibrium.

For simplicity, below we limit ourselves to the case of isothermal plasma with equal temperature components ($T_\alpha = T$). The case of the two-fluid and two-temperature plasma will be dealt with later.

The local Maxwellian is conveniently written in a somewhat different way by introducing the definition of the chemical potential [3]

$$\mu_\alpha = kT \ln \frac{p_\alpha}{(kT)^{5/2}} = kT \ln \frac{n_\alpha}{(kT)^{3/2}}. \tag{2.5.1}$$

Then Eq. (2.3.12) can be rewritten as

$$f_\alpha^{(0)} = \exp\left[\frac{\mu_\alpha}{kT} - \frac{m_\alpha}{2kT}(\mathbf{v}_\alpha - \mathbf{u})^2\right]. \tag{2.5.2}$$

Substituting Eq. (2.5.2) into the definition of entropy density (2.3.1), we have

$$ns = -k\sum_\alpha \int f_\alpha^{(0)}\left[\frac{\mu_\alpha}{kT} - \frac{m_\alpha}{2kT}(\mathbf{v}_\alpha - \mathbf{u})^2\right]dv + kn \tag{2.5.3}$$

with an accuracy up to the constant.

Using the well-known definitions of the number density n_α and kinetic energy density (i.e. energy per plasma unit volume) $nE = \sum_\alpha n_\alpha E_\alpha$, let us represent Eq. (2.5.3) as

$$ns = -\frac{1}{T}\left(\sum_\alpha n_\alpha \mu_\alpha - nE - p\right). \tag{2.5.4}$$

Introducing the specific volume $v = 1/n$ and relative molar concentration of species $y_\alpha = n_\alpha/n$, Eq. (2.5.4) may be rewritten as

$$Ts = E + pv - \sum_\alpha y_\alpha \mu_\alpha. \tag{2.5.5}$$

Differentiating this relationship and using the familiar thermodynamic identity [3]

$$sdT = -vdp - \sum_\alpha y_\alpha d\mu_\alpha$$

we obtain the Gibbs relationship [2]

$$Tds = dE + pdv - \sum_\alpha \mu_\alpha dy_\alpha. \tag{2.5.6}$$

Taking into account that $E = (3/2)\,kT$ and $v = n^{-1}$, rewrite Eq. (2.5.6) as

$$\frac{ds}{dt} = \frac{3}{2}\frac{k}{T}\frac{dT}{dt} - \frac{k}{n}\frac{dn}{dt} - \frac{1}{T}\sum_\alpha \mu_\alpha \frac{dy_\alpha}{dt}. \tag{2.5.7}$$

Now let us transform the entropy balance equation (2.3.3). Using the conservation equation of the total number of particles in a plasma (2.2.13), it is convenient to express (2.3.3) as[1]

$$n\frac{ds}{dt} + \nabla\cdot\mathbf{J}_s - s\nabla\cdot\sum_\alpha n_\alpha \mathbf{w}_\alpha = \sigma. \tag{2.5.8}$$

Now let us use the conservation equations for particle number density and temperature. By virtue of Eqs. (2.2.2) and (2.2.13), we have

$$n\frac{dy_\alpha}{dt} = y_\alpha \nabla\cdot\sum_\alpha n_\alpha \mathbf{w}_\alpha - \nabla\cdot n_\alpha \mathbf{w}_\alpha. \tag{2.5.9}$$

---

[1]Note that the occurrence of the third term on the left-hand side of Eq. (2.5.8) is related to the use of the chosen definitions of entropy density and other additional thermodynamic variables as referred to the number of particles, i.e. per particle rather than per mass unit as is done in most cases [2,9]. However, this does not change the final result of the expressions of interest such as the entropy flux and local entropy production.

Substitution of dT/dt from Eq. (2.2.18), dn/dt from Eq. (2.2.13) and dy_α/dt from Eq. (2.5.9) into the expression for ds/dt (2.5.7) and a number of transformations with allowance of Eq. (2.5.2) give

$$\begin{aligned} n\frac{ds}{dt} &= -\nabla\cdot\left(\frac{\mathbf{q}-\sum_\alpha \mu_\alpha n_\alpha \mathbf{w}_\alpha}{T}\right)+s\nabla\cdot\sum_\alpha n_\alpha \mathbf{w}_\alpha-\frac{1}{T^2}\mathbf{q}\cdot\nabla T \\ &-\frac{1}{T}\pi_{rs}\frac{\partial u_r}{\partial x_s}-\frac{1}{T}\sum_\alpha n_\alpha \mathbf{w}_\alpha\cdot\left[T\nabla\left(\frac{\mu_\alpha}{T}\right)-\mathbf{X}_\alpha\right]+\frac{1}{T}\mathbf{j}\cdot\mathbf{E}'. \end{aligned} \tag{2.5.10}$$

Comparison with the entropy balance equation (2.5.8) leads to [2,8]

$$\begin{aligned} \mathbf{J}_s &= \frac{1}{T}\left(\mathbf{q}-\sum_\alpha \mu_\alpha n_\alpha \mathbf{w}_\alpha\right), \\ \sigma &= -\frac{\mathbf{q}}{T^2}\cdot\nabla T-\frac{1}{T}\pi_{rs}\frac{\partial u_r}{\partial x_s}-\frac{1}{T}\sum_\alpha n_\alpha \mathbf{w}_\alpha \\ &\times\left[T\nabla\left(\frac{\mu_\alpha}{T}\right)-\mathbf{X}_\alpha\right]+\frac{1}{T}\mathbf{j}\cdot\mathbf{E}'. \end{aligned} \tag{2.5.11}$$

Now let us transform the expression for local entropy production, using the explicit expression of the chemical potential (2.5.1). Introducing the reduced heat flux

$$\mathbf{h}=\mathbf{q}-\frac{5}{2}\sum_\alpha p_\alpha \mathbf{w}_\alpha, \tag{2.5.12}$$

we can find

$$\begin{aligned} \sigma &= -\frac{\mathbf{h}}{T^2}\cdot\nabla T-\frac{1}{T}\pi_{rs}\frac{\partial u_r}{\partial x_s}-\frac{1}{T}\sum_\alpha \mathbf{w}_\alpha\cdot(\nabla p_\alpha-n_\alpha\mathbf{X}_\alpha) \\ &+\frac{1}{T}\mathbf{j}\cdot\mathbf{E}'. \end{aligned} \tag{2.5.13}$$

The expression (2.5.13) can be suitably presented in the form which illustrates its relation to the results of the kinetic theory of dilute gas mixtures to be considered in Chapter 5. Using the condition $\sum_\alpha \rho_\alpha \mathbf{w}_\alpha = 0$ and $\sum_\alpha n_\alpha e_\alpha = 0$, we obtain

$$\sigma=-\frac{1}{T}\left(\mathbf{h}\cdot\nabla\ln T+\pi_{rs}\varepsilon_{rs}+p\sum_\alpha \mathbf{w}_\alpha\cdot\mathbf{d}_\alpha\right). \tag{2.5.14}$$

Here

$$\mathbf{d}_\alpha = \nabla y_\alpha + \left(y_\alpha - \frac{\rho_\alpha}{\rho}\right)\nabla \ln p$$
$$-\frac{1}{p}\left(n_\alpha \mathbf{X}_\alpha - \frac{\rho_\alpha}{\rho}\sum_\alpha n_\alpha \mathbf{X}_\alpha\right) - \frac{1}{p} n_\alpha e_\alpha \mathbf{E}' \quad (2.5.15)$$

and

$$\varepsilon_{rs} = \left\{\frac{\partial u_r}{\partial x_s}\right\} = \frac{1}{2}\left(\frac{\partial u_r}{\partial x_s} + \frac{\partial u_s}{\partial x_r}\right) - \frac{1}{3}\delta_{rs}\frac{\partial u_l}{\partial x_l}. \quad (2.5.16)$$

Let us transform the last term in Eq. (2.5.14) taking into account that only $N-1$ diffusion velocities of species are independent and using the evident requirement $\sum_\alpha \mathbf{d}_\alpha = 0$. The result is

$$\sigma = -\frac{1}{T}\left[\mathrm{h} \cdot \nabla \ln \nabla T + \pi_{rs}\varepsilon_{rs} + p\sum_{\alpha=1}^{N-1}(\mathbf{w}_\alpha - \mathbf{w}_N)\cdot \mathbf{d}_\alpha\right]. \quad (2.5.17)$$

As is known, one of the postulates of nonequilibrium thermodynamics is the statement that with a slight deviation of the system from equilibrium the fluxes can be represented as linear functions of the generalized forces [2]

$$J_m = \sum L_{mn} X_n. \quad (2.5.18)$$

In the general case, the generalized fluxes and forces can be scalars, vectors and tensors. The values L_{mn} are called phenomenological coefficients. The diagonal coefficients L_{mm} are the values describing direct effects while nondiagonal ones L_{mn} $(m \neq n)$ are phenomenological coefficients of the crossed effects. They are usually scalar, but for the system with a pronounced anisotropy and with the magnetic field available they turn out to be tensor values.

Another significant assumption of non-equilibrium thermodynamics is the statement that the change of entropy in time for an isolated system can be represented as a bilinear combination of generalized forces and flows. If the equation of entropy balance is considered in the local form (2.5.8), then this statement regards the structure of the expression for local entropy production σ.

Calculation of entropy production gives in fact an advice as to the correct choice of generalized fluxes and forces inherent in the linear phenomenological relationships (2.5.17). Such a choice of fluxes and forces makes the reciprocity relationships of Onsager for crossed coefficients look valid

$$L_{mn} = L_{nm}. \tag{2.5.19}$$

It is to be reminded that with the magnetic field present the reciprocity relationships accept a different more complicated form [2]

$$L_{mn}(\mathbf{B}) = L_{nm}(-\mathbf{B}), \tag{2.5.20}$$

valid on the assumption that both of the forces X_m and X_n are odd functions of particle velocities or

$$L_{mn}(\mathbf{B}) = -L_{nm}(-\mathbf{B}), \tag{2.5.20'}$$

if one of the forces is odd whereas the other is even.

Turning back to the expression derived above for Eq. (2.5.17), we can see that σ is a bilinear combination of fluxes (the difference of diffusion velocities $(\mathbf{w}_\alpha - \mathbf{w}_N)$, reduced heat flux $\mathbf{h}$ and non-diagonal part of the pressure tensor or viscous-stress tensor π_{rs}) and thermodynamic forces $(p\mathbf{d}_\alpha, \nabla \ln T, \varepsilon_{rs})$ under the action of which the fluxes are generated. The linear phenomenological relationships in this representation are of the form

$$\begin{aligned}
\mathbf{h} &= -\frac{L_{00}}{T}\nabla T - \sum_{\beta=1}^{N-1} L_{0\beta} p\mathbf{d}_\beta, \\
\mathbf{w}_\alpha - \mathbf{w}_N &= -\frac{L_{\alpha 0}}{T}\nabla T - \sum_{\beta=1}^{N-1} L_{\alpha\beta}\mathbf{d}_\beta, \\
\pi_{rs} &= -\Lambda_{00}\varepsilon_{rs}.
\end{aligned} \tag{2.5.21}$$

Notice that without the magnetic field $(\mathbf{B}=0)$ the Onsager reciprocity relationships (2.5.19) meet the requirement $L_{0\alpha} = L_{\alpha 0}$ and $L_{\alpha\beta} = L_{\beta\alpha}$

Being phenomenological, the nonequilibrium thermodynamics theory permits one to establish just a general structure of equations describing nonequilibrium phenomena as well as a certain linkage (symmetry relations) between the coefficients of the equations. Independent check of these relationships as well as direct calculation of the kinetic (transport) coefficients is a prerogative of the kinetic theory. We intend to discuss the results of the nonequilibrium thermodynamics and kinetic theory in Chapter 5, when the relevant expressions for the fluxes and kinetic coefficients are calculated in terms of the moment method. A way to generalize to some extent the results of the conventional linear nonequilibrium thermodynamics accounting for spatial derivatives of heat flux and viscous-stress tensor will also be given.

Concluding the section, let us treat the case of a two-fluid and two-temperature plasma [6,7]. The entropy balance equation can be written in this case for each of the plasma species, i.e. separately for electrons and ions in the case of the fully ionized plasma, as

$$\frac{\partial n_\alpha s_\alpha}{\partial t} + \nabla\cdot(n_\alpha s_\alpha \mathbf{u}_\alpha + \mathbf{J}_{s\alpha}) = \sigma_\alpha \tag{2.5.22}$$

or, using the continuity equations (2.4.8), as

$$n_\alpha \frac{d_\alpha s_\alpha}{dt} + \nabla\cdot\mathbf{J}_{s\alpha} = \sigma_\alpha . \tag{2.5.22$'$}$$

Using the local Maxwellian in the form (2.4.5), it is easy to verify that the entropy density (per particle) for the species particle can be represented accurate to the constant by the expression

$$s_\alpha = \frac{3}{2} k \ln T_\alpha^* - k \ln n_\alpha . \tag{2.5.23}$$

Calculating $d_\alpha s_\alpha/dt$ and eliminating $d_\alpha n_\alpha/dt$ and $d_\alpha T_\alpha^*/dt$ with the help of Eqs. (2.2.2) and (2.4.8), we obtain the entropy balance equation (2.5.22′) in which the entropy flux density and local entropy production of species particles are described by

$$\begin{aligned} \mathbf{J}_{s\alpha} &= \frac{\mathbf{q}_\alpha^*}{T_\alpha^*}, \\ \sigma_\alpha &= -\frac{\mathbf{q}_\alpha^*}{T_\alpha^*}\cdot\nabla \ln T_\alpha^* - \frac{1}{T_\alpha^*}\pi_{\alpha rs}^* \varepsilon_{\alpha rs}^* + \frac{1}{T_\alpha^*} Q_\alpha^*, \end{aligned} \tag{2.5.24}$$

where $\varepsilon_{\alpha rs}^* = \{\partial u_{\alpha r}/\partial x_s\}$.

Now consider a special case of a two-component fully ionized plasma ($\alpha = e, i$). The electronic heat exchange term Q_e^* is related to the corresponding ionic quantity by

$$Q_e^* = -(\mathbf{u}_e - \mathbf{u}_i)\cdot\mathbf{R}_e - Q_i^* = \frac{1}{n_e e}\mathbf{j}\cdot\mathbf{R}_e - Q_i^* . \tag{2.5.25}$$

Then the local entropy production for the electrons and ions can be written as

$$\sigma_e = -\frac{\mathbf{q}_e^*}{T_e^*}\cdot\nabla\ln T_e^* - \frac{1}{T_e^*}\pi_{ers}^*\varepsilon_{ers}^* + \frac{1}{n_e e T_e^*}\mathbf{j}\cdot\mathbf{R}_e - \frac{1}{T_e^*}Q_i^*,$$

$$\sigma_i = -\frac{\mathbf{q}_i^*}{T_i^*}\cdot\nabla\ln T_i - \frac{1}{T_i^*}\pi_{irs}^*\varepsilon_{irs}^* + \frac{1}{T_i}Q_i^*. \tag{2.5.26}$$

The two balance equations (2.5.22) can also be put together into a global entropy balance

$$\frac{\partial ns}{\partial t} + \sum_\alpha \nabla\cdot\left(n_\alpha s_\alpha \mathbf{u}_\alpha + \frac{\mathbf{q}_\alpha^*}{T_\alpha^*}\right) = \sigma, \qquad \alpha = e, i, \tag{2.5.27}$$

where

$$\begin{aligned}\sigma \;=\; & -\sum_\alpha\left(\frac{\mathbf{q}_\alpha^*}{T_\alpha^*}\cdot\nabla\ln T_\alpha^* + \frac{1}{T_\alpha^*}\pi_{\alpha rs}^*\varepsilon_{\alpha rs}^*\right) + \frac{1}{n_e e T_e^*}\mathbf{j}\cdot\mathbf{R}_e \\ & + Q_i^*\left(\frac{1}{T_i^*} - \frac{1}{T_e^*}\right).\end{aligned} \tag{2.5.28}$$

It follows from Eq. (2.5.26) that the conjugated quantities between which the phenomenological relationships can be established for the electron transport are the generalized fluxes $\mathbf{q}_e^*, \pi_{ers}^*, \mathbf{R}_e$, and generalized forces $\nabla\ln T_e, \varepsilon_{ers}$, and $(\mathbf{u}_e - \mathbf{u}_i)$. The application of Eq. (2.5.26) allows one, in terms of the Onsager principle, to obtain non-trivial relations of symmetry between different electron kinetic coefficients for a two-temperature fully ionized plasma in the magnetic field [6,7]. For heavy particles (ions) the conclusions derived from the relevant expression of entropy production are quite common and were analyzed previously for the case of an isothermal plasma. As it follows from Eq. (2.5.28), the new additional phenomenological relation between the generalized flux Q_i^* and generalized force $(T_e^* - T_i^*)/T_e^*T_i^*$ appears in the case of the two-temperature plasma.

Chapter 3

Quasihydrodynamic approach

3.1 Average transfer of momentum and energy in collisions

Consider the problem of calculating the right-hand sides of motion and energy equations for species α which correspond to the average transfer of momentum and energy during the collisions of particles. It will be supposed that the distribution function of particles of species α be taken in the form of a local Maxwellian displaced relative to the own mean velocity of the species $\mathbf{u}_\alpha$ (2.4.5).

Assuming the diffusion velocities of the plasma species small compared with their mean thermal speed

$$|\mathbf{w}_\alpha| \ll \left(\frac{2kT_\alpha}{m_\alpha}\right)^{1/2} \tag{3.1.1}$$

and neglecting terms squared with respect to $\mathbf{w}_\alpha$, we can represent (2.4.5) as

$$f_\alpha = f_\alpha^{(0)}\left(1 + \gamma_\alpha \mathbf{w}_\alpha \cdot \mathbf{c}_\alpha\right), \qquad \gamma_\alpha = \frac{m_\alpha}{kT_\alpha}, \tag{3.1.2}$$

where $f_\alpha^{(0)}$ corresponds to the usual local Maxwellian (2.4.2) shifted with respect to the mass-average velocity $\mathbf{u}$ and defined at temperature T_α.

It is noted that in the approximation considered

$$\pi_{\alpha rs} = 0, \qquad \mathbf{q}_\alpha = \frac{5}{2} p_\alpha \mathbf{w}_\alpha. \tag{3.1.3}$$

Now let us calculate the right-hand sides of Eqs. (2.2.20) and (2.2.21) using the approach of (3.1.2). In the expressions (2.2.5) and (2.2.6) we can integrate over angles if the variables are introduced in the centre of mass of colliding particles (1.2.4) and (1.2.5). In terms of relationships (1.2.6), we have

$$\begin{aligned} m_\alpha(\mathbf{c}_\alpha' - \mathbf{c}_\alpha) &= -\mu_{\alpha\beta}(\mathbf{g}' - \mathbf{g}), \\ \frac{m_\alpha}{2}(\mathbf{c}_\alpha'^2 - \mathbf{c}_\alpha^2) &= -\mu_{\alpha\beta}\mathbf{U}\cdot(\mathbf{g}' - \mathbf{g}), \end{aligned} \tag{3.1.4}$$

where $\mathbf{U} = \mathbf{G} - \mathbf{u}$.

Using the relations of $\mathbf{g}'$ and $\mathbf{g}$ (1.2.8) and after integrating over φ, we obtain

$$\mathbf{R}_\alpha = -\sum_\beta \mu_{\alpha\beta} \iint \mathbf{g} g Q_{\alpha\beta}^{(1)}(g)\, f_\alpha f_\beta d\mathbf{c}_\alpha d\mathbf{c}_\beta, \tag{3.1.5}$$

$$Q_\alpha = -\sum_\beta \mu_{\alpha\beta} \iint (\mathbf{g}\cdot\mathbf{U})\, g Q_{\alpha\beta}^{(1)}(g)\, f_\alpha f_\beta d\mathbf{c}_\alpha d\mathbf{c}_\beta, \tag{3.1.6}$$

where $Q_{\alpha\beta}^{(1)}(g)$ is the momentum transfer cross section defined by Eq. (1.3.1).

Using the representation (3.1.2) and leaving the linear terms with respect to $\mathbf{w}_\alpha$ and $\mathbf{w}_\beta$ in the product of $f_\alpha f_\beta$, we have

$$\begin{aligned} f_\alpha f_\beta &= n_\alpha n_\beta \left[\frac{\gamma_\alpha\gamma_\beta}{(2\pi)^2}\right]^{3/2} \exp\left(-\frac{\gamma_\alpha}{2}c_\alpha^2 - \frac{\gamma_\beta}{2}c_\beta^2\right) \\ &\quad \times \left(1 + \gamma_\alpha \mathbf{w}_\alpha\cdot\mathbf{c}_\alpha + \gamma_\beta \mathbf{w}_\beta\cdot\mathbf{c}_\beta\right). \end{aligned}$$

Further it would be convenient to pass from the variable $\mathbf{U}$ to the variable $\mathbf{X}$ by means of the relation

$$\mathbf{X} = \frac{\gamma_\alpha \mathbf{c}_\alpha + \gamma_\beta \mathbf{c}_\beta}{\gamma_\alpha + \gamma_\beta} = \mathbf{U} + \frac{\mu_{\alpha\beta}}{\gamma_\alpha + \gamma_\beta}\left(\frac{\gamma_\alpha}{m_\alpha} - \frac{\gamma_\beta}{m_\beta}\right)\mathbf{g}. \tag{3.1.7}$$

Then

$$\begin{aligned} f_\alpha f_\beta = n_\alpha n_\beta &\left(\frac{\gamma_\alpha + \gamma_\beta}{2\pi}\right)^{3/2} \left(\frac{\gamma_{\alpha\beta}}{2\pi}\right)^{3/2} \exp\left(-\frac{\gamma_\alpha + \gamma_\beta}{2}X^2\right) \\ \times \exp\left(-\frac{\gamma_{\alpha\beta}}{2}g^2\right) &\left[1 + (\gamma_\alpha \mathbf{w}_\alpha + \gamma_\beta \mathbf{w}_\beta)\cdot\mathbf{X} + \gamma_{\alpha\beta}(\mathbf{w}_\alpha - \mathbf{w}_\beta)\cdot\mathbf{g}\right], \end{aligned} \tag{3.1.8}$$

where

$$\gamma_{\alpha\beta} = \frac{\gamma_\alpha \gamma_\beta}{(\gamma_\alpha + \gamma_\beta)}. \tag{3.1.9}$$

Substituting (3.1.8) for Eqs. (3.1.5) and (3.1.6) and taking into account that $d\mathbf{c}_\alpha d\mathbf{c}_\beta = d\mathbf{X}d\mathbf{g}$, we can integrate the appropriate expressions over $\mathbf{X}$. In doing so, some integrals cancel because of the odd subintegral expressions over $\mathbf{X}$, and the integrals left give

$$\mathbf{R}_\alpha = -\sum_\beta n_\alpha \mu_{\alpha\beta} \tau_{\alpha\beta}^{-1} (\mathbf{w}_\alpha - \mathbf{w}_\beta), \tag{3.1.10}$$

$$Q_\alpha = -\sum_\beta 3kn_\alpha \left(\frac{\mu_{\alpha\beta}}{m_\alpha + m_\beta}\right) \tau_{\alpha\beta}^{-1} (T_\alpha - T_\beta). \tag{3.1.11}$$

The averaged momentum transfer collisional frequency of species α and β $\tau_{\alpha\beta}^{-1}$ in these expressions is defined as

$$\tau_{\alpha\beta}^{-1} = \frac{4}{3} n_\beta \langle v_{\alpha\beta}\rangle \overline{Q}_{\alpha\beta}, \qquad \langle v_{\alpha\beta}\rangle = \left(\frac{8}{\pi\gamma_{\alpha\beta}}\right)^{1/2},$$
$$\overline{Q}_{\alpha\beta} = \int_0^\infty \zeta^5 \exp(-\zeta)^2 Q_{\alpha\beta}^{(1)}(g)\, d\zeta, \tag{3.1.12}$$

where $\langle v_{\alpha\beta}\rangle$ is the mean relative speed of the particles, $\overline{Q}_{\alpha\beta}$ is the average momentum transfer cross section, $\zeta = (\gamma_{\alpha\beta}/2)^{1/2} g$.

For electron–atom collisions, at $\gamma_e \ll \gamma_\beta$ ($\beta \neq e$) the value $\tau_{e\beta}^{-1}$ is coincident with the effective collisional frequency of the electron commonly used in the kinetics of plasma [1,2]

$$\nu_{eff} = \frac{1}{3}\left(\frac{2}{\pi}\right)^{1/2} \gamma_e^{5/2} \int_0^\infty \nu_{e\beta}(v)\, v^4 \exp\left(-\frac{\gamma_e v^2}{2}\right) dv, \tag{3.1.13}$$

where

$$\nu_{e\beta}(v) = n_\beta v \int_\Omega \sigma_{e\beta}(v,\chi)(1-\cos\chi)\, d\Omega \tag{3.1.14}$$

is the electron momentum transfer collisional frequency which generally depends on velocity.

When the temperatures of all species in the plasma are the same, the average collisional frequency $\tau_{\alpha\beta}^{-1}$ is related to the integral $\Omega_{\alpha\beta}^{11}$, which is a special case of the familiar integrals of Chapman–Cowling ($\Omega^{lr}_{\alpha\beta}$-integrals), by a simple relationship

$$\tau^{-1}_{\alpha\beta} = \frac{16}{3} n_\beta \Omega^{11}_{\alpha\beta}. \tag{3.1.15}$$

For collisions of heavy particles (e.g. of ions and atoms or atoms and atoms) it is convenient to express $\tau^{-1}_{\alpha\beta}$ via the binary coefficient diffusion $[\mathfrak{D}_{\alpha\beta}]_1$. Since $[\mathfrak{D}_{\alpha\beta}]_1 = 3kT/16n\mu_{\alpha\beta}\,\Omega^{11}_{\alpha\beta}$ (the first Chapman–Cowling approximation of [3]), we have

$$\mu_{\alpha\beta}\tau^{-1}_{\alpha\beta} = \frac{n_\beta kT}{n\,[\mathfrak{D}_{\alpha\beta}]_1}. \tag{3.1.16}$$

A point of interest is the interaction of an ion and an atom of the same kind for which the main process is the resonant charge transfer. Unlike a purely elastic collision when the mean transfer of the momentum in the collision is proportional to $\mu_{\alpha\beta}\mathbf{g}\,(1-\cos\chi)$, a transfer of the whole momentum proportional to the ion mass m_i does occur in this case. In fact, to preserve the earlier used structure of the expressions for $\mathbf{R}_\alpha$ and Q_α with $\mu_{i\alpha} = m_i/2$, it would be quite sufficient to replace $\overline{Q}_{\alpha\beta}$ in the expression for $\tau^{-1}_{\alpha\beta}$ (3.1.12) with $\overline{Q}'_{\alpha\beta} = \overline{Q}_{\alpha\beta} + 2Q^*_{\alpha\beta}$, where $Q^*_{\alpha\beta}$ is the total effective cross section of the resonant charge transfer.

In Coulomb interactions of particles, by using the expression of the diffusion cross section (1.3.23) obtained by cutting off the integration with respect to the impact parameter at a distance of the order of the Debye length, we have

$$\tau^{-1}_{\alpha\beta} = \frac{16\pi^{1/2}}{3} n_\beta \left(\frac{\gamma_{\alpha\beta}}{2}\right)^{3/2} \left(\frac{e_\alpha e_\beta}{4\pi\varepsilon_0\mu_{\alpha\beta}}\right)^2 \ln\Lambda_{\alpha\beta}, \tag{3.1.17}$$

where $\Lambda_{\alpha\beta}$ is defined by Eq. (1.3.32).

The applicability of the expression (3.1.17) is limited, strictly speaking, by the requirement $\ln\Lambda \gg 1$. It was obtained on the basis of unified theory [4,5], where the shielding effect is taken into account more correctly, that the similar expression is valid for a more soft condition $\Lambda \gg 1$ if we substitute $\ln\Lambda - 1.37$ for $\ln\Lambda$.

With allowance for (2.2.8) the motion equation for species α (2.2.20) is written in the approximate form as

$$\rho_\alpha \frac{d_\alpha \mathbf{u}_\alpha}{dt} + \nabla p_\alpha - n_\alpha \mathbf{X}_\alpha - n_\alpha e_\alpha (\mathbf{E} + \mathbf{u}_\alpha \times \mathbf{B}) = \mathbf{R}_\alpha, \tag{3.1.18}$$

where a small difference between p_α and p^*_α is neglected. Note that on the left-hand side of Eq. (3.1.18) the full derivative $d_\alpha \mathbf{u}_\alpha/dt$ can be substituted for $d\mathbf{u}/dt$ which corresponds to neglecting the terms of the order of $d\mathbf{w}_\alpha/dt$ as compared with $\tau_{\alpha\beta}^{-1}\mathbf{w}_\alpha$ on the right-hand side of the equation. This is quite in agreement with the assumption made further as to the macroscopic parameters of the plasma weakly changing at distances and in time of the order of mean free path and mean time between collisions of particles. As a result, instead of Eq. (3.1.18) we get

$$n_\alpha \sum_\beta \mu_{\alpha\beta}\tau_{\alpha\beta}^{-1}(\mathbf{w}_\alpha - \mathbf{w}_\beta) = n_\alpha e_\alpha \left(\mathbf{E}' + \mathbf{w}_\alpha \times \mathbf{B}\right) + n_\alpha \mathbf{X}_\alpha - \mathbf{G}_\alpha, \tag{3.1.19}$$

where

$$\mathbf{G}_\alpha = \nabla p_\alpha + \rho_\alpha \frac{d\mathbf{u}}{dt}. \tag{3.1.20}$$

If the derivative $d\mathbf{u}/dt$ is excluded by means of the general motion equation of plasma (2.2.17), where $\pi_{rs} = 0$, then instead of (3.1.19) we have

$$n_\alpha \sum_\beta \mu_{\alpha\beta}\tau_{\alpha\beta}^{-1}(\mathbf{w}_\alpha - \mathbf{w}_\beta) = n_\alpha e_\alpha \left(\mathbf{E}' + \mathbf{w}_\alpha \times \mathbf{B}\right)$$
$$+ \left(n_\alpha \mathbf{X}_\alpha - \frac{\rho_\alpha}{\rho}\sum_\alpha n_\alpha \mathbf{X}_\alpha\right) - \left(\nabla p_\alpha - \frac{\rho_\alpha}{\rho}\nabla p\right) - \frac{\rho_\alpha}{\rho}\mathbf{j} \times \mathbf{B}. \tag{3.1.21}$$

The energy equation for species α, using (2.2.21) and eliminating $d\mathbf{u}/dt \approx d_\alpha \mathbf{u}_\alpha/dt$ with (3.1.18), in the same approximation can be rewritten as

$$\frac{3}{2} n_\alpha k \frac{d_\alpha T_\alpha}{dt} + p_\alpha \nabla \cdot \mathbf{u}_\alpha - \frac{5}{2}\nabla \cdot (p_\alpha \mathbf{w}_\alpha) + \mathbf{w}_\alpha \cdot \mathbf{R}_\alpha = Q_\alpha. \tag{3.1.22}$$

Here the terms squared relative to the diffusion velocity under the derivative sign are omitted.

Equations (3.1.18) and (3.1.22), along with the continuity equation (2.2.1), form a closed system of equations to define the parameters of $\rho_\alpha, \mathbf{u}_\alpha, T_\alpha$ in the approximation under study.

For a spatially homogeneous plasma and with the lack of diffusion transport of particles ($\mathbf{w}_\alpha = 0$), Eq. (3.1.22) describes the relaxation of temperature T_α to the equilibrium temperature T with the characteristic time of energy relaxation

$$\tau_{\alpha\beta}^E = \frac{(m_\alpha + m_\beta)^2}{2 m_\alpha m_\beta}\tau_{\alpha\beta}. \tag{3.1.23}$$

For the interaction of electrons and heavy particles (ions or atoms) $\tau_{e\beta}^{E} = (m_\beta/m_e)\,\tau_{e\beta}$. This time is substantially greater than the characteristic time of energy exchange inside both the electrons and any heavy species of the plasma. As was described earlier, this lets us introduce various temperatures for electron and heavy species and, if the characteristic time of energy exchange markedly exceeds the characteristic macroscopic time in the problem under investigation, to consider this difference to be nearly constant in time. The difference between the electron temperature and the temperature of the other species can also be due to the electron current in the plasma (or, to be more precise, due to the energy which is transferred by the electrical field to electrons in the unit plasma volume).

Indeed, if Eq. (2.2.21) is written for electron species ($\alpha = e$), then for the stationary homogeneous plasma we have

$$\mathbf{j}_e \cdot \mathbf{E}' = \frac{3}{2} k n_e \sum_\beta \left(\tau_{e\beta}^{E}\right)^{-1} (T_e - T_\beta), \tag{3.1.24}$$

$$\mathbf{j}_e = -\, n_e e \mathbf{w}_e. \tag{3.1.25}$$

3.2 Generalized Ohm's law for a partially ionized three-component plasma

Equations (3.1.19) or (3.1.21) can be solved with respect to diffusion velocities of species if the condition (1.1.4) is taken into account. In practice, Eq. (3.1.19) becomes usable if $d\mathbf{u}/dt$ is given (as in the problems of plasma rotation in cross fields $\mathbf{E}$ and $\mathbf{B}$), or if the inertial terms can be neglected. In the general case, it is more common to solve Eq. (3.1.21), having previously substituted $\mathbf{j} = \sum n_\alpha e_\alpha \mathbf{w}_\alpha$ into them. The formal solution of these equations and derivation of the generalized Ohm's law for the case of arbitrary multicomponent plasma will be dealt with in the next sections. Here we intend to consider a traditional derivation of the generalized Ohm's law for a special case of a three-component partially ionized gas [6,7]. The limiting transition to the case of a two-component fully ionized gas is ensured when the density of the neutral component is assumed to be zero.

Let the indices e, i, a correspond to electrons, ions and neutral particles (atoms or molecules), the charges of electrons and ions be equal to $e_e = -e$ and $e_i = Ze$, respectively. The masses of ions and neutral particles are supposed to be equal ($m_i = m_a = m$).

Note that, thanks to the condition of quasineutrality,

$$\sum_{\alpha} n_\alpha e_\alpha = 0 \tag{3.2.1}$$

the expression for the electric current density (2.2.14) can be generally written as

$$\mathbf{j} = \sum_{\beta \neq k} n_\beta e_\beta \left(\mathbf{w}_\beta - \mathbf{w}_k \right), \tag{3.2.2}$$

where k is the index of an arbitrarily chosen species of the plasma. For the plasma with a single ion species $(n_i Z = n_e)$,

$$\mathbf{j} = n_e e \left(\mathbf{w}_i - \mathbf{w}_e \right). \tag{3.2.3}$$

Because of the condition $\sum \rho_\alpha \mathbf{w}_\alpha = 0$, the diffusion velocity $\mathbf{w}_\alpha$ can be also expressed in terms of difference of the species velocities

$$\mathbf{w}_\alpha = \sum_{\beta \neq \alpha} \frac{\rho_\beta}{\rho} \left(\mathbf{w}_\alpha - \mathbf{w}_\beta \right). \tag{3.2.4}$$

Let us use Eq. (3.1.21) and neglect the forces of non-electromagnetic character for simplicity. The equation for the electron component $(\alpha = e)$ can be simplified due to the small parameter $\delta = m_e/m_i$ and, in particular, the terms proportional to ρ_e/ρ can be omitted. As a result,

$$n_e m_e \left[\tau_{ei}^{-1} \left(\mathbf{w}_e - \mathbf{w}_i \right) + \tau_{ea}^{-1} \left(\mathbf{w}_e - \mathbf{w}_a \right) \right] = - n_e e \mathbf{E}^* - n_e m_e \omega_e \mathbf{w}_e \times \mathbf{k}, \tag{3.2.5}$$

where

$$\omega_e = \frac{e}{m_e} |\mathbf{B}| \tag{3.2.6}$$

has the physical meaning of the cyclotron frequency for electrons, $\mathbf{k} = \mathbf{B}/B$ and the field $\mathbf{E}^*$ is defined by

$$\mathbf{E}^* = \mathbf{E}' + \frac{1}{n_e e} \nabla p_e. \tag{3.2.7}$$

Introducing the variables $\mathbf{w} = \mathbf{w}_e - \mathbf{w}_i$ and $\mathbf{s} = \mathbf{w}_i - \mathbf{w}_a$ we shall re-write Eq. (3.2.5) as

$$n_e m_e \left[\tau_0^{-1} \mathbf{w} + \tau_{ea}^{-1} \mathbf{s} \right] = - n_e e \mathbf{E}^* - n_e m_e \omega_e \mathbf{w}_e \times \mathbf{k}, \tag{3.2.8}$$

where τ_0^{-1} is defined by

$$\tau_0^{-1} = \tau_{ei}^{-1} + \tau_{ea}^{-1}. \tag{3.2.9}$$

Let us introduce the degree of plasma ionization $\alpha = n_i/(n_i + n_a)$. Then, neglecting the terms of the order of $\delta = m_e/m_i$, on the basis of the relationship (3.2.4), we have

$$\mathbf{w}_e = \mathbf{w} + (1-\alpha)\,\mathbf{s}. \tag{3.2.10}$$

Ultimately, to define the current density $\mathbf{j} = -\,n_e e \mathbf{w}$ by means of Eq. (3.2.8), it would be necessary to have only one additional relationship connecting the "ion-slip" velocity $\mathbf{s}$ and the value $\mathbf{w}$. To this end, Eq. (3.1.21) written for the neutral component ($\alpha = a$) will be used. Considering that $\rho_a/\rho \approx 1-\alpha$, it can be presented as

$$\begin{aligned} & -n_e m_e \tau_{ea}^{-1}\mathbf{w} - \left(\frac{n_i m_i}{2}\tau_{ia}^{-1} + n_e m_e \tau_{ea}^{-1}\right)\mathbf{s} \\ = & -\left[\nabla p_a - (1-\alpha)\,\nabla p\right] + (1-\alpha)\,n_e m_e \omega_e \mathbf{w}\times\mathbf{k}. \end{aligned}$$

Taking $\mathbf{s}$ from it, we find

$$\mathbf{s} = -\frac{\varepsilon}{1+\varepsilon}\left\{\mathbf{w} + (1-\alpha)\,\omega_e\tau_{ea}\mathbf{w}_e\times\mathbf{k}\right\} + \frac{2\tau_{ia}}{n_i m\,(1+\varepsilon)}\mathbf{P}, \tag{3.2.11}$$

where

$$\mathbf{P} = \alpha\nabla p - \nabla\,(p_e + p_i)$$

and

$$\varepsilon = \frac{2n_e m_e \tau_{ia}}{n_i m \tau_{ea}} \sim Z\,(m_e T/m T_e)^{1/2} \ll 1.$$

This estimation assumes that $\overline{Q}_{ea} \sim \overline{Q}_{ia}$.

Substituting $\mathbf{s}$ for Eq. (3.2.8) on account of (3.2.10), we can see that in the expression obtained the terms proportional to the small value of ε could be neglected when they are not multiplied by ω_e which can take arbitrary values. The resulting expression for the generalized Ohm's law will look like

$$\sigma_0\left[\mathbf{E}^* + \frac{\delta_0}{n_e e\,(1-\alpha)}\mathbf{P}\times\mathbf{k}\right] = \mathbf{j} + \beta_0\mathbf{j}\times\mathbf{k} + s\mathbf{k}\times(\mathbf{j}\times\mathbf{k}), \tag{3.2.12}$$

where

$$\sigma_0 = \frac{n_e e^2}{m_e}\tau_0 \tag{3.2.13}$$

corresponds to the electrical conductivity of the plasma in the absence of the magnetic field, $\beta_0 = \omega_e \tau_0$ is the Hall parameter for electrons and

$$s = (1-\alpha)^2 \beta_0 \beta_i = \delta_0 \beta_0^2 \tag{3.2.14}$$

is the ion-slip factor. Here $\beta_i = \omega_i \tau_{ia}$ is the Hall parameter for ion-neutral collisions, where

$$\omega_i = \frac{Ze}{\mu_{ia}} |\mathbf{B}| . \tag{3.2.15}$$

The projection of Eq. (3.2.12) in the direction of the magnetic field is obtained by forming a scalar product of its left- and right-hand sides with the vector $\mathbf{k}$. The resulting parallel constituent of the current density is

$$\mathbf{j}_{\parallel} = \sigma_{\parallel} \mathbf{E}^*_{\parallel}, \qquad \sigma_{\parallel} = \sigma_0,$$

where $\mathbf{E}^*_{\parallel} = \mathbf{k} \cdot (\mathbf{E}^* \cdot \mathbf{k})$.

The full current density is defined as

$$\mathbf{j} = \mathbf{j}_{\parallel} + \mathbf{j}_{\perp}.$$

The constituents of the current perpendicular to $\mathbf{B}$ are easily found if one forms a vector product of Eq. (3.2.12) with the vector $\mathbf{k}$. As a result, the expression for $\mathbf{j}$ takes the form of

$$\mathbf{j} = \sigma_{\parallel} \mathbf{E}^*_{\parallel} + \sigma_{\perp} \mathbf{E}^*_{\perp} + \sigma_{\wedge} \mathbf{k} \times \mathbf{E}^* + \beta_{\perp} \mathbf{P}_{\perp} + \beta_{\wedge} \mathbf{k} \times \mathbf{P}, \tag{3.2.16}$$

where $\mathbf{E}^*_{\perp} = \mathbf{k} \times (\mathbf{E}^* \times \mathbf{k})$, $\mathbf{P}_{\perp} = \mathbf{k} \times (\mathbf{P} \times \mathbf{k})$ and $\sigma_{\perp}$, $\sigma_{\wedge}$ are the perpendicular and Hall electrical conductivities of the plasma which are defined by the expressions

$$\sigma_{\perp} = \sigma_0 \frac{1+s}{(1+s)^2 + \beta_0^2}, \qquad \sigma_{\wedge} = \sigma_0 \frac{\beta_0}{(1+s)^2 + \beta_0^2}. \tag{3.2.17}$$

In a fully ionized plasma $\delta_0 = 0$ and $\tau_0 = \tau_{ei}$. Then

$$\sigma_{\parallel} = \sigma_0 = \frac{n_e e^2}{m_e} \tau_{ei}, \qquad \sigma_{\perp} = \sigma_0 \frac{1}{1+\omega_e^2 \tau_{ei}^2}, \qquad \sigma_{\wedge} = \sigma_0 \frac{\omega_e \tau_{ei}}{1+\omega_e^2 \tau_{ei}^2}. \tag{3.2.18}$$

In accordance with the definition of (3.1.17), when the condition $\gamma_e \ll \gamma_i$ or $(m_e T_i / m_i T_e)^{1/2} \ll 1$ is met, we have

$$\tau_{ei}^{-1} = \frac{1}{3} \frac{Z^2 e^4}{m_e^{1/2} \varepsilon_0 (2\pi k T_e)^{3/2}} n_i \ln \Lambda_{ei}. \tag{3.2.19}$$

The formal need to take into account the terms with the vector $\mathbf{P}$ in Eq. (3.2.16) arises only in the case of a partially ionized plasma, where

$$\beta_\perp = \sigma_\perp \frac{1}{n_e e (1-\alpha)} \frac{s}{1+s},$$
$$\beta_\wedge = -\sigma_\perp \frac{s}{n_e e (1-\alpha)}. \tag{3.2.20}$$

In practice (cf. the relevant estimates in [7]) these terms can be neglected in comparison with $\nabla p_e / n_e e$ in the expression for $\mathbf{E}^*$.

Note that for a partially ionized plasma satisfying the requirement $\tau_{ea}^{-1} \geq \tau_{ei}^{-1}$

$$\delta_0 = (1-\alpha)^2 \frac{\beta_0}{\beta_i} \sim \varepsilon \ll 1,$$

therefore, the effect of ion-slip becomes telltale on the parameters of $\sigma_\perp$ and $\sigma_\wedge$ much later than the influence of the Hall parameter for electrons.

3.3 The formal solution of the equations of multicomponent diffusion in a plasma

To derive the generalized Ohm's law for the case of the arbitrary N-component plasma (including the plasma with several species of ions) requires the general solution of a set of equations for multicomponent diffusion (3.1.21). These equations are linearly dependent, hence, the actual number of equations necessary for defining $\mathbf{w}_\alpha$ is a unit less than the number of the components.

Substitute the expression for $\mathbf{j}$ in the form of (3.2.2) into the right-hand side of Eqs. (3.1.21), and the expression for $\mathbf{w}_\alpha$ (3.2.4) into the term with the magnetic field $\mathbf{w}_\alpha \times \mathbf{B}$. Then, introducing the variable $\mathbf{x}_\beta = \mathbf{w}_\beta - \mathbf{w}_k$, we can represent Eqs. (3.1.21) as

$$\sum_{\beta \neq k}^{N} c_{\alpha\beta} \mathbf{x}_\beta = \boldsymbol{\varepsilon}_\alpha + \sum_{\beta \neq k}^{N} b_{\alpha\beta} \mathbf{x}_\beta \times \mathbf{k}, \tag{3.3.1}$$

where

$$\varepsilon_\alpha = n_\alpha e_\alpha \mathbf{E}' - \left(\nabla p_\alpha - \frac{\rho_\alpha}{\rho}\nabla p\right) + \left(n_\alpha \mathbf{X}_\alpha - \frac{\rho_\alpha}{\rho}\sum_\alpha n_\alpha \mathbf{X}_\alpha,\right) \tag{3.3.2}$$

and factors $c_{\alpha\beta}$ and $b_{\alpha\beta}$ are defined as

$$\begin{gathered} c_{\alpha\alpha} = \sum_{\gamma\neq\alpha}^{N} \lambda_{\alpha\gamma}, \qquad c_{\alpha\beta} = -\lambda_{\alpha\beta}, \\ b_{\alpha\alpha} = n_\alpha e_\alpha \left(1 - \frac{2\rho_\alpha}{\rho}\right)|\mathbf{B}|, \\ b_{\alpha\beta} = -\frac{n_\alpha n_\beta}{\rho}\left(e_\alpha m_\beta - e_\beta m_\alpha\right)|\mathbf{B}|. \end{gathered} \tag{3.3.3}$$

and

$$\lambda_{\alpha\beta} = \lambda_{\beta\alpha} = n_\alpha \mu_{\alpha\beta} \tau_{\alpha\beta}^{-1} = n_\beta \mu_{\alpha\beta} \tau_{\beta\alpha}^{-1}. \tag{3.3.4}$$

Let us form the vector product of the left- and right-hand sides of (3.3.1) and the vector $\mathbf{k}$. As a result, we get

$$\sum_{\beta\neq k}^{N} c_{\alpha\beta}\mathbf{x}_\beta \times \mathbf{k} = \varepsilon_\alpha \times \mathbf{k} - \sum_{\beta\neq k}^{N} b_{\alpha\beta}\left(\mathbf{x}_\beta - \mathbf{x}_{\beta||}\right), \tag{3.3.5}$$

where $\mathbf{x}_{\beta||} = \mathbf{k}\cdot(\mathbf{x}_\beta \cdot \mathbf{k})$.

The equation for $\mathbf{x}_{\beta||}$ is obtained if we form the double scalar product of Eqs. (3.3.1) and the vector $\mathbf{k}$

$$\sum_{\beta\neq k}^{N} c_{\alpha\beta}\mathbf{x}_{\beta||} = \varepsilon_{\alpha||}, \qquad \varepsilon_{\alpha||} = \mathbf{k}\cdot(\varepsilon_\alpha \cdot \mathbf{k}). \tag{3.3.6}$$

The solution of Eqs. (3.3.6) is

$$\mathbf{x}_{\alpha||} = \left|c'\right|^{-1} \sum_{\beta\neq k}^{N} \left|c'\right|_{\beta\alpha} \varepsilon_{\beta||}, \tag{3.3.7}$$

where $\left|c'\right|$ is the determinant of the matrix of the elements $c_{\alpha\beta}$ where every kth line and kth column are crossed out and $\left|c'\right|_{\beta\alpha}$ is the cofactor of the element $\beta\alpha$ of the determinant.

Solving Eqs. (3.3.5) with respect to $\mathbf{x}_\alpha \times \mathbf{k}$, we obtain

$$\mathbf{x}_\alpha \times \mathbf{k} = \left|c'\right|^{-1}\left[\sum_{\beta\neq k}^{N} \left|c'\right|_{\beta\alpha} \varepsilon_\beta \times \mathbf{k} - \sum_{\beta,\gamma\neq k}^{N} \left|c'\right|_{\beta\alpha} b_{\beta\gamma}\left(\mathbf{x}_\gamma - \mathbf{x}_{\gamma||}\right)\right].$$

Substituting the solution into Eq. (3.3.1), replacing the terms with $\mathbf{x}_\beta$ into the left-hand side and applying the expression for $\mathbf{x}_{\gamma||}$, we get

$$\sum_{\beta\neq k}^{N} a_{\alpha\beta}\mathbf{x}_\beta = \boldsymbol{\varepsilon}_\alpha + \sum_{\beta\neq k}^{N} k_{\alpha\beta}\boldsymbol{\varepsilon}_\beta \times \mathbf{k} + \sum_{\beta\neq k}^{N} \chi_{\alpha\beta}\boldsymbol{\varepsilon}_{\beta||}, \tag{3.3.8}$$

where

$$\begin{aligned} a_{\alpha\beta} &= c_{\alpha\beta} + \left|c'\right|^{-1} \sum_{\gamma,\delta\neq k}^{N} \left|c'\right|_{\gamma\delta} b_{\alpha\delta} b_{\gamma\beta}, \\ k_{\alpha\beta} &= \left|c'\right|^{-1} \sum_{\gamma\neq k}^{N} \left|c'\right|_{\beta\gamma} b_{\alpha\gamma}, \\ \chi_{\alpha\beta} &= \left|c'\right|^{-2} \sum_{\gamma,\delta,\varepsilon\neq k}^{N} \left|c'\right|_{\gamma\varepsilon} \left|c'\right|_{\beta\delta} b_{\alpha\varepsilon} b_{\gamma\delta}. \end{aligned} \tag{3.3.9}$$

The solution of Eqs. (3.3.8) is written in the form

$$\begin{aligned} \mathbf{x}_\alpha \;=\; & \left|a'\right|^{-1} \left(\sum_{\beta\neq k}^{N} \left|a'\right|_{\beta\alpha} \boldsymbol{\varepsilon}_\beta \right. \\ & \left. + \sum_{\beta,\gamma\neq k}^{N} \left|a'\right|_{\beta\alpha} \kappa_{\beta\gamma}\boldsymbol{\varepsilon}_\gamma \times \mathbf{k} + \sum_{\beta,\gamma\neq k}^{N} \left|a'\right|_{\beta\alpha} \chi_{\beta\gamma}\boldsymbol{\varepsilon}_{\gamma||} \right). \end{aligned}$$

Now let us use the relationship which follows from Eqs. (3.3.8) if the double scalar products of both its left- and right-hand sides are formed with the vector $\mathbf{k}$

$$\left|a'\right|^{-1} \left(\sum_{\beta\neq k}^{N} \left|a'\right|_{\beta\alpha} \boldsymbol{\varepsilon}_{\beta||} + \sum_{\beta,\gamma\neq k}^{N} \left|a'\right|_{\beta\alpha} \chi_{\beta\gamma}\boldsymbol{\varepsilon}_{\gamma||} \right) = \left|c'\right|^{-1} \sum_{\beta\neq k}^{N} \left|c'\right|_{\beta\alpha} \boldsymbol{\varepsilon}_{\beta||}.$$

Then, taking into account that $\boldsymbol{\varepsilon}_\beta = \boldsymbol{\varepsilon}_{\beta||} + \boldsymbol{\varepsilon}_{\beta\perp}$, where $\boldsymbol{\varepsilon}_{\beta\perp} = \mathbf{k}\times(\boldsymbol{\varepsilon}_\beta \times \mathbf{k})$, we arrive at the final expression

$$\begin{aligned} \mathbf{x}_\alpha \;=\; & \left|c'\right|^{-1} \sum_{\beta\neq k}^{N} \left|c'\right|_{\beta\alpha} \boldsymbol{\varepsilon}_{\beta||} + \left|a'\right|^{-1} \sum_{\beta\neq k}^{N} \left|a'\right|_{\beta\alpha} \boldsymbol{\varepsilon}_{\beta\perp} \\ & + \left|a'\right|^{-1} \sum_{\beta,\gamma\neq k}^{N} \kappa_{\gamma\beta} \left|a'\right|_{\gamma\alpha} \boldsymbol{\varepsilon}_\beta \times \mathbf{k}. \end{aligned} \tag{3.3.10}$$

To define the particular diffusion velocities of species α, we may use the relation (3.2.4) from which it follows that

$$\mathbf{w}_\alpha = \mathbf{x}_\alpha - \sum_{\gamma=1}^{N} \left(\frac{\rho_\gamma}{\rho}\right) \mathbf{x}_\gamma. \tag{3.3.11}$$

We shall not analyse the expressions for the diffusion velocities in the general case in detail and apply the solutions obtained to derive the generalized Ohm's law for the arbitrary multicomponent plasma. Some special problems (e.g. diffusion in a weakly ionized gas and diffusion across a strong magnetic field in the fully ionized multicomponent plasma) will be considered at length in this chapter later.

3.4 Generalized Ohm's law for a multicomponent plasma

Let us make use of the fact that, following from the relation (3.2.2),

$$\mathbf{j} = \sum_{\alpha \neq k}^{N} n_\alpha e_\alpha \mathbf{x}_\alpha. \tag{3.4.1}$$

Substituting the solutions derived above for $\mathbf{x}_\alpha$ (3.3.10) for (3.4.1), we come to the expression of the generalized Ohm's law, which correlates the current density $\mathbf{j}$ and components of vector $\boldsymbol{\varepsilon}_\beta$ parallel and perpendicular with respect to the vector of the magnetic field $\mathbf{B}$. Without the pressure gradients of species ($\nabla p_\alpha = \nabla p = 0$) and the forces of non-electromagnetic nature, when $\boldsymbol{\varepsilon}_\beta = n_\beta e_\beta \mathbf{E}'$, the expression for $\mathbf{j}$ can be written as

$$\mathbf{j} = \sigma_{||} \mathbf{E}'_{||} + \sigma_\perp \mathbf{E}'_\perp + \sigma_\wedge \mathbf{k} \times \mathbf{E}'. \tag{3.4.2}$$

The parallel, perpendicular and Hall electrical conductivities of the plasma are defined by

$$\begin{aligned}
\sigma_{||} &= \left|c'\right|^{-1} \sum_{\alpha,\beta \neq k}^{N} n_\alpha e_\alpha n_\beta e_\beta \left|c'\right|_{\beta\alpha}, \\
\sigma_\perp &= \left|a'\right|^{-1} \sum_{\alpha,\beta \neq k}^{N} n_\alpha e_\alpha n_\beta e_\beta \left|a'\right|_{\beta\alpha}, \\
\sigma_\wedge &= -\left|a'\right|^{-1} \sum_{\alpha,\beta,\gamma \neq k}^{N} n_\alpha e_\alpha n_\beta e_\beta \kappa_{\beta\gamma} \left|a'\right|_{\gamma\alpha}.
\end{aligned} \tag{3.4.3}$$

For a two-component fully ionized plasma ($e, i = 1, 2$, $e_1 = -e$, $e_2 = Ze$) it follows from the expressions (3.4.3) that

$$\sigma_{||} = \sigma_0 = \frac{(n_1 e_1)^2}{c_{11}}, \qquad \sigma_\perp = \frac{(n_1 e_1)^2}{a_{11}},$$

$$\sigma_\wedge = -\frac{(n_1 e_1)^2 \kappa_{11}}{a_{11}}.$$

It is easily seen that

$$c_{11} = n_e m_e \tau_{ei}^{-1}, \qquad a_{11} = c_{11}\left(1 + \omega_e^2 \tau_{ei}^2\right),$$

$$\kappa_{11} = \frac{b_{11}}{c_{11}} = -\omega_e \tau_{ei},$$

and thus we obtain the expressions for $\sigma_{||}, \sigma_\perp$ and $\sigma_\wedge$ (3.2.18) derived in Section 3.2.

When treating the plasma of a more complex structure, the expressions (3.4.3) can be simplified to a certain extent due to the small parameter $\delta = m_e/m_\beta$ ($\beta \neq e$). For the case of a partially ionized three-component plasma ($e, i, a = 1, 2, 3$, $e_1 = -e$, $e_2 = Ze$, $e_3 = 0$, $m_2 = m_3 = m$), assuming that $k = 2$ and neglecting the terms $c_{13}/c_{33} \sim \delta^{1/2}$, we get the results familiar as the expressions (3.2.17).

Another example can be provided by the case of a fully ionized plasma with two species of ions ($e_1 = -e$, $e_2 = Z_2 e$, $e_3 = Z_3 e$, $m_2 \neq m_3$). Simplification analogous to that used above results in

$$\sigma_{||} = \sigma_0 = \frac{n_e e^2}{m_e}\tau_0,$$

$$\sigma_\perp = \sigma_0 \frac{1+b}{(1+a)^2 + \beta_0^2}, \tag{3.4.4}$$

$$\sigma_\wedge = \sigma_0 \frac{(1+f)\beta_0}{(1+a)^2 + \beta_0^2},$$

where

$$\tau_0^{-1} = \tau_{e2}^{-1} + \tau_{e3}^{-1}, \qquad \beta_0 = \omega_e \tau_0,$$

$$a = n_e \frac{\left(\rho_2^2 n_3 Z_3 + \rho_3^2 n_2 Z_2\right) e^2 B^2}{\rho^2 c_{11} c_{33}},$$

$$b = \frac{(\rho_2 n_3 Z_3 - \rho_3 n_2 Z_2)^2 e^2 B^2}{\rho^2 c_{11} c_{33}},$$

$$f = \frac{n_2 Z_2 n_3 Z_3}{n_e} \frac{\left(\rho_2^2 n_3 Z_3 + \rho_3^2 n_2 Z_2\right) e^2 B^2}{\rho^2 c_{33}^2}, \tag{3.4.5}$$

$$c_{11} = n_e m_e \tau_0^{-1}, \qquad c_{33} = n_3 \mu_{32} \tau_{32}^{-1} + n_e m_e \tau_{e3}^{-1}.$$

At $Z_2 = Z_3 = 1$, these results agree with the expressions derived in a different way in [8,9].

In the case of an arbitrary multicomponent plasma, the calculation of electrical conductivity turns out to be more complicated, though the application of the small parameter $\delta = m_e/m_\beta$ does help make such simplifications. In particular, for the longitudinal electrical conductivity if the terms $\sim c_{e\beta}/c_{\alpha\beta}$ $(\alpha, \beta \neq e)$, of the order of $\delta^{1/2}$ compared to unity, are neglected when calculating the determinants, we have

$$\sigma_{||} = \sigma_0 = \frac{n_e e^2}{m_e} \tau_0, \tag{3.4.6}$$

$$\tau_0^{-1} = \sum_\beta \tau_{e\beta}^{-1}, \tag{3.4.7}$$

where β is related either to ions or atoms (molecules) of any species.

From the analysis of (3.2.17) and (3.4.4) it is evident that on condition that $s \ll 1$ and $s_{ik} \ll 1$, where $s = (\rho_a/\rho)^2 \beta_0 \beta_{ia}$ and $s_{ik} \sim (\rho_k/\rho)^2 \beta_0 \beta_{ik}$ ($i \neq k$ are the indices of ion species) the values of the perpendicular and Hall electrical conductivities cease to depend on the ion-neutral or ion–ion frequency of collisions. This implies that under certain conditions the current density $\mathbf{j}$ is determined by electron transport alone, i.e. it can be defined from the independent equations (3.1.21) written for a singular electron species of the plasma ($\alpha = e$) whereas the velocities of the heavy species are taken to be approximately equal ($\mathbf{w}_\beta = 0$). Such on approximation is sometimes called "quasi-two-species plasma approximation". Neglecting the terms proportional to (ρ_e/ρ) on the right-hand side and omitting the forces of non-electromagnetic nature for simplicity, we obtain the equation for electrons in the following form:

$$n_e m_e \left(\tau_0^{-1} \mathbf{w}_e + \omega_e \mathbf{w}_e \times \mathbf{k}\right) = - n_e e \mathbf{E}^*. \tag{3.4.8}$$

where $\mathbf{E}^*$ is determined by (3.2.7).

Since $\mathbf{j} = - n_e e \mathbf{w}_e$, then when solving Eq. (3.4.8), we can find

$$\mathbf{j} = \sigma_{||} \mathbf{E}_{||}^* + \sigma_\perp \mathbf{E}_\perp^* + \sigma_\wedge \mathbf{k} \times \mathbf{E}^*, \tag{3.4.9}$$

where $\sigma_{||}, \sigma_{\perp}$ and $\sigma_{\wedge}$ for an arbitrary multicomponent plasma is defined by the expressions analogous to the case of the two-component plasma (Eq. 3.2.18) with the simultaneous replacement of τ_{ei} by τ_0 defined by (3.4.7).

Let us consider in greater detail the expression for the longitudinal electrical conductivity of a multicomponent plasma $\sigma_{||} = \sigma_0$.

If the condition $\gamma_e \ll \gamma_\beta$ or $(m_e T_\beta / m_\beta T_e)^{1/2} \ll 1$ is fulfilled, then, following Eqs. (3.4.7) and (3.1.12), we have

$$\tau_0^{-1} = \frac{16}{3}\left(\frac{kT_e}{2\pi m_e}\right)^{1/2} \sum_{\beta \neq e} n_\beta \overline{Q}_{e\beta}. \tag{3.4.10}$$

The resulting expression for σ_0 can be written as

$$\sigma_0 = 0.471 \frac{n_e e^2}{(m_e k T_e)^{1/2}} \frac{1}{\sum\limits_{\beta \neq e} n_\beta \overline{Q}_{e\beta}}. \tag{3.4.11}$$

For electron-neutral interaction, $\overline{Q}_{e\beta}$ is defined by expression (3.1.12) and can be found if the differential effective cross section of elastic scattering of the electron by an atom (or molecule) is known. For the Coloumb interaction of the electron and the ions with $Z_i e$ charge, the use of the expression (3.1.17) gives rise to

$$\tau_0^{-1} = \frac{1}{3} \frac{e^4}{m_e^{1/2} \varepsilon_0^2 (2\pi k T_e)^{3/2}} \sum_i n_i Z_i^2 \ln \Lambda_{ei}, \tag{3.4.12}$$

where the summation includes the ion species of the plasma. When the difference of ions charges is not so pronounced, then, taking into account the weaker dependence upon Z_i under the sign of the logarithm it is quite justifiable to take approximate $\ln \Lambda_{ei} = \ln \Lambda$, where Λ is determined at a mean value $Z = Z_{eff}$. Then

$$\begin{aligned} \Lambda &= \frac{12\pi\varepsilon_0^{3/2} kT}{Z_{eff} e^2} \left[\frac{kT}{n_e e^2 (1 + Z_{eff})}\right]^{1/2} \quad \text{for} \quad T_e = T, \\ \Lambda &= \frac{12\pi\varepsilon_0^{3/2} kT_e}{Z_{eff} e^2} \left[\frac{kT}{n_e e^2 Z_{eff}}\right]^{1/2} \quad \text{for} \quad T_e \gg T. \end{aligned} \tag{3.4.13}$$

In this case, the expression of the electrical conductivity of a fully ionized multicomponent plasma is written as

$$\sigma_0 = 0.298 \frac{(kT_e)^{3/2}}{m_e^{1/2}} \frac{(4\pi\varepsilon_0)^2}{Z_{eff} e^2 \ln \Lambda}, \tag{3.4.14}$$

where

$$Z_{eff} = \frac{\sum_i n_i Z_i^2}{\sum n_i Z_i} = \sum_i \frac{n_i}{n_e} Z_i^2. \tag{3.4.15}$$

The value Z_{eff} corresponds to a certain effective multiplicity charge of the ion species.

In terms of accuracy the expressions (3.4.11) and (3.4.14) agree with the so called first Chapman–Cowling approximation which corresponds to taking into consideration only one Sonine polynomials in the expansion of the nonequilibrium correction to the distribution function [3]. In the case of a weakly ionized gas for the model of particles as hard elastic spheres ($Q_{e\beta}^{(1)}$ = const), the numerical coefficient in (3.4.11) differs from the exact value obtained in the Lorentz approximation by 13% (0.471 instead of 0.532). If we take into account the next approximation, the difference is reduced to 4%. For a fully ionized plasma the first approximation is considerably less accurate, although the second approximation, by increasing the numerical coefficient at $Z_i = 1$ by nearly two, proves to be practically close to the exact result [10]. In the general case of a multicomponent plasma when the dependence of σ_0 on other parameters still remains, the dependence of σ_0 on Z_{eff} turns out to be more intricate.

More accurate expressions for both the longitudinal and transverse components of the electrical conductivity of plasma will be considered in the chapters to follow.

3.5 Diffusion in a weakly ionized gas

Let us investigate a weakly ionized gas mixture, when the degree of ionization is so small that we can neglect the interaction of charged particles compared with their interaction with neutral particles. Estimation of the ionization level required in this case can be made by treating the relative contribution of the interaction into the electrical conductivity of plasma (3.4.10). Thus, for a conventional three-component weakly ionized plasma an obvious condition of weak ionization is the requirement

$$\tau_{ei}^{-1} \ll \tau_{ea}^{-1} \qquad \text{or} \qquad n_i \overline{Q}_{ei} \ll n_a \overline{Q}_{ea}. \tag{3.5.1}$$

Independent diffusion of charged particles of various species in a weakly ionized gas mixture can be treated as a process relative to which the neutral species act as the medium of common hydrodynamic velocity approximately coincident with the mass-average velocity of the mixture. This lets us assume that the macroscopic velocities of neutral species $\mathbf{u}_k$ are equal to $\mathbf{u}$ (i.e. $\mathbf{w}_k = 0$) in the diffusion equations (3.1.21) written for charged particles. It is natural that in the set of diffusion equations for the neutral species proper the relative velocity of the neutral particles of various species be conserved without any assumptions.

We shall consider a mixture with an arbitrary number of ion species. With allowance for assumptions made earlier, Eq. (3.1.21) for electrons and ions ($\alpha = e, i_1, i_2, ...i_s$) acquires the form

$$n_\alpha \mathbf{w}_\alpha \sum_{\beta=1}^{k} \mu_{\alpha\beta}\tau_{\alpha\beta}^{-1} = n_\alpha e_\alpha \mathbf{E}' + n_\alpha e_\alpha \mathbf{w}_\alpha \times \mathbf{B} - \mathbf{G}_\alpha, \qquad (3.5.2)$$

where

$$\mathbf{G}_\alpha = \nabla p_\alpha + \rho_\alpha \frac{d\mathbf{u}}{dt} = \nabla p_\alpha - \frac{\rho_\alpha}{\rho}\left(\nabla p - \mathbf{j} \times \mathbf{B}\right), \qquad (3.5.3)$$

and summation is carried out with respect to the indices of the neutral species from 1 to k.

Solving Eqs. (3.5.2) with respect to $\mathbf{w}_\alpha$, we get

$$\begin{aligned} \mathbf{w}_\alpha &= Z_\alpha \left(b_{\alpha\|}\mathbf{E}'_{\|} + b_{\alpha\perp}\mathbf{E}'_{\perp} + b_{\alpha\wedge}\mathbf{E}' \times \mathbf{k}\right) \\ &\quad - p_\alpha^{-1}\left(D_{\alpha\|}\mathbf{G}_{\alpha\|} + D_{\alpha\perp}\mathbf{G}_{\alpha\perp} + D_{\alpha\wedge}\mathbf{G}_\alpha \times \mathbf{k}\right), \qquad (3.5.4) \end{aligned}$$

where

$$b_{\alpha\|} = \frac{e}{\sum\limits_{\beta=1}^{k} \mu_{\alpha\beta}\tau_{\alpha\beta}^{-1}}, \qquad b_{\alpha\perp} = \frac{b_{\alpha\|}}{1+\beta_\alpha^2},$$
$$b_{\alpha\wedge} = \frac{b_{\alpha\|}\beta_\alpha}{1+\beta_\alpha^2}, \qquad \beta_\alpha = Z_\alpha b_{\alpha\|}\left|\mathbf{B}\right|. \qquad (3.5.5)$$

In this expression, $Z_\alpha = e_\alpha/e$ is the charge multiplicity (charge number) of particles; for electrons $Z_e = -1$. In the absence of the magnetic field and for the flows established ($d\mathbf{u}/dt = 0$)

$$\mathbf{w}_\alpha = Z_\alpha b_\alpha \mathbf{E} - D_\alpha \nabla \ln p_\alpha. \qquad (3.5.6)$$

The diffusion coefficient D_α is related to the mobility coefficient b_α by the Einstein relationship

$$D_\alpha = \frac{kT_\alpha}{e} b_\alpha. \qquad (3.5.7)$$

In the presence of the magnetic field this relation is also valid for the transverse coefficients.

Using the relation (3.1.16) between $\tau_{\alpha\beta}^{-1}$ and the binary diffusion coefficient $[\mathfrak{D}_{\alpha\beta}]_1$, the ion mobility coefficient can be defined by the relation ($T_\alpha = T$ for $\alpha \neq e$)

$$\frac{1}{b_i} = \frac{kT}{e} \sum_k \frac{y_k}{[\mathfrak{D}_{ik}]_1}, \tag{3.5.8}$$

where $y_k = n_k/n$ is the relative concentration of neutral species in the mixture. If $b_{ik} = (e/kT)\,[\mathfrak{D}_{ik}]_1$ are the ion mobility coefficients of species i in each pure gas which constitute the mixture under investigation, then from (3.5.8) it follows that

$$\frac{1}{b_i} = \sum_k \frac{y_k}{b_{ik}}, \tag{3.5.8$'$}$$

which is well known as Blank's law [11].

Now we shall derive the expression of the current density in a weakly ionized multicomponent plasma. If we use the definition of $\mathbf{j} = \sum n_\alpha e_\alpha \mathbf{w}_\alpha$ and take account of (3.5.4), we find

$$\begin{aligned} \mathbf{j} \quad &= \sigma_{\parallel}\mathbf{E}'_{\parallel} + \sigma_{\perp}\mathbf{E}'_{\perp} + \sigma_{\wedge}\mathbf{k}\times\mathbf{E}' - \sum_\alpha Z_\alpha b_{\alpha\parallel}\mathbf{G}_{\alpha\parallel} \\ &- \sum_\alpha Z_\alpha b_{\alpha\perp}\mathbf{G}_{\alpha\perp} - \sum_\alpha Z_\alpha b_{\alpha\wedge}\mathbf{G}_\alpha \times \mathbf{k}, \end{aligned} \tag{3.5.9}$$

where the electrical conductivities are expressed as

$$\begin{gathered} \sigma_{\parallel} = e\sum_\alpha n_\alpha Z_\alpha^2 b_{\alpha\parallel}, \qquad \sigma_{\perp} = e\sum_\alpha n_\alpha Z_\alpha^2 b_{\alpha\perp}, \\ \sigma_{\wedge} = -e\sum_\alpha n_\alpha Z_\alpha^2 b_{\alpha\wedge}. \end{gathered} \tag{3.5.10}$$

When $\overline{Q}_{ek} \sim \overline{Q}_{ik}$, the ratio b_i/b_e is approximately equal to $(m_e T/\mu_{ik} T_e)^{1/2}$, therefore when calculating the longitudinal electrical conductivity, the ion mobility contributed to it as compared with the electron mobility can be neglected. This statement fully agrees with the conclusions made in Section 3.4 and the expression for $\sigma_{\parallel}$ is of the kind (3.4.6). To estimate the contribution of the ion and electron constituents in $\sigma_\perp$ and $\sigma_\wedge$, note that at $Z_i = 1$ and $n_e = \sum_{i=1}^{s} n_i$, where the index i corresponds to ions of various species, the expressions of the transverse electrical conductivity at $b_e \gg b_i$ can be written as

$$\begin{aligned}\sigma_{\perp} &= e\sum_{i=1}^{s} n_i\left(\frac{b_e}{1+\beta_e^2}+\frac{b_i}{1+\beta_i^2}\right)\\ &\approx e\sum_{i=1}^{s} n_i\frac{b_e\left(1+\beta_0\beta_i\right)}{\left(1+\beta_0\beta_i\right)^2+\beta_0^2},\\ \sigma_{\wedge} &= -e\sum_{i=1}^{s} n_i\left(\frac{b_e\beta_e}{1+\beta_e^2}+\frac{b_i\beta_i}{1+\beta_i^2}\right)\\ &\approx e\sum_{i=1}^{s} n_i\frac{b_e\beta_0}{\left(1+\beta_0\beta_i\right)^2+\beta_0^2},\end{aligned} \tag{3.5.11}$$

where

$$b_e = \frac{e\tau_0}{m_e}, \qquad \beta_0 = |\boldsymbol{\beta}_e| = \omega_e\tau_0, \qquad \omega_e = \frac{e}{m_e}|\mathbf{B}|,$$

$$\beta_i = b_i|\mathbf{B}| = \frac{e|\mathbf{B}|}{\sum\limits_{\beta=1}^{k}\mu_{i\beta}\tau_{i\beta}^{-1}}.$$

These expressions when applied to a weakly ionized gas generalize the results obtained earlier for a three-component plasma with a single ion species for the case of several ion species in the mixture. Note that we can neglect the ion's contribution to $\sigma_{\perp}$ and $\sigma_{\wedge}$ when the condition $\beta_0\beta_i \ll 1$ is fulfilled. In this case we come to an approximation, when the current density is defined by the electron transport alone. Vice versa, under the condition of $\beta_0\beta_i \gg 1$ the transverse electrical conductivities are determined by the contribution of the ion species alone and in this case the expressions for $\sigma_{\perp}$ and $\sigma_{\wedge}$ (3.5.10), in which summation includes only ions, are valid.

Now let us consider the ambipolar diffusion in a weakly ionized plasma with some species of ions. Charged particles diffuse under the action of the gradients of their own partial pressure due to the higher mobility of electrons compared with that of ion. This causes separation of the charge. The electrical field in this case (the field of the volume charge) causes electrons and ions to diffuse simultaneously, and this diffusion is termed ambipolar [12].

In the absence of the magnetic field ($\mathbf{B} = 0$), the diffusion of the charged particles of the plasma is described by Eq. (3.5.6). If the current in the plasma is zero which is possible, for example, when there is no contact of the plasma with the external conductors, then it follows from the definition of $\mathbf{j}$ that

$$\sum_{\alpha} n_\alpha e_\alpha \mathbf{w}_\alpha = 0. \tag{3.5.12}$$

Here, substituting $\mathbf{w}_\alpha$ of (3.5.6), we can express the field of the volume charge $\mathbf{E}$ in terms of the gradients of partial pressure of the charged particles

$$\mathbf{E} = \sigma^{-1} \sum_{\alpha} n_\alpha e_\alpha D_\alpha \nabla \ln p_\alpha,$$

where $\sigma = e \sum_{\alpha} n_\alpha Z_\alpha^2 b_\alpha$ is the electrical conductivity of the plasma in the absence of the magnetic field. Eliminating $\mathbf{E}$ in (3.5.6), we obtain

$$n_\alpha \mathbf{w}_\alpha = -n_\alpha D_\alpha \left(\nabla \ln p_\alpha - \frac{Z_\alpha}{T_\alpha} \frac{\sum_{\beta} n_\beta Z_\beta T_\beta b_\beta \nabla \ln p_\beta}{\sum_{\beta} n_\beta Z_\beta^2 b_\beta} \right). \tag{3.5.13}$$

In particular, for electrons ($\alpha = e$) separating $\beta = e$ from summation over β, we get

$$n_e \mathbf{w}_e = -n_e D_e \frac{\nabla \ln p_e \sum_{i} n_i Z_i^2 b_i + \sum_{i} n_i Z_i b_i \frac{T_i}{T_e} \nabla \ln p_e}{n_e b_e + \sum_{i} n_i Z_i^2 b_i}, \tag{3.5.13'}$$

where summation is carried out only with respect to ion species. It is to be noted that in the general case the diffusion flux of any species depends on all the gradients of the partial pressure of the charged particles.

The expressions obtained can be easily generalized for the case of diffusion of the charged particles across the magnetic field. Consider as an example a cylindrical column of plasma in the transverse magnetic field. Assuming that all the gradients and the electrical field exist only in the radial direction, we have

$$n_\alpha w_{\alpha r} = n_\alpha Z_\alpha b_{\alpha\perp} E_r - n_\alpha D_{\alpha\perp} \frac{\partial \ln p_\alpha}{\partial r}, \tag{3.5.14}$$

where

$$D_{\alpha\perp} = \frac{kT_\alpha}{e} b_{\alpha\perp} = \frac{D_\alpha}{1 + \beta_\alpha^2}. \tag{3.5.15}$$

Again, if E_r is expressed with the help of Eq. (3.5.12), the formula for $n_\alpha w_{\alpha r}$ is obtained which fully agrees with (3.5.13) if D_α and b_β are substituted for $D_{\alpha\perp}$ and $b_{\beta\perp}$.

As it is common practice to assume that the condition $b_e \gg b_i$ is fulfilled, the expression (3.5.13′) may be written as

$$n_e \mathbf{w}_e = -\sum_i n_i Z_i D_i \left(Z_i \frac{T_e}{T_i} \nabla \ln p_e + \nabla \ln p_i \right), \qquad (3.5.16)$$

i.e. the diffusion transport of electrons similarly to that of ions is determined by the ion diffusion coefficient. For the case of diffusion across the magnetic field the requirement $b_{e\perp} \gg b_{i\perp}$ is not satisfied in the general case. Moreover, in the strong magnetic field, when $\beta_\alpha^2 \gg 1$, the reverse condition $b_{i\perp} \gg b_{e\perp}$ holds true. Then

$$n_e w_{er} = -n_e D_{e\perp} \frac{\partial \ln p_e}{\partial r} + \frac{1}{\sum_i n_i Z_i^2 b_{i\perp}} \sum_i n_i Z_i b_{i\perp} \frac{T_i}{T_e} \frac{\partial \ln p_i}{\partial r}. \qquad (3.5.17)$$

For the plasma with a single species of ions of charge Ze, when $n_e = Zn_i$ and under the assumption that $\nabla T_e = \nabla T_i = 0$, it follows from Eq. (3.5.13) that

$$n_e \mathbf{w}_e = Zn_i \mathbf{w}_i = -D_A \nabla n_e, \qquad (3.5.18)$$

where D_A is the coefficient of ambipolar diffusion determined by

$$D_A = D_i \frac{b_e}{b_e + Zb_i} \left(1 + Z\frac{T_e}{T_i} \right). \qquad (3.5.19)$$

Respectively, for the diffusion across the magnetic field we have

$$n_e \mathbf{w}_{er} = Zn_i \mathbf{w}_{ir} = -D_{A\perp} \frac{\partial n_e}{\partial r}, \qquad (3.5.20)$$

where $D_{A\perp}$ is obtained from Eq. (3.5.19) by replacing all the coefficients of b_α and D_α by the transverse coefficients $b_{\alpha\perp}$ and $D_{\alpha\perp}$. Then

$$D_A = D_i \left(1 + Z\frac{T_e}{T_i} \right) \quad \text{at} \quad b_e \gg b_i,$$
$$D_{A\perp} = D_{e\perp} \left(1 + \frac{1}{Z}\frac{T_i}{T_e} \right) \quad \text{at} \quad b_{e\perp} \ll b_{i\perp}. \qquad (3.5.21)$$

For $Z = 1$ and $T_e = T_i$ the well-known result [12] follows from Eqs. (3.5.21). In accordance with this result the coefficient of ambipolar diffusion in the absence of (or along with) the magnetic field equals the double coefficient of ion diffusion ($D_A = 2D_i$), whereas across the strong magnetic field this value is equal to twice the coefficient of the transverse diffusion of electrons ($D_{A\perp} = 2D_{e\perp} = 2D_e/\beta_e^2$).

3.6 Transverse transport of particles in a magnetized plasma

Consider the diffusion of the charged particles in a fully ionized multicomponent plasma across a strong magnetic field when the requirement

$$|\omega_\alpha|\,\tau_{\alpha\beta} \gg 1 \qquad (3.6.1)$$

is satisfied for all the constituent species of the charged particles, where $\omega_\alpha = e_\alpha |\mathbf{B}|/m_\alpha$ is the cyclotron frequency of species α, $\tau_{\alpha\beta}$ is the effective time between the collisions defined by (3.1.17). The plasma whose particles obey the requirement of (3.6.1) is commonly termed magnetized.

The starting point of the analysis will be based on the system of Eqs. (3.1.19) for the species diffusion velocities which, accounting for the definition of ω_α, can be written as

$$\sum_\beta (\omega_\alpha \tau_{\alpha\beta})^{-1} \frac{\mu_{\alpha\beta}}{m_\alpha} (\mathbf{w}_\alpha - \mathbf{w}_\beta)$$
$$= \mathbf{w}_\alpha \times \mathbf{k} + (\rho_\alpha \omega_\alpha)^{-1} \left(n_\alpha e_\alpha \mathbf{E}' + n_\alpha \mathbf{X}_\alpha - \mathbf{G}_\alpha\right). \qquad (3.6.2)$$

The small parameter $(\omega\tau)^{-1} \ll 1$ on the left-hand side of these equations allows one to use the technique of successive approximations for their solution. In the zero approximation neglecting the collisions (i.e. the left-hand side of (3.6.2)), we have

$$\mathbf{w}_{\alpha\perp}^{(0)} = \frac{1}{B}\mathbf{E}' \times \mathbf{k} + \frac{1}{\rho_\alpha \omega_\alpha} \mathbf{k} \times \nabla p_\alpha + \frac{1}{\omega_\alpha} \mathbf{k} \times \frac{d\mathbf{u}}{dt} - \frac{1}{m_\alpha \omega_\alpha} \mathbf{k} \times \mathbf{X}_\alpha \qquad (3.6.3)$$

for the diffusion velocity of particles across the magnetic field.

The first order correction for $\mathbf{w}_{\alpha\perp}$ can be found by substituting (3.6.3) on the left-hand side of (3.6.2)

$$\mathbf{w}_{\alpha\perp}^{(1)} = -\frac{1}{m_\alpha^2 \omega_\alpha^2} \sum_\alpha \mu_{\alpha\beta} \tau_{\alpha\beta}^{-1} \left[\left(\frac{\nabla_\perp p_\alpha}{n_\alpha} - \frac{Z_\alpha}{Z_\beta} \frac{\nabla_\perp p_\beta}{n_\beta} \right) \right.$$
$$\left. + \left(m_\alpha - \frac{Z_\alpha}{Z_\beta} m_\beta \right) \frac{d\mathbf{u}_\perp}{dt} - \left(\mathbf{X}_{\alpha\perp} - \frac{Z_\alpha}{Z_\beta} \mathbf{X}_{\beta\perp} \right) \right]. \qquad (3.6.4)$$

Now let us analyse some sequels from the formulas obtained.

Summing the diffusion velocities of species $\mathbf{w}_{\alpha\perp} = \mathbf{w}^{(0)}_{\alpha\perp} + \mathbf{w}^{(1)}_{\alpha\perp}$ with the weight $n_\alpha Z_\alpha e$ (for the electrons $Z_e = -1$), we get the expression of the transverse constituent of the current density in the plasma

$$\mathbf{j}_\perp = \frac{1}{B}\left(\mathbf{k}\times\nabla p + \rho\mathbf{k}\times\frac{d\mathbf{u}}{dt} - \mathbf{k}\times\sum_\alpha n_\alpha \mathbf{X}_\alpha\right). \tag{3.6.5}$$

The same expression results from the full motion equation of the plasma (2.2.17) if it is solved with respect to $\mathbf{j}$, and the force of the viscous friction is disregarded. To put down the expression (3.6.5) in the form close to the conventional generalized Ohm's law, we can express $d\mathbf{u}/dt$ in terms of $\mathbf{E}'$, using the obvious requirement $\sum \rho_\alpha \mathbf{w}_{\alpha\perp} = 0$ and substituting $\mathbf{w}_{\alpha\perp}$ into it. The final expression for $\mathbf{j}_\perp$ takes the form

$$\begin{aligned}
\mathbf{j}_\perp &= \frac{1}{B}\Bigg\{\mathbf{k}\times\nabla\left(p - \langle p\rangle\right) + \rho e\left\langle\frac{m}{Z}\right\rangle^{-1}\mathbf{k}\times\mathbf{E}' \\
&+ \frac{1}{eB}\left\langle\frac{m}{Z}\right\rangle^{-1}\sum_\alpha \frac{\rho_\alpha \mu_{\alpha\beta}}{Z_\alpha^2}\tau_{\alpha\beta}^{-1}\Bigg[\left(\frac{\nabla_\perp p_\alpha}{n_\alpha} - \frac{Z_\alpha}{Z_\beta}\frac{\nabla_\perp p_\beta}{n_\beta}\right) \\
&+ \left(m_\alpha - \frac{Z_\alpha}{Z_\beta}m_\beta\right)\left(\left\langle\frac{m}{Z}\right\rangle^{-1} e\mathbf{E}'_\perp - \frac{1}{\rho}\nabla_\perp\langle p\rangle\right)\Bigg]\Bigg\},
\end{aligned} \tag{3.6.6}$$

where

$$\left\langle\frac{m}{Z}\right\rangle = \sum_\alpha \frac{m_\alpha}{Z_\alpha}\frac{\rho_\alpha}{\rho}, \qquad \langle p\rangle = \left\langle\frac{m}{Z}\right\rangle^{-1}\sum_\alpha \frac{m_\alpha}{Z_\alpha}p_\alpha. \tag{3.6.7}$$

Note that due to the small parameter of m_e/m_α $(\alpha \neq e)$ the electrons do not contribute to both $\langle m/Z\rangle$ and $\langle p\rangle$ at all.

The coefficients at $\mathbf{E}'_\perp$ and $\mathbf{k}\times\mathbf{E}'$ in (3.6.6) correspond to the transverse electrical conductivities in the magnetized plasma $\sigma_\perp$ and $\sigma_\wedge$. It is easy to make sure that in the case of plasma with two ion species these coefficients coincide with the expressions following from Eqs. (3.4.4)–(3.4.5) when the limit $\omega_e\tau_0 \gg 1$, $\omega_i\tau_{23} \gg 1$ is considered.

Now let us consider the diffusion of a plasma species across the magnetic field. When the inertial terms in the motion equation are negligibly small ($d\mathbf{u}/dt = 0$), the equilibrium of plasma as a whole is described by the equation of the momentum balance

$$\nabla p = \mathbf{j}\times\mathbf{B}. \tag{3.6.8}$$

Combined with the requirement imposed on $\mathbf{B}$ by the Maxwell equations, Eq. (3.6.8) determines the so called equilibrium magnetohydrodynamic configuration of the plasma. Let us analyse diffusion in the cylindrical column of the plasma under this condition, assuming that the magnetic field B_Z is directed along the axes of the cylinder. For a two-component plasma of electrons and an ion species, allowing for the quasineutral state of the plasma, it follows from Eqs. (3.6.3) and (3.6.4) that the radial electron and ion transports occur at the same velocity

$$u_{ir} = u_{er} = \frac{1}{\sigma_0 B^2}\frac{\partial p}{\partial r} \tag{3.6.9}$$

sometimes termed the "velocity of percolation" of the plasma [13]. For a plasma with two ion species, the estimate of the radial ion transport according to Eq. (3.6.4) shows that their velocities are $(m_i/m_e)^{1/2}$ times greater than the velocity of percolation of the simple plasma (3.6.9). As a result during the time, which is markedly less than the typical time of percolation the equilibrium distribution of ion densities of various species is established along the radius; it can be found by making the radial constituents of ion velocities and radial current density equal to zero. For the radial velocities of ion of i and I species, neglecting the contribution of electron–ion collisions in expressions (3.6.4), we obtain

$$Z_i n_i w_{ir} = -Z_I n_I w_{Ir} = \frac{\mu_{iI}\tau_{iI}}{Z_i B^2}\left(\frac{\partial p_i}{\partial r} - \frac{Z_i n_i}{Z_I n_I}\frac{\partial p_I}{\partial r}\right). \tag{3.6.10}$$

Setting these expressions equal to zero leads to the well-known result [13,14]

$$\frac{n_I}{n_i} = A n_i^{\frac{Z_I}{Z_i}-1}, \qquad A = \text{const}, \tag{3.6.11}$$

whose consequence is that ions with charge multiplicity of greater numbers concentrate in the field of greater density. The generalization of this result, which will take into account different ion charge states of each species as well as the influence of the radial temperature gradient in the plasma, will be presented in Chapter 9.

Chapter 4

Grad's method

4.1 Expansion of the distribution function

The approximation to f_α in the form (3.1.2) considered in the preceding chapter is a special case of a more general expansion of the distribution function in series in orthogonal tensorial polynomials of the particle velocities. The use of such expansions is an essential feature of the so called moment methods of solving the kinetic equation when defining $f_\alpha(\mathbf{v}_\alpha, \mathbf{r}, t)$ or the moments corresponding to it amounts to the solution of an infinite set of equations for the expansion coefficients [1–3]. The character of the expansion and type of the polynomials are uniquely dictated by the choice of the weight function or the distribution function of zero approximation [4]. It is clear that the practical realization of the moment method is confined to a finite number of the expansion terms or the truncated set of equations for the expansion coefficients.

When treating the slightly non-equilibrium gas states, it would be natural to choose the local Maxwellian as the distribution function of the zero approximation. After writing it down in the Cartesian system of coordinates and forming a number of polynomials out of the ordered full series of monomials $1, \xi_r, \xi_r\xi_s, \ldots$, where $\boldsymbol{\xi} = (m/kT)^{1/2}\mathbf{c}$ is the dimensionless velocity (for simplicity the species index α is omitted for the time being), we can easily find [4] that the orthogonalization procedure for the series terms with the weight function $\omega(|\boldsymbol{\xi}|) = (2\pi)^{-3/2}\exp(-\xi^2/2)$ results in the tensorial Hermite polynomials in the expansion defined as

$$H^n_{r_1\ldots r_n}(\boldsymbol{\xi}) = (-1)^n \exp\left(\frac{\xi^2}{2}\right)\frac{\partial^n \exp\left(-\frac{\xi^2}{2}\right)}{\partial\xi_{r_1}\ldots\partial\xi_{r_n}}. \qquad (4.1.1)$$

Then the expansion of the distribution function in series over $H^n_{r_1...r_n}(\boldsymbol{\xi})$ is represented as [1,2]

$$f(\boldsymbol{\xi},\mathbf{r},t) = f^{(0)} \sum_{n=0}^{\infty} \frac{1}{n!} a^n_{r_1...r_n}(\mathbf{r},t)\, H^n_{r_1...r_n}(\boldsymbol{\xi})$$

$$f^{(0)} = n\left(\frac{m}{kT}\right)^{3/2} \omega(\xi) = n\left(\frac{m}{2\pi kT}\right)^{3/2} \exp\left(-\frac{\xi^2}{2}\right), \tag{4.1.2}$$

where the Hermite polynomials meet the orthogonality conditions

$$\int \omega(\xi)\, H^n_{r_1...r_n}(\xi)\, H^m_{s_1...s_m}(\xi)\, d\xi = 0, \qquad m \neq n;$$

$$\int \omega(\xi)\, H^n_{r_1...r_n}(\xi)\, H^m_{s_1...s_m}(\xi)\, d\xi = \delta^{(n)}_{rs}, \quad m = n; \tag{4.1.3}$$

with $\delta^{(1)}_{rs} = \delta_{rs}$, $\delta^{(2)}_{rs} = \delta_{r_1 s_1}\delta_{r_2 s_2} + \delta_{r_1 s_2}\delta_{r_2 s_1} + \delta_{r_1 r_2}\delta_{s_1 s_2}$ etc. The components of tensor $\delta^{(n)}$ involve $(2n)!/2^n n!$ terms which are in fact possible but various products of second-rank unit tensors obtained by permutation of $2n$ indices (each product emerges once in every operation). Therefore, the polynomials $H^m_{r_1...r_n}$ and $H^n_{s_1...s_n}$ are orthogonal if the $r_1...r_n$ indices are not permutations of $s_1...s_n$.

It should be noted that the first tensorial polynomials are

$$H^0 = 1, \qquad H^1_r = \xi_r, \qquad H^2_{rs} = \xi_r\xi_s - \delta_{rs},$$

$$H^3_{rst} = \xi_r\xi_s\xi_t - (\xi_r\delta_{st} + \xi_s\delta_{rt} + \xi_t\delta_{rs}), \dots . \tag{4.1.4}$$

Taking into account (4.1.3) and with the help of (4.1.2), we can readily obtain the following relations between the expansion coefficients $a^n_{r_1...r_n}$ and the moments of the distribution function

$$na^n_{r_1...r_n} = \int H^n_{r_1...r_n}(\xi)\, f dc. \tag{4.1.5}$$

Grad [1,2] was the first to suggest the expansion of f in the tensorial Hermite polynomials (4.1.2). A variation of the original Grad's procedure, which in particular helps us to follow the relation of the procedure and the expansion used in the Chapman–Enskog–Burnett theory [5,6] comes in the application of the irreducible Hermite polynomials $H^{mn}_{r_1...r_n}(\boldsymbol{\xi})$ instead of the polynomials $H^n_{r_1...r_m}(\boldsymbol{\xi})$. The introduction of these polynomials is easily justified if for the distribution function of the zero approximation the variables are taken in the spherical (ξ, Θ, φ) rather than Cartesian system of coordinates [7]. The orthogonalization procedure with respect to the series $\xi^m, \xi^{m+2}, \xi^{m+4}, \ldots$ $(0 \leq m \leq \infty)$ with the Gaussian weight function over $\xi^2/2$ brings about the polynomials $L_m^n(\xi^2/2)$, which are directly related to the Sonine polynomials $S_{m+1/2}^n(\xi^2/2)$ used in the Chapman–Enskog–Burnett theory

$$L_m^n(\xi^2/2) = (-2)^n n! S_{m+1/2}^n(\xi^2/2),$$

where

$$S_{m+1/2}^n(u^2) = \sum_{p=0}^{\infty} \frac{(-u^2)^p (m+n+\frac{1}{2})!}{p!(n-p)!(m+p+\frac{1}{2})!}. \tag{4.1.6}$$

In particular, at any m

$$S_{m+1/2}^0(u^2) = 1, \qquad S_{m+1/2}^1(u^2) = m + \frac{3}{2} - u^2.$$

The orthogonalization of the series on the unit sphere with the weight function $\sin\Theta$ over Θ and the unit weight function over φ generates tensorial spherical harmonics $Y_m^k(u, \Theta, \varphi)$ in the expansion, that are the product of u^m and the common spherical harmonics

$$Y_m^k(u, \Theta, \varphi) = u^m (-1)^m \frac{(\sin\Theta)^k}{2^m m!} \frac{d^{m+k}(\sin\Theta)^{2m}}{d(\cos\Theta)^{m+k}} \exp(ik\varphi).$$

It is natural that the expansions in both the Cartesian and spherical coordinate systems should be fully equivalent. In fact, this implies that the tensorial Hermite polynomials of any order can be represented as a linear combination of irreducible tensorial polynomials $H_{r_1 \ldots r_m}^{mn}(\boldsymbol{\xi})$ which are the products of polynomials $L_m^n(\xi^2/2)$ and irreducible harmonic polynomials $P_{r_1 \ldots r_m}^{(m)}(\boldsymbol{\xi})$ or [2,7]

$$H^{mn}(\boldsymbol{\xi}) = (-2)^n n! S_{m+1/2}^n(\xi^2/2) P^{(m)}(\boldsymbol{\xi}). \tag{4.1.7}$$

It is to be pointed out that the tensor $P^{(m)}(\boldsymbol{\xi})$ is equivalent to $2m+1$ linearly independent functions with the same m [8]. A few first harmonic polynomials are [7]

$$P^{(0)} = 1, \qquad P_r^{(1)} = \xi_r, \qquad P_{rs}^{(2)} = \xi_r\xi_s - \frac{1}{3}\delta_{rs}\xi^2,$$
$$P_{rst}^{(3)} = \xi_r\xi_s\xi_t - \frac{1}{5}\xi^2(\xi_r\delta_{st} + \xi_s\delta_{rt} + \xi_t\delta_{rs}) \tag{4.1.8}$$

Now we shall turn our attention to the multicomponent plasma. Let us use the local Maxwellian taken at the species temperature T_α and shifted with respect to the mass-average velocity $\mathbf{u}$ as the zero approximation. Then the expansion for $f_\alpha(\boldsymbol{\xi}_\alpha, \mathbf{r}, t)$, accounting for the tensor decomposition into its irreducible components, is represented as

$$f_\alpha = f_\alpha^{(0)} \sum_{n=0}^{\infty} \sum_{m=0}^{\infty} \sigma_{mn} a_{\alpha r_1 \dots r_m}^{mn}(\mathbf{r},t) H_{\alpha r_1 \dots r_m}^{mn}(\boldsymbol{\xi}_\alpha), \tag{4.1.9}$$

where

$$f_\alpha^{(0)} = n_\alpha \left(\frac{\gamma_\alpha}{2\pi}\right)^{3/2} \exp\left(-\frac{\xi_\alpha^2}{2}\right), \quad \boldsymbol{\xi}_\alpha = \gamma_\alpha^{1/2} \mathbf{c}_\alpha, \quad \gamma_\alpha = \frac{m_\alpha}{kT_\alpha}, \tag{4.1.10}$$

and

$$\sigma_{mn} = \frac{(2m+1)!\,(m+n)!}{n!\,(m!)^2\,(2m+2n+1)!}$$

is the normalization coefficient due to the fact that the internal product $a^{mn}H^{mn}$ implicitly comprises tensor contractions.

Because of the orthogonality of the polynomials, the expansion coefficients (4.1.9) are defined by

$$n_\alpha a_{\alpha r_1 \dots r_m}^{mn} = \int H_{\alpha r_1 \dots r_m}^{mn}(\boldsymbol{\xi}_\alpha) f_\alpha d\mathbf{c}. \tag{4.1.11}$$

The set of equations for the coefficients $a_{\alpha r_1 \dots r_m}^{mn}$ is obtained by multiplying the kinetic equation (1.2.18) by polynomials $H_{\alpha r_1 \dots r_m}^{mn}(\boldsymbol{\xi}_\alpha)$, and their subsequent integration over particle velocities. In fact, these equations are coincident with the transport equation (2.1.9) in which polynomials $H_\alpha^{mn}(\boldsymbol{\xi}_\alpha)$ should work instead of ψ_α. The equations for the coefficients $a_{\alpha r_1 \dots r_m}^{mn}$ form an infinite hierarchy of the coupling equations since every equation of the kth order includes the coefficients of the $(k+1)$th and $(k-1)$th orders. Another step is to truncate the moment hierarchy at a certain level so that any given number of equations should be solvable, i.e. the number of unknown values should correspond to the number of equations. It is at this step that the main discrepancy between the results obtained by the canonical Chapman–Enskog procedure and the Grad's method arises.

The results of the former technique follow from the moment equations when a certain iteration procedure is used in which a small parameter ε is the so called Knudsen number, equal in the order of value to the ratio of the mean free path of particles λ and the typical dimension of the problem L, or the ratio of the mean time between the collisions

τ and the characteristical time of the problem τ_L. In doing so all the coefficients or the moments of the higher orders are expressed via the functions of the lower moments ρ_α, $\mathbf{u}$,T_α and their spatial derivatives.

In the Grad's method the equations of the coefficients are truncated on the assumption that for every given set of equations of the kth order the distribution function can be approximated with a series in which the coefficients of higher than kth order are assumed to be zero which allows one to make the system closed.

The difference between the results disappears if we consider a gas state in which the macroscopic parameters slightly change at the distances of the order of λ and the times of the order of τ.

Along with it, for some kinds of gas flows even within this range there are peculiarities and differences in the results obtained with the help of two approaches which will be considered in Chapter 5.

4.2 13N-moment approximation

The conservation equations derived in the previous chapter contain the moments of higher order $\pi_{\alpha rs}$ and $\mathbf{q}_\alpha$ besides the variables ρ_α, $\mathbf{u}_\alpha$ and T_α. As the tensor $\pi_{\alpha rs}$ is symmetrical and its trace is zero, it is defined by five independent components. The total number of variables determining the state of the gas mixture (plasma) in this case is equal to $13N$ (where N is the number of plasma species) [9–11].

Using the explicit expressions for a few first polynomials H^{mn}

$$\begin{aligned} &H^{00} = 1, \qquad H^{01} = \xi^2 - 3, \qquad H^{10}_r = \xi_r, \\ &H^{20}_{rs} = \xi_r\xi_s - \frac{1}{3}\delta_{rs}\xi^2, \quad H^{11}_r = \xi_r\left(\xi^2 - 5\right), \end{aligned} \tag{4.2.1}$$

we have the expansion

$$\begin{aligned} f_\alpha \;=\; & f^{(0)}_\alpha\Big[a^{00}_\alpha + \frac{1}{6}a^{01}_\alpha\left(\xi^2 - 3\right) + a^{10}_{\alpha r}\xi_r \\ & + \frac{1}{2}a^{20}_{\alpha rs}\left(\xi_r\xi_s - \frac{1}{3}\delta_{rs}\xi^2\right) + \frac{1}{10}a^{11}_{\alpha r}\xi_r\left(\xi^2 - 5\right)\Big]. \end{aligned} \tag{4.2.2}$$

Let us write the expansion coefficients in terms of moments of the distribution function with the help of (4.1.11)

$$\begin{aligned} &a^{00}_\alpha = 1, \qquad a^{01}_\alpha = 0, \qquad a^{10}_{\alpha r} = \gamma_\alpha^{1/2} w_{\alpha r}, \\ &a^{20}_{\alpha rs} = p_\alpha^{-1}\pi_{\alpha rs}, \quad a^{11}_\alpha = 2\gamma_\alpha^{1/2} p_\alpha^{-1} h_{\alpha r}, \end{aligned} \tag{4.2.3}$$

where

$$\mathbf{h}_\alpha = \mathbf{q}_\alpha - \frac{5}{2} p_\alpha \mathbf{w}_\alpha \tag{4.2.4}$$

is the reduced heat flux of species α.

Limiting ourselves to these coefficients in the expansion, which are related to macroscopic quantities of clear physical meaning, we have

$$\begin{aligned} f_\alpha &= f_\alpha^{(0)} \left[1 + \gamma_\alpha^{1/2} w_{\alpha r} \xi_r + \frac{1}{2} p_\alpha^{-1} \pi_{\alpha rs} \left(\xi_r \xi_s - \frac{1}{3} \delta_{rs} \xi^2 \right) \right. \\ &\quad \left. + \frac{1}{5} \gamma_\alpha^{1/2} p_\alpha^{-1} h_{\alpha r} \xi_r \left(\xi^2 - 5 \right) \right] \end{aligned} \tag{4.2.5}$$

for the distribution function in the approximation of $13N$-moments.

The closed system of equations correspondent to this approximation includes equations for the parameters ρ_α, $\mathbf{u}$ and T_α contained in the distribution function of the zero approximation (the weight function) as well as equations for the coefficients $a_{\alpha r}^{10}$, $a_{\alpha rs}^{20}$ and $a_{\alpha r}^{11}$ or the quantities $\mathbf{w}_\alpha$, $\pi_{\alpha rs}$ and $\mathbf{h}_\alpha$. Equations for $a_{\alpha r}^{10}$, $a_{\alpha rs}^{20}$ and $a_{\alpha r}^{11}$ follow from the general transport equation (2.1.9) if $\psi_\alpha = H_{\alpha r}^{10}$, $H_{\alpha rs}^{20}$ and $H_{\alpha r}^{11}$ is substituted into it whereas the moments of higher orders in the equations are expressed in terms of the assumed set $13N$-moments with the help of (4.2.5).

Sometimes it is more convenient to deal with the equations of the moments of the distribution function rather than the equations for the coefficients a_α^{mn}. To this end, let us introduce the polynomials $G_\alpha^{mn}(\mathbf{c}_\alpha, \gamma_\alpha)$ related to $H_\alpha^{mn}(\boldsymbol{\xi}_\alpha)$ via

$$G_\alpha^{mn}(\mathbf{c}_\alpha, \gamma_\alpha) = 2^{-n} m_\alpha \gamma_\alpha^{-\left(n + \frac{m}{2}\right)} H_\alpha^{mn}\left(\gamma_\alpha^{1/2} \mathbf{c}_\alpha \right), \tag{4.2.6}$$

and the coefficients b_α^{mn} defined as

$$n_\alpha b_\alpha^{mn} = \int G_\alpha^{mn} f_\alpha d\mathbf{c}.$$

In the approximation tested

$$G_\alpha^{00} = m_\alpha, \quad G_{\alpha r}^{10} = m_\alpha c_{\alpha r}, \quad G_\alpha^{01} = \frac{m_\alpha}{2} \left(c_\alpha^2 - \frac{3}{\gamma_\alpha} \right),$$

$$G_{\alpha rs}^{20} = m_\alpha \left(c_{\alpha r} c_{\alpha s} - \frac{1}{3} \delta_{rs} c_\alpha^2 \right), \quad G_{\alpha r}^{11} = \frac{m_\alpha}{2} c_{\alpha r} \left(c_\alpha^2 - \frac{5}{\gamma_\alpha} \right) \tag{4.2.7}$$

and, respectively,

$$n_\alpha b_\alpha^{00} = \rho_\alpha, \quad n_\alpha b_{\alpha r}^{10} = \rho_\alpha w_{\alpha r}, \quad n_\alpha b_\alpha^{01} = 0,$$
$$n_\alpha b_{\alpha rs}^{20} = \pi_{\alpha rs}, \qquad n_\alpha b_{\alpha r}^{11} = h_{\alpha r}. \tag{4.2.8}$$

Derivation of the general infinite set of equations for coefficients $n_\alpha b_\alpha^{mn}$ resulting from (2.1.9) with the substitution of $\psi_\alpha = G_\alpha^{mn}(\mathbf{c}_\alpha, \gamma_\alpha)$ is given in Appendix 1.

The associated closed system of equations in the $13N$-moment approximation, which includes, together with the equations for $\mathbf{w}_\alpha$, $\pi_{\alpha rs}$ and $\mathbf{h}_\alpha$, the equations for the basic thermodynamic variables ρ_α, $\mathbf{u}$ and T_α, becomes [9–13]

$$\frac{d\rho_\alpha}{dt} + \rho_\alpha \nabla \cdot \mathbf{u} + \nabla \cdot \rho_\alpha \mathbf{w}_\alpha = 0, \tag{4.2.9}$$

$$\rho \frac{d\mathbf{u}}{dt} + \nabla p + \nabla \overset{\wedge}{\pi} - \sum_\alpha n_\alpha \langle \mathbf{F}_\alpha \rangle = 0, \tag{4.2.10}$$

$$\frac{3}{2}\frac{d}{dt}(n_\alpha k T_\alpha) + \frac{5}{2} n_\alpha k T_\alpha \nabla \cdot \mathbf{u} + \nabla \cdot \mathbf{q}_\alpha + \pi_{\alpha rs}\frac{\partial u_r}{\partial x_s} - n_\alpha \mathbf{w}_\alpha \cdot \langle \mathbf{F}_\alpha^* \rangle = R_\alpha^{01}, \tag{4.2.11}$$

$$\frac{d}{dt}\rho_\alpha w_{\alpha r} + \rho_\alpha w_{\alpha s}\frac{\partial u_r}{\partial x_s} + \rho_\alpha w_{\alpha r}\nabla \cdot \mathbf{u} + \frac{\partial p_\alpha}{\partial x_r} + \frac{\partial \pi_{\alpha rs}}{\partial x_s} - n_\alpha \langle F_{\alpha r}^* \rangle = R_{\alpha r}^{10}, \tag{4.2.12}$$

$$\frac{d\pi_{\alpha rs}}{dt} + \pi_{\alpha rs}\nabla \cdot \mathbf{u} + 2\left\{\pi_{\alpha rl}\frac{\partial u_s}{\partial x_l}\right\} + \frac{4}{5}\left\{\frac{\partial q_{\alpha r}}{\partial x_s}\right\} + 2p_\alpha \varepsilon_{rs} - 2\left\{n_\alpha w_{\alpha r}\langle F_{\alpha s}^* \rangle\right\} - 2\frac{e_\alpha}{m_\alpha}\left\{\left[\overset{\wedge}{\pi}_\alpha \times \mathbf{B}\right]_{rs}\right\} = R_{\alpha rs}^{20}, \tag{4.2.13}$$

$$\frac{dh_{\alpha r}}{dt} + \frac{7}{5}h_{\alpha s}\frac{\partial u_r}{\partial x_s} + \frac{2}{5}h_{\alpha s}\frac{\partial u_s}{\partial x_r} + \frac{7}{5}h_{\alpha r}\nabla \cdot \mathbf{u} + \frac{5}{3}p_\alpha w_{\alpha r}\nabla \cdot \mathbf{u} + 2p_\alpha w_{\alpha s}\varepsilon_{rs} + \frac{1}{\gamma_\alpha}\frac{\partial \pi_{\alpha rs}}{\partial x_s} + \frac{7}{2}\frac{1}{\gamma_\alpha}\pi_{\alpha rs}\frac{1}{T_\alpha}\frac{\partial T_\alpha}{\partial x_s} + \frac{5}{2}\frac{1}{\gamma_\alpha}\frac{p_\alpha}{T_\alpha}\frac{\partial T_\alpha}{\partial x_r} + \frac{5}{2}\frac{p_\alpha}{T_\alpha}w_{\alpha r}\frac{dT_\alpha}{dt} - \frac{1}{m_\alpha}\pi_{\alpha rs}\langle F_{\alpha s}^* \rangle - \frac{e_\alpha}{m_\alpha}(\mathbf{h}_\alpha \times \mathbf{B})_r = R_{\alpha r}^{11}, \tag{4.2.14}$$

where the reductions are made

$$\{K_r L_s\} = \frac{1}{2}(K_r L_s + L_r K_s) - \frac{1}{3}K_l L_l \delta_{rs},$$

$$\varepsilon_{rs} = \left\{\frac{\partial u_r}{\partial x_s}\right\}, \quad \left[\hat{\pi}_\alpha \times \mathbf{B}\right]_{rs} = \pi_{\alpha lr}\varepsilon_{slm}B_m,$$

and ε_{slm} is the antisymmetrical unit tensor of the third rank or the permutation tensor.

On the right-hand sides of (4.2.11)–(4.2.14) the "moments with respect to the collision integral" appear

$$R^{mn}_{\alpha r_1 \ldots r_m} = \sum_\beta \iint \left[G^{mn}_{\alpha r_1 \ldots r_m}\left(\mathbf{c}'_\alpha\right) - G^{mn}_{\alpha r_1 \ldots r_m}\left(\mathbf{c}_\alpha\right)\right] \times f_\alpha f_\beta g \sigma_{\alpha\beta}(g, \chi)\, d\Omega d\mathbf{c}_\alpha d\mathbf{c}_\beta. \tag{4.2.15}$$

It should be pointed out that (4.2.11) and (4.2.12) are conservation equations of energy and momentum for species α of the mixture considered earlier (Eqs. (2.2.4) and (2.2.3)) with $R^{10}_{\alpha r} = R_{\alpha r}$, $R^{01}_\alpha = Q_\alpha$.

The values of $R^{mn}_{\alpha r_1 \ldots r_m}$ can be calculated by the same technique which was used in finding $\mathbf{R}_\alpha$ and Q_α in the previous chapter. Passing over to the variables in the centre-of-mass system of the colliding particles and using the relationship between the relative velocities before and after the collision, we can calculate the integrals with respect to angles. Then the expressions for f_α and f_β are substituted into (4.2.15) in the form of (4.2.5) in which the transition to $\mathbf{X}$ and $\mathbf{g}$ (3.1.7) is performed. Ignoring the moment-squared terms and after integrating with respect to $\mathbf{X}$, we obtain the following general expressions [11–13]

$$R^{10}_{\alpha r} = \sum_\beta G^{(1)}_{\alpha\beta}(w_{\alpha r} - w_{\beta r}) + \sum_\beta \gamma_{\alpha\beta} G^{(2)}_{\alpha\beta}\left(\frac{h_{\alpha r}}{\gamma_\alpha p_\alpha} - \frac{h_{\beta r}}{\gamma_\beta p_\beta}\right),$$

$$R^{01}_\alpha = \sum_\beta G^{(1)}_{\alpha\beta}\frac{3k}{m_\alpha + m_\beta}(T_\alpha - T_\beta),$$

$$R^{20}_{\alpha rs} = \sum_\beta \frac{1}{\gamma_\alpha + \gamma_\beta}\left(G^{(3)}_{\alpha\beta}\frac{\pi_{\alpha rs}}{p_\alpha} + G^{(4)}_{\alpha\beta}\frac{\pi_{\beta rs}}{p_\beta}\right),$$

$$R^{11}_{\alpha r} = \frac{1}{\gamma_\alpha}\sum_\beta \left[G^{(5)}_{\alpha\beta}\frac{h_{\alpha r}}{p_\alpha} + G^{(6)}_{\alpha\beta}\frac{h_{\beta r}}{p_\beta} + \frac{5}{2}\frac{\gamma_{\alpha\beta}}{\gamma_\alpha}G^{(7)}_{\alpha\beta}(w_{\alpha r} - w_{\beta r}) + 5\Theta_{\alpha\beta}G^{(1)}_{\alpha\beta}w_{\alpha r}\right]. \tag{4.2.16}$$

The coefficients $G^{(n)}_{\alpha\beta}$ turn out to be linear combinations of the values $\Omega^{lr}_{\alpha\beta}$ which are the generalized well-known Chapman–Cowling integrals [5] for the case of different temperatures of the species.

The expressions of the coefficients are [13]

$$
\begin{aligned}
&G^{(1)}_{\alpha\beta} = B^{(1)}_{\alpha\beta}, \qquad\qquad G^{(2)}_{\alpha\beta} = B^{(2)}_{\alpha\beta},\\
&G^{(3)}_{\alpha\beta} = B^{(3)}_{\alpha\beta} - \frac{\gamma_\beta}{\gamma_\alpha}\Theta_{\alpha\beta}C^{(1)}_{\alpha\beta}, \quad G^{(4)}_{\alpha\beta} = B^{(4)}_{\alpha\beta} - \Theta_{\alpha\beta}C^{(1)}_{\alpha\beta},\\
&G^{(5)}_{\alpha\beta} = B^{(5)}_{\alpha\beta} + \frac{1}{2}\frac{\gamma_\beta}{\gamma_\alpha}\Theta_{\alpha\beta}\left(C^{(2)}_{\alpha\beta} - C^{(3)}_{\alpha\beta} - 2\frac{\gamma_\alpha}{\gamma_\beta}C^{(4)}_{\alpha\beta}\right)\\
&\qquad - \frac{\gamma_\beta}{\gamma_\alpha}\Theta^2_{\alpha\beta}\left(C^{(2)}_{\alpha\beta} - \frac{1}{2}C^{(3)}_{\alpha\beta}\right),\\
&G^{(6)}_{\alpha\beta} = B^{(6)}_{\alpha\beta} - \frac{1}{2}\Theta_{\alpha\beta}\left(C^{(2)}_{\alpha\beta} - C^{(3)}_{\alpha\beta} + 2C^{(4)}_{\alpha\beta}\right) + \Theta^2_{\alpha\beta}\left(C^{(2)}_{\alpha\beta} - \frac{1}{2}C^{(3)}_{\alpha\beta}\right),\\
&G^{(7)}_{\alpha\beta} = B^{(2)}_{\alpha\beta} + \Theta_{\alpha\beta}C^{(5)}_{\alpha\beta} - \Theta^2_{\alpha\beta}C^{(6)}_{\alpha\beta},
\end{aligned}
\tag{4.2.17}
$$

where

$$
\begin{aligned}
B^{(1)}_{\alpha\beta} &= -\frac{16}{3}\mu_{\alpha\beta}n_\alpha n_\beta\Omega^{11}_{\alpha\beta},\\
B^{(2)}_{\alpha\beta} &= -\frac{16}{3}\mu_{\alpha\beta}n_\alpha n_\beta\left(\frac{2}{5}\Omega^{12}_{\alpha\beta} - \Omega^{11}_{\alpha\beta}\right),\\
B^{(3)}_{\alpha\beta} &= -\frac{16}{5}\mu_{\alpha\beta}n_\alpha n_\beta\left(\frac{\gamma_\beta}{\gamma_\alpha}\Omega^{22}_{\alpha\beta} + \frac{10}{3}\Omega^{11}_{\alpha\beta}\right),\\
B^{(4)}_{\alpha\beta} &= -\frac{16}{5}\mu_{\alpha\beta}n_\alpha n_\beta\left(\Omega^{22}_{\alpha\beta} - \frac{10}{3}\Omega^{11}_{\alpha\beta}\right),\\
B^{(5)}_{\alpha\beta} &= -\frac{64}{15}\mu_{\alpha\beta}\kappa_{\alpha\beta}n_\alpha n_\beta\left[\Omega^{22}_{\alpha\beta} + \left(\frac{15}{4}\frac{\gamma_\alpha}{\gamma_\beta} + \frac{25}{8}\frac{\gamma_\beta}{\gamma_\alpha}\right)\Omega^{11}_{\alpha\beta}\right.\\
&\qquad\left. - \frac{1}{2}\frac{\gamma_\beta}{\gamma_\alpha}\left(5\Omega^{12}_{\alpha\beta} - \Omega^{13}_{\alpha\beta}\right)\right],\\
B^{(6)}_{\alpha\beta} &= -\frac{64}{15}\mu_{\alpha\beta}\kappa_{\alpha\beta}n_\alpha n_\beta\left[\Omega^{22}_{\alpha\beta} - \frac{55}{8}\Omega^{11}_{\alpha\beta} + \frac{1}{2}\left(5\Omega^{12}_{\alpha\beta} - \Omega^{13}_{\alpha\beta}\right)\right],
\end{aligned}
\tag{4.2.18}
$$

$$
\begin{aligned}
C_{\alpha\beta}^{(1)} &= -\frac{16}{5}\mu_{\alpha\beta}n_\alpha n_\beta\left(\Omega_{\alpha\beta}^{22}-\frac{4}{3}\Omega_{\alpha\beta}^{12}\right),\\
C_{\alpha\beta}^{(2)} &= -\frac{64}{15}\mu_{\alpha\beta}\kappa_{\alpha\beta}n_\alpha n_\beta\left(5\Omega_{\alpha\beta}^{12}-2\Omega_{\alpha\beta}^{13}\right),\\
C_{\alpha\beta}^{(3)} &= -\frac{64}{15}\mu_{\alpha\beta}\kappa_{\alpha\beta}n_\alpha n_\beta\left(5\Omega_{\alpha\beta}^{22}-2\Omega_{\alpha\beta}^{23}\right),\\
C_{\alpha\beta}^{(4)} &= -\frac{64}{15}\mu_{\alpha\beta}\kappa_{\alpha\beta}n_\alpha n_\beta\left(\Omega_{\alpha\beta}^{22}-\frac{11}{2}\Omega_{\alpha\beta}^{12}+\frac{25}{4}\Omega_{\alpha\beta}^{11}\right),\\
C_{\alpha\beta}^{(5)} &= -\frac{64}{15}\mu_{\alpha\beta}n_\alpha n_\beta\left(\Omega_{\alpha\beta}^{22}-\Omega_{\alpha\beta}^{12}-\frac{5}{2}\Omega_{\alpha\beta}^{11}\right),\\
C_{\alpha\beta}^{(6)} &= -\frac{64}{15}\mu_{\alpha\beta}n_\alpha n_\beta\left(\Omega_{\alpha\beta}^{22}-2\Omega_{\alpha\beta}^{12}\right).
\end{aligned}
\tag{4.2.19}
$$

Here

$$
\Omega_{\alpha\beta}^{lr} = \left(\frac{2\pi}{\gamma_{\alpha\beta}}\right)^{1/2}\int_0^\infty\int_0^\pi \zeta^{2r+3}e^{-\zeta^2}\left(1-\cos^l\chi\right)\sigma_{\alpha\beta}(\zeta,\chi)\sin\chi d\chi d\zeta.
\tag{4.2.20}
$$

and $\zeta = (\gamma_{\alpha\beta}/2)^{1/2}g$. The values $\kappa_{\alpha\beta}, \gamma_{\alpha\beta}$ and $\Theta_{\alpha\beta}$ are defined by the expressions

$$
\kappa_{\alpha\beta} = \frac{\gamma_{\alpha\beta}}{\gamma_\alpha+\gamma_\beta}, \qquad \gamma_{\alpha\beta} = \frac{\gamma_\alpha\gamma_\beta}{\gamma_\alpha+\gamma_\beta},
$$

$$
\Theta_{\alpha\beta} = \frac{1-\dfrac{T_\beta}{T_\alpha}}{1+\dfrac{m_\beta}{m_\alpha}}.
\tag{4.2.21}
$$

4.3 Linear transport relations

The equations in the previous section form a system of quasilinear (i.e. linear with respect to the derivatives) differential equations of the first order which must be solved in the general case on account of specific initial and boundary conditions. However, the equations for $\mathbf{w}_\alpha$, $\pi_{\alpha rs}$ and $\mathbf{h}_\alpha$ can be essentially simplified assuming that the macroscopic parameters of both the species and plasma as a whole slightly change at distances and times of the order of the mean free path and time between collisions of the particles, i.e.

$$\lambda/L \ll 1, \qquad \tau/\tau_L \ll 1. \tag{4.3.1}$$

Then the small gradients of the basic variables ρ_α, u and T_α and small values of $\gamma_\alpha^{1/2} w_{\alpha r}$, $p_\alpha^{-1}\pi_{\alpha rs}$ and $\gamma_\alpha^{1/2} p_\alpha^{-1} h_{\alpha r}$, due to the vicinity of the system to the local equilibrium, allow us to disregard the products of small terms (λ/L squared) in Eqs. (4.2.12)–(4.2.14). Note that the coefficients $G_{\alpha\beta}^{(n)}$ on the right-hand sides of the equations are of the same order of values as $G_{\alpha\beta}^{(1)} = -n_\alpha \mu_{\alpha\beta}\tau_{\alpha\beta}^{-1}$. Thus, because of the second requirement (4.3.1) the time derivatives of the moments on the left-hand sides compared with the right-hand ones can be neglected. This means that in the case of a slow change in the time parameters considered there arises a quasiequilibrium after a time interval equal to some mean times between the collisions in the plasma. Instead of Eqs. (4.2.12)–(4.2.14), a system of equations can be approximately applied [12,13]

$$\frac{\partial p_\alpha}{\partial x_r} + \frac{\partial \pi_{\alpha rs}}{\partial x_s} - n_\alpha \langle F_{\alpha r}^* \rangle = R_{\alpha r}^{10}, \tag{4.3.2}$$

$$2p_\alpha \varepsilon_{rs} + \frac{4}{5}\left\{\frac{\partial q_{\alpha r}}{\partial x_s}\right\} - 2\left\{n_\alpha w_{\alpha r}\langle F_{\alpha s}^*\rangle\right\} - 2\left\{\pi_{\alpha l r} e_{slm}\omega_{\alpha m}\right\} = R_{rs}^{20}, \tag{4.3.3}$$

$$\frac{5}{2}\frac{k}{m_\alpha} p_\alpha \frac{\partial T_\alpha}{\partial x_r} + \frac{kT_\alpha}{m_\alpha}\frac{\partial \pi_{\alpha rs}}{\partial x_s} - \frac{1}{m_\alpha}\pi_{\alpha rs}\langle F_{\alpha s}^*\rangle - (\mathbf{h}_\alpha \times \boldsymbol{\omega}_\alpha)_r = R_{\alpha r}^{11}, \tag{4.3.4}$$

where $\boldsymbol{\omega}_\alpha = \omega_\alpha \mathbf{k}$ and $\omega_\alpha = (e_\alpha/m_\alpha)|\mathbf{B}|$ is the cyclotron frequency of the charged particles of species α.

The terms of the form $\partial\pi_{\alpha rs}/\partial x_s$ and $\partial q_{\alpha r}/\partial x_s$ are sometimes of the order of $(\lambda/L)^2$. However, they should be taken into account in the problems when the longitudinal and transverse scales of changing macroscopic plasma parameters are essentially different. Some examples (in particular, the contribution of the pressure gradient to the heat flux, the effect of the viscous transport of the momentum on the diffusion in a steady viscous flow of the mixture [11]) will be considered later. As to the terms depending on $\langle F_{\alpha s}^*\rangle$ in Eqs. (4.3.3)–(4.3.4), their influence manifests itself only in rather strong electric fields [10].

These specific cases apart and with the corresponding terms on the left-hand side of (4.3.2)–(4.3.4) omitted, the final simplified system of equations for $\mathbf{w}_\alpha$, $\pi_{\alpha rs}$ and $\mathbf{h}_\alpha$ is written as[1]

$$\left(\nabla p_\alpha - \frac{\rho_\alpha}{\rho}\nabla p\right) - \left(n_\alpha \mathbf{X}_\alpha - \frac{\rho_\alpha}{\rho}\sum_\alpha n_\alpha \mathbf{X}_\alpha\right) - n_\alpha e_\alpha \mathbf{E}'$$
$$-\rho_\alpha \mathbf{w}_\alpha \times \boldsymbol{\omega}_\alpha + \frac{\rho_\alpha}{\rho}\mathbf{j}\times\mathbf{B} = \sum_\beta G^{(1)}_{\alpha\beta}(\mathbf{w}_\alpha - \mathbf{w}_\beta)$$
$$+\sum_\beta \gamma_{\alpha\beta} G^{(2)}_{\alpha\beta}\left(\frac{\mathbf{h}_\alpha}{\gamma_\alpha p_\alpha} - \frac{\mathbf{h}_\beta}{\gamma_\beta p_\beta}\right), \tag{4.3.5}$$

$$2p_\alpha \varepsilon_{rs} - 2\{\pi_{\alpha lr} e_{slm}\omega_{\alpha m}\} = \sum_\beta \frac{1}{\gamma_\alpha + \gamma_\beta}\left(G^{(3)}_{\alpha\beta}\frac{\pi_{\alpha rs}}{p_\alpha} + G^{(4)}_{\alpha\beta}\frac{\pi_{\beta rs}}{p_\beta}\right), \tag{4.3.6}$$

$$\frac{5}{2}\frac{k}{m_\alpha}p_\alpha \nabla T_\alpha - \mathbf{h}_\alpha \times \boldsymbol{\omega}_\alpha = \frac{1}{\gamma_\alpha}\sum_\beta \left[G^{(5)}_{\alpha\beta}\frac{\mathbf{h}_\alpha}{p_\alpha} + G^{(6)}_{\alpha\beta}\frac{\mathbf{h}_\beta}{p_\beta}\right.$$
$$\left. + \frac{5}{2}\frac{\gamma_{\alpha\beta}}{\gamma_\alpha}G^{(7)}_{\alpha\beta}(\mathbf{w}_\alpha - \mathbf{w}_\beta) + 5\Theta_{\alpha\beta}G^{(1)}_{\alpha\beta}\mathbf{w}_\alpha\right]. \tag{4.3.7}$$

4.4 One-temperature plasma

It is appropriate to consider cases when the set of Eqs. (4.3.5)–(4.3.7) allows further simplifications. Let us discuss an isothermal multicomponent plasma, e.g., when $T_\alpha = T$ for all α. In this case, due to $\Theta_{\alpha\beta} = 0$ the coefficients $G^{(n)}_{\alpha\beta}$ on the right-hand sides of the equations are substantially simplified and the last term on the right-hand side of (4.3.7) can be omitted. As a result, the system of equations for $\mathbf{w}_\alpha$, $\pi_{\alpha rs}$ and $\mathbf{h}_\alpha$ is written as [11,12]

$$p\mathbf{d}_\alpha - \rho_\alpha \mathbf{w}_\alpha \times \boldsymbol{\omega}_\alpha + \frac{\rho_\alpha}{\rho}\mathbf{j}\times\mathbf{B}$$
$$= \sum_\beta B^{(1)}_{\alpha\beta}(\mathbf{w}_\alpha - \mathbf{w}_\beta) + \sum_\beta \mu_{\alpha\beta}B^{(2)}_{\alpha\beta}\left(\frac{\mathbf{h}_\alpha}{m_\alpha p_\alpha} - \frac{\mathbf{h}_\beta}{m_\beta p_\beta}\right), \tag{4.4.1}$$

---

[1]We have excluded the term $d\mathbf{u}/dt$ in $\langle \mathbf{F}^*_\alpha \rangle$ with the help of the motion equation of the plasma as a whole (2.2.17).

$$2p_\alpha\varepsilon_{rs} - 2\{\pi_{\alpha lr}e_{slm}\omega_{\alpha m}\}$$
$$= \sum_\beta \frac{kT}{m_\alpha + m_\beta}\left(B^{(3)}_{\alpha\beta}\frac{\pi_{\alpha rs}}{p_\alpha} + B^{(4)}_{\alpha\beta}\frac{\pi_{\beta rs}}{p_\beta}\right), \qquad (4.4.2)$$

$$\frac{5}{2}\frac{k}{m_\alpha}p_\alpha\nabla T - \mathbf{h}_\alpha\times\omega_\alpha$$
$$= \frac{kT}{m_\alpha}\sum_\beta\left[B^{(5)}_{\alpha\beta}\frac{\mathbf{h}_\alpha}{p_\alpha} + B^{(6)}_{\alpha\beta}\frac{\mathbf{h}_\beta}{p_\beta} + \frac{5}{2}\frac{\mu_{\alpha\beta}}{m_\alpha}B^{(2)}_{\alpha\beta}(\mathbf{w}_\alpha - \mathbf{w}_\beta)\right],$$
$$(4.4.3)$$

where

$$\mathbf{d}_\alpha = \nabla y_\alpha + \left(y_\alpha - \frac{\rho_\alpha}{\rho}\right)\nabla\ln p$$
$$-\frac{1}{p}\left(n_\alpha\mathbf{X}_\alpha - \frac{\rho_\alpha}{\rho}\sum_\alpha n_\alpha\mathbf{X}_\alpha\right) - \frac{1}{p}n_\alpha e_\alpha\mathbf{E}' \qquad (4.4.4)$$

and $y_\alpha = n_\alpha/n$ is the relative concentration of species α. In the expressions for $B^{(n)}_{\alpha\beta}$ (4.2.18) $\gamma_\beta/\gamma_\alpha = m_\beta/m_\alpha$, $\kappa_{\alpha\beta} = \mu_{\alpha\beta}/(m_\alpha + m_\beta)$ should be taken.

Note that the assumption of the rigorous equality of temperatures of the plasma species at the gradients of thermodynamic values and fields which perturb the equilibrium state is, to be precise, not absolutely correct. In fact, when formally describing such a mixture within the framework of the moment method, the Maxwellian written down at T rather than T_α should be used. Here T is the general temperature of the mixture described by the relationship $nT = \sum_\alpha n_\alpha T_\alpha$. In doing so, the expression of the distribution function of the species α in the mixture in the $13N$-moment approximation acquires the form

$$f_\alpha = f^{(0)}_\alpha(T)\left[1 + \frac{1}{6}a^{01}_\alpha(\xi^2 - 3) + a^{10}_{\alpha r}\xi_r\right.$$
$$\left. + \frac{1}{2}a^{20}_{\alpha rs}\left(\xi_r\xi_s - \frac{1}{3}\delta_{rs}\xi^2\right) + \frac{1}{10}a^{11}_r\xi_r(\xi^2 - 5)\right],$$

where all the coefficients a^{mn}_α are defined at temperature T. An additional term of the form $(1/6)\,a^{01}_\alpha(\xi^2 - 3)$ with $a^{01}_\alpha = 3(T_\alpha - T)/T$ also develops in the expansion as $a^{01}_\alpha \neq 0$. Let the relative difference of temperatures $(T_\alpha - T)/T$ be a small perturbation parameter as the other expansion coefficient. If we neglect the moment-squared terms, the new term available in the expression will take effect only in calculating the value of Q_α on the right-hand side of the energy equation for species α. In fact, the value $Q_\alpha = R_\alpha^{01}$ retains its previous form (4.2.16) with the only difference that $G_{\alpha\beta}^{(1)} = B_{\alpha\beta}^{(1)}$ is defined at temperature T.

Let us use the energy equation for particles of species α (2.2.21) and exclude $dT/dt + (2/3)T\nabla \cdot \mathbf{u}$ with the help of the energy equation for the plasma as a whole (2.2.18). The result will be

$$\frac{3}{2}kn_\alpha \frac{d}{dt}(T_\alpha - T) + kn_\alpha (T_\alpha - T)\nabla \cdot \mathbf{u} + \left(\pi_{\alpha rs} - \frac{n_\alpha}{n}\pi_{rs}\right)\frac{\partial u_r}{\partial x_s}$$

$$+\left(\nabla \cdot \mathbf{q}_\alpha - \frac{n_\alpha}{n}\nabla \cdot \mathbf{q}\right) - \left(n_\alpha \mathbf{w}_\alpha \cdot \mathbf{X}_\alpha - \frac{n_\alpha}{n}\sum_\alpha n_\alpha \mathbf{w}_\alpha \cdot \mathbf{X}_\alpha\right)$$

$$-(n_\alpha e_\alpha \mathbf{w}_\alpha - \mathbf{j}) \cdot \mathbf{E}' - \rho_\alpha \mathbf{w}_\alpha \cdot \frac{d\mathbf{u}}{dt} = -\frac{3}{2}kn_\alpha \sum_\beta \left(\tau_{\alpha\beta}^E\right)^{-1}(T_\alpha - T_\beta). \tag{4.4.5}$$

For the plasma species whose particle masses do not substantially differ, the characteristic time of energy exchange $\tau_{\alpha\beta}^E$ is of the order of the mean characteristic time $\tau_{\alpha\beta}$. Then in Eq. (4.4.5), written for particles of this type because of the requirement (4.3.1), the derivative $d(T_\alpha - T)/dt$ can be neglected on the left-hand side compared with the right-hand side. Eliminating $d\mathbf{u}/dt$ by means of the motion equation, (2.2.17), one can easily see that in the absence of external fields the temperature differences defined from (4.4.5) turn out to be proportional to the term combinations representing the products of flux and gradients of thermodynamical values or divergences of the fluxes proper. As these terms are as a rule of the order of $(\lambda/L)^2$, then in the approximation linear in the Knudsen number $Kn = \lambda/L$, the assumption of nearly equal temperatures of the heavy components of the plasma look quite justified. Clearly, this assumption is not fulfilled for both electrons and light mass particles noticeably differing from the mass of the rest of the atoms or ions as $\tau_{\alpha\beta}^E \gg \tau_{\alpha\beta}$. For the isothermal state of the plasma to be realized in this case, a more rigid requirement $\tau_{\alpha\beta}^E \ll \tau_L$ must be satisfied.

The application of the one-temperature approach in the plasma theory allows one not only to simplify the form of the equations used to define the non-equilibrium parameters $\mathbf{w}_\alpha$, $\pi_{\alpha rs}$ and $\mathbf{h}_\alpha$ but it also permits one decrease markedly the complete number of equations necessary for a full macroscopic description of plasma. For this purpose, it is practical to include the conservation equations for the plasma as a whole

$$\frac{d\rho}{dt} + \rho\nabla\cdot\mathbf{u} = 0,$$

$$\rho\frac{d\mathbf{u}}{dt} + \nabla p + \nabla\,\hat{\pi} - \sum_{\alpha} n_\alpha \mathbf{X}_\alpha - \mathbf{j}\times\mathbf{B} = 0,$$

$$\frac{3}{2}\frac{dp}{dt} + \frac{5}{2}p\nabla\cdot\mathbf{u} + \pi_{rs}\frac{\partial u_r}{\partial x_s} + \nabla\cdot\mathbf{q} - \sum_{\alpha} n_\alpha \mathbf{w}_\alpha\cdot\mathbf{X}_\alpha - \mathbf{j}\cdot\mathbf{E}' = 0 \tag{4.4.6}$$

and the $N-1$ continuity equations for n_α and the state equation

$$\frac{dn_\alpha}{dt} + n_\alpha\nabla\cdot\mathbf{u} + \nabla\cdot n_\alpha\mathbf{w}_\alpha = \dot{n}_\alpha, \quad p = kT\sum_{\alpha} n_\alpha. \tag{4.4.7}$$

The constituents of this system of equations such as the diffusion velocity $\mathbf{w}_\alpha$, the current density $\mathbf{j} = \sum_{\alpha} n_\alpha e_\alpha \mathbf{w}_\alpha$, the viscous-stress tensor $\pi_{rs} = \sum_{\alpha} \pi_{\alpha rs}$ and the total heat flux $\mathbf{q} = \sum_{\alpha}[\mathbf{h}_\alpha + (5/2)p_\alpha\mathbf{w}_\alpha]$ are defined from the solutions of the algebraic system of Eqs. (4.4.1)–(4.4.3) to be added by the condition $\sum_{\alpha} \rho_\alpha \mathbf{w}_\alpha = 0$.

To close the given system fully, the Maxwell equations describing the changes of the electric and magnetic fields must be also added

$$\nabla\times\mathbf{B} = \mu_0\mathbf{j}, \quad \nabla\times\mathbf{E} = -\frac{\partial\mathbf{B}}{\partial t}, \quad \nabla\cdot\mathbf{B} = 0. \tag{4.4.8}$$

When writing these equations, we neglect the replacement current which is valid when considering rather slow processes in the plasma. Besides, the Poisson equation $\varepsilon_0\nabla\cdot\mathbf{E} = \rho^e$ including explicitly the space change is omitted since the quasi-neutrality condition $\rho_e = \sum_{\alpha} n_\alpha e_\alpha = 0$ is used. In practice, this equation will be needed only to determine the deviation of plasma from the quasineutral state on the basis of the field $\mathbf{E}$ derived from the rest of the system equations which must be taken into account either in the vicinity of the boundaries of plasma (in the layer of the thickness of the order of the Debye radius) or in the analysis of high frequency oscillations of plasma.

Along with Eqs. (4.4.1)–(4.4.3) and electrodynamics equations (4.4.8) the system (4.4.6) underlies the so called magnetohydrodynamic description of plasma [14]. Note that due to (4.3.1) it is generally suitable for considering a slowly varying state of the plasma. Further simplifications of this system lead to various forms of the so called magnetohydrodynamics used to describe the interactions of the moving plasma and electromagnetic fields, with a wide range of practical applications [14–16]. For example, consider the case when the forces of non-electromagnetic nature correspond to the conventional gravitation ($\mathbf{X}_\alpha = -m_\alpha \mathbf{g}$). Then the equation system (4.4.6) may be rewritten as

$$\frac{d\rho}{dt} + \rho \nabla \cdot \mathbf{u} = 0,$$

$$\rho \frac{d\mathbf{u}}{dt} + \nabla p + \nabla \overset{\wedge}{\pi} + \rho \mathbf{g} - \mathbf{j} \times \mathbf{B} = 0, \tag{4.4.9}$$

$$\frac{3}{2}\frac{dp}{dt} + \frac{5}{2} p \nabla \cdot \mathbf{u} + \pi_{rs} \frac{\partial u_r}{\partial x_s} + \nabla \cdot \mathbf{q} - \mathbf{j} \cdot \mathbf{E}' = 0.$$

If the generation and disappearance of particles of species α per plasma unit volume is characterized by high process speeds ($\tau_{react} \ll \tau_L$), then the continuity equations (4.4.7) for the species are reduced to the system $\dot{n}_\alpha = 0$. In practice that means that the densities of particles of species be found from the equations of local chemical equilibrium. In the case when the chemical reactions are the ionization and recombination, an example of such an equation is the well-known Sacha equation [17]. The state equation acquires the form

$$p = \rho \frac{kT}{\langle m \rangle}, \qquad \langle m \rangle = \sum_\alpha m_\alpha y_\alpha, \tag{4.4.10}$$

where $\langle m \rangle$ is expressed in terms of parameters T, p with the help of the equations of chemical equilibrium $y_\alpha = f(T, p)$. The functions of these parameters are found to be the coefficients $B_{\alpha\beta}^{(n)}$ of the right-hand sides of Eqs. (4.4.1)–(4.4.3). As a result, to close the system of Eqs. (4.4.9)–(4.4.10) it is necessary to have the relation of the current density $\mathbf{j}$ to be found from the generalized Ohm's law and the expressions for π_{rs} and $\mathbf{q}$ relating the fluxes to the gradients of hydrodynamic parameters and the values of electric and magnetic fields.

We have considered special cases of the generalized Ohm's law earlier. A fuller statement of the law with regard for thermal diffusion and corrections of higher approximations to the electrical conductivity will be made later. The topics of further discussions will also be expressions of the viscous-stress tensor and heat flux vector.

4.5 Two-temperature plasma

When an electric field is applied to a plasma, it mostly affects the mobile electron component of the plasma. As was mentioned, electrons lose only a small amount of their energy in collisions with heavy particles and, because of the constant energy supply by the field, that can lead to a marked difference between the electron temperature and the temperature of heavy particles. For the stationary homogeneous plasma the order of value of the field $\mathbf{E}$ necessary in this case is easily estimated with the help of (3.1.24) drawn from the energy conservation equation for particles of species α. Substituting $\mathbf{j}_e \cong \sigma_0 \mathbf{E} = \left(n_e e^2 \tau_0 / m_e\right) \mathbf{E}$, we find that $(T_e - T)/T$ is of the order of unity at $E = E_p$ defined by the expression

$$E_p^2 = \frac{3}{2} \frac{m_e k T}{e^2 \tau_0} \sum_\beta \delta_\beta \tau_{e\beta}^{-1},$$

where $\tau_0^{-1} = \sum_\beta \tau_{e\beta}^{-1}$ and $\delta_\beta = 2m_e/m_\beta$ is the mean proportion of the energy that is transferred in an individual elastic collision of an electron and a heavy particle of species β.

In collisions of heavy particles in the plasma, the mean energy transferred in a collision proves to be of the same order as the eigen energy of the particles. Due to the intense energy transfer resulting from collisions the magnitude of the field to ensure the noticeable difference in the temperatures of the heavy particles is much greater than the field E_p. It is this consideration that allows us to introduce the model of two-temperature plasma in which the temperature of heavy particles is taken to be approximately equal ($T_\beta \cong T$ at $\beta \neq e$) and the temperature of electrons T_e can be appreciably different from T. Clearly, at $E \ll E_p$ the electron temperature is practically the same as the general plasma temperature, i.e. the equations of one-temperature approximation become valid.

It is appropriate to make a remark here. The estimates of the field magnitude necessary for the marked difference between T_e and T given here are based on the equation of the energy balance for the electrons whose right-hand side is calculated on the assumption that the distribution function of the electrons corresponds to the Maxwell distribution at temperature T_e. This assumption can be shown to be fulfilled in the case of a highly ionized plasma if the effective frequency of electron–electron collisions satisfies the requirement [17,18]

$$\tau_{ee}^{-1} \gg \sum_{\beta \neq e} \delta_\beta \tau_{e\beta}^{-1}.$$

For a weakly ionized plasma, the Maxwell distribution function for the electrons at temperature T_e and the correspondent expression for the mean transfer of energy in collisions are formally retained only in the case when the effective collision frequency of electrons and neutrals does not depend on velocity ($\nu_{e\beta}(v) = vQ^{(1)}_{e\beta}(v) = \text{const}$). Under the arbitrary dependence $\nu_{e\beta}(v)$ the distribution function $f^{(0)}_e$ can turn out to be essentially non-Maxwellian, in particular for the model of hard elastic spheres when $Q^{(1)}_{e\beta}(v) = \text{const}$, at $E \gg E_p$, the Druyvestein distribution takes place and at high velocities it declines faster than the Maxwell distribution [18].

Another remark is pertinent in the case when there are both atomic and molecular species in the plasma. The mean part of the transferred energy in an inelastic collision can exceed the value of $\delta_\beta = 2m_e/m_\beta$ as there are losses of electron energy because of exciting the rotational and vibrational degrees of freedom of the molecules. In such a plasma the difference between T_e and T turns out to be less than in the case of only elastic collisions, all factors being the same.

Now we shall examine what equations for $\mathbf{w}_\alpha$, $\pi_{\alpha rs}$ and $\mathbf{h}_\alpha$ will look like for the case of a two-temperature plasma. As the temperatures of heavy particles were assumed to be the same, the left-hand sides of Eqs. (4.3.5)–(4.3.7) for $\alpha \neq e$ can be written as was done in the equations of the one-temperature approach (4.4.1)–(4.4.3). For the electron component ($\alpha = e$) it is appropriate to keep the left-hand sides in the form of (4.3.5)–(4.3.7), since $T_\alpha = T_e$ and $p_\alpha = n_e kT_e$ in this case. Looking at the coefficients in the right-hand sides, it is seen that if the requirement $\gamma_e \ll \gamma_\beta$ (or $m_e T/m_\beta T_e \ll 1$) is fulfilled, then $\gamma_{e\beta} \cong \gamma_e$. Besides, in the expressions for $G^{(n)}_{\alpha\beta}$ the terms of the orders $\Theta_{e\beta} = (m_e/m_\beta)\left[1 - (T/T_e)\right]$ can be neglected compared with unity, therefore, $G^{(n)}_{e\beta} \cong B^{(n)}_{e\beta}$. As a result, the right-hand sides of the equations for electrons are written in the form similar to that for the equal component temperatures, but with the substitute of T for T_e in the coefficients $\gamma_e = m_e/kT_e$, $\gamma_{e\beta} = \gamma_e$, $B^{(n)}_{ee}$ and $B^{(n)}_{e\beta}$. The right-hand sides of equations for the heavy particles are written in the same way as in the equations of the one-temperature approximations (i.e. at temperature T) except for the terms in these equations, which are written with coefficients $B_{\alpha e}$ ($\alpha \neq e$) and must be written at temperature T_e. A detailed analysis proves that the latter terms in the equations for heavy particles can be neglected as a rule.

It is essential that the peculiar features of the equations just discussed also remain for the equations of higher approximations of the moment method. This allows us to avoid rather complex calculations of $R^{mn}_{\alpha r_1 \ldots r_m}$ of higher orders at different species temperatures but to use instead the fact that these values can be represented by means of the so called bracket integrals of the Sonine polynomials which were repeatedly calculated for the case of an isothermal gas mixture ($T_\alpha = T$)

(ref. [5,6] and Appendix 2).

The general system of Eqs. (4.4.6) suitable to describe the magnetohydrodynamic state of plasma remains also valid for the two-temperature plasma, though the state equation acquires another form

$$p = n_e kT_e + \sum_{\beta \neq e} n_\beta kT, \tag{4.5.1}$$

where T is the temperature of the heavy components of the plasma. Besides, the equation of energy conservation for electrons which follows from (2.2.21) should be added to the set of defining equations. It can be written as

$$\frac{3}{2}\frac{d}{dt}(n_e kT_e) + \frac{5}{2} n_e kT_e \nabla \cdot \mathbf{u} + \nabla \cdot \mathbf{q}_e + \pi_{ers} \frac{\partial u_r}{\partial x_s}$$
$$-n_e \mathbf{w}_e \cdot \mathbf{X}_e - \mathbf{j}_e \cdot \mathbf{E}' = -\frac{3}{2} k n_e \sum_\beta \frac{2m_e}{m_\beta} \tau_{e\beta}^{-1} (T_e - T). \tag{4.5.2}$$

We have neglected the term $\rho_e \mathbf{w}_e \cdot d\mathbf{u}/dt$ in this equation because of its small value.

When accounting for inelastic processes in the plasma in the energy equation (4.5.2) certain changes should be introduced which take into consideration the contribution of ionization energy ε_i [17]. On the left-hand side of the equation the values $(3/2)\, p_e + \varepsilon_i$ and $(5/2)\, p_e + \varepsilon_i$ are used instead of $(3/2)\, p_e$ and $(5/2)\, p_e$ whereas the electron heat flux is defined as $(\mathbf{q}_e - n_e \mathbf{w}_e \varepsilon_i)$. The right-hand side of (4.5.2) includes the electron losses due to inelastic collisions with atoms with different levels of electron excitation as well as energy losses because of exciting the internal degree of freedom of molecules in the molecular plasma. It is to be mentioned that the energy losses in inelastic collisions with atoms may be expressed in terms of the local rate of energy losses at the expense of radiation on assumption that the atoms interact energetically only with free electrons and they change their energy at the expense of emitting and absorbing a photon alone [17].

4.6 Higher approximations of the moment method

Higher approximations of the Grad's moment method are consistent with accounting for a greater number of polynomials in the expansion (4.1.9) than the set (4.2.1) corresponding to the $13N$-moment approximation. The new coefficients a_α^{mn} corresponding to these polynomials are related to the moment of the distribution function which have no physical meaning; however, if we limit ourselves to a finite number of the expansion terms, then the equations of the coefficients are part of a full closed system of equations together with the equations for ρ_α, $\mathbf{u}$, T_α, $\mathbf{w}_\alpha$, $\pi_{\alpha rs}$, and $\mathbf{h}_\alpha$. The solution of such a system would permit us to find expressions of the macroscopic plasma parameters of interest as well as the relevant kinetic coefficients when the contribution of the expansion terms of higher orders is taken into account.

The general system of the moment equations (or equations for the coefficients $n_\alpha b_\alpha^{mn}$) in any order of approximation is given in Appendix 1. Calculation of the right-hand sides of the equations with neglecting of the coefficient-squared expansion terms allows us to write them in the form of linear sums b_β^{mn}. In turn, the coefficients at b_β^{mn} in the sums are expressed in terms of the bracket integrals of the Sonine polynomials [5,6] and are of the order of $n_\beta \tau_{\alpha\beta}^{-1}$. When the requirement of (4.3.1) is fulfilled, i.e. when plasma parameters slowly vary at characteristic lengths and at time of the order of the mean free path and the time between the particle collisions, the left-hand sides of the equations for $n_\alpha b_\alpha^{mn}$ can be substantially simplified along the same lines which were suggested in Section 4.3.

The simplifying procedure becomes more illustrative if the coefficients $n_\alpha b_\alpha^{mn}$ are represented by the formal expansion in series with respect to the parameter $\varepsilon = \lambda/L$, where $\lambda \sim \langle v_{\alpha\beta}\rangle \tau_{\alpha\beta}$. Omitting the derivatives $dn_\alpha b_\alpha^{mn}/dt$ compared with the right-hand side of Eqs. (A.1.7) owing to the second requirement of (4.3.1) and reducing the equations to the dimensionless form, it is easy to find that in front of the linear combination of the coefficients the factor of the order of ε^{-1} ($\varepsilon \ll 1$) appears on the right-hand side. This prompts us to apply the technique of successive approximations in which the expansion of the form of $b_\alpha^{mn} = \sum_{k=1} \varepsilon^k b_\alpha^{mn}(k)$ is substituted into the original equations and the terms with the same degrees of ε are made equal. Compared with the equations of the $13N$-moment approximation (4.2.12)–(4.2.14), the application of such an approach brings about the linear transport relations (4.3.2.)–(4.3.4) for the values $\mathbf{w}_\alpha$, $\pi_{\alpha rs}$, and $\mathbf{h}_\alpha$ in the first approximation with respect to ε. Using the expressions obtained as a first approximation in ε and substituting them into the left-hand side of the equations for $n_\alpha b_\alpha^{mn}$, we can get equations valid up to the accuracy of the $\sim \varepsilon^2$ terms which include the dependence on the gradient products and the second derivatives of the hydrodynamic parameters that corresponds to the so called Burnett approximation [5,6].

The procedure being described is to a certain extent equivalent to that used in the Chapman–Enskog method. The only difference is that in the latter the expansion in series with respect to ε is applied to the distribution function and to the kinetic equation proper whose reduction to the dimensionless form results in the factor ε^{-1} on its right-hand side, before the collision integral.

It is clear that the linear transport relations at a first approximation to ε can be also obtained by using the original kinetic equation linearized near the local Maxwell distribution rather than the complete nonlinear Boltzmann equation. Indeed, if the distribution function is represented as

$$f_\alpha = f_\alpha^{(0)}\left(1+\Phi_\alpha\right), \qquad |\Phi_\alpha| \ll 1, \tag{4.6.1}$$

then after linearizing the kinetic equation in the form of (2.1.7) near $f_\alpha^{(0)}$ and omitting the Φ_α squared terms, one can find [19]

$$\begin{aligned} & f_\alpha^{(0)}\left\{\frac{d\Phi_\alpha}{dt} + (\mathbf{c}_\alpha\cdot\nabla)\,\Phi_\alpha + \frac{1}{m_\alpha}\left(e_\alpha\mathbf{E}' + \mathbf{X}_\alpha\right)\cdot\nabla_c\Phi_\alpha\right. \\ & + \left[\frac{e_\alpha}{m_\alpha}\left(\mathbf{c}_\alpha\times\mathbf{B}\right)\cdot\nabla_c\right]\Phi_\alpha + \frac{1}{2T_\alpha}\left(\gamma_\alpha c_\alpha^2 - 5\right)\mathbf{c}_\alpha\cdot\nabla T_\alpha \\ & \left. + \gamma_\alpha\left(c_{\alpha r}c_{\alpha s} - \frac{1}{3}\delta_{rs}c_\alpha^2\right)\varepsilon_{rs} + \frac{1}{p_\alpha}\left(\mathbf{c}_\alpha\cdot\nabla p_\alpha - n_\alpha\mathbf{c}_\alpha\cdot\mathbf{F}_\alpha^*\right)\right\} \\ & = \sum_\beta\left[J_{\alpha\beta}\left(f_\alpha^{(0)}\Phi_\alpha, f_\beta^{(0)}\right) + J_{\alpha\beta}\left(f_\alpha^{(0)}, f_\beta^{(0)}\Phi_\beta\right)\right]. \end{aligned} \tag{4.6.2}$$

The linearized equations for the coefficients $n_\alpha b_\alpha^{mn}$ are obtained by multiplying the left- and right-hand sides of Eqs. (4.6.2) by the polynomials G_α^{mn} and integrating with respect to velocities. It is clear that they are in perfect agreement with the linearized moment equations which follow from the complete system of moment equations given in Appendix 1 if all the nonlinear terms including the products of moments (or the coefficients $n_\alpha b_\alpha^{mn}$) and small gradients of thermodynamic variables as well as the small gradients of the potential are neglected on the left-hand side. For the states of multicomponent plasma satisfying the condition (4.3.1), the term with $d\Phi_\alpha/dt$ can be omitted in Eq. (4.6.2) since in the corresponding moment equations we can neglect the terms of the order of $dn_\alpha b_\alpha^{mn}/dt$ compared with the right-hand side of the order of $n_\alpha\tau_{\alpha\beta}^{-1}b_{\alpha\beta}^{mn}$. For the time being the term with $\mathbf{c}_\alpha\cdot\nabla\Phi_\alpha$ is omitted because accounting for it would demand saving the derivatives in the coordinates of the coefficients of $(m+1)$th and $(m-1)$th order in the moment equations (cf. the terms in Eqs. (A.1.7)). In the $13N$-moment approximation considered above the derivatives $\partial\pi_{\alpha rs}/\partial x_s$ and $\{\partial q_{\alpha r}/\partial x_s\}$ in Eqs. (4.3.3)–(4.3.4) correspond to these terms. We have already mentioned the cases when the account of these terms may turn out to be essential. A more detailed discussion will be given in the next chapter. When the terms resulting from integrating the terms with $\langle \mathbf{F}_\alpha^* \rangle \cdot \nabla \Phi_\alpha$, whose effect manifests itself only in strong fields, are omitted, the original linearized kinetic equation can be written as

$$\begin{aligned} D\left(f_\alpha^{(0)}\right) \equiv f_\alpha^{(0)} \Big\{ & \mathbf{c}_\alpha \cdot [\frac{1}{p_\alpha}(\nabla p_\alpha - n_\alpha \mathbf{F}_\alpha^*)] \\ & + \frac{1}{2T_\alpha}[(\gamma_\alpha c_\alpha^2 - 5)\, \mathbf{c}_\alpha \cdot \nabla T_\alpha] \\ & + \gamma_\alpha \left(c_{\alpha r} c_{\alpha s} - \frac{1}{3}\delta_{rs} c_\alpha^2 \right) \varepsilon_{rs} \Big\} + f_\alpha^{(0)} \left[(\mathbf{c}_\alpha \times \boldsymbol{\omega}_\alpha) \cdot \nabla_c \right] \Phi_\alpha \\ = \sum_\beta & \left[J_{\alpha\beta}\left(f_\alpha^{(0)} \Phi_\alpha,\, f_\beta^{(0)} \right) + J_{\alpha\beta}\left(f_\alpha^{(0)},\, f_\beta^{(0)} \Phi_\beta \right) \right]. \end{aligned} \tag{4.6.3}$$

Note that polynomials $G_{\alpha r}^{10}$, $G_{\alpha r}^{11}$ and $G_{\alpha rs}^{20}$ occur on the left-hand side of the equation. By multiplying (4.6.3) by G_α^{mn} and integrating with respect to velocities, with account for the orthogonality condition for the polynomials H_α^{mn} and the expansion of (4.1.9) for Φ_α we arrive to an infinite system of linear algebraic equations for the coefficients b_α^{mn}

$$\begin{aligned} & p\mathbf{d}_\alpha + \frac{\rho_\alpha}{\rho} \mathbf{j} \times \mathbf{B} \delta_{m1}\delta_{n0} + \frac{5}{2}\frac{k}{m_\alpha} p_\alpha \nabla T_\alpha \delta_{m1}\delta_{n1} + 2p_\alpha \varepsilon_{rs}\delta_{m2}\delta_{n0} \\ = & -\sum_\beta \sum_k \left(A_\alpha^{mnk} b_\alpha^{mk} + B_{\alpha\beta}^{mnk} b_\beta^{mk} \right) + n_\alpha \langle b_{\alpha s}^{mn} \delta_r \rangle\, e_{rsl} \omega_{\alpha l}. \end{aligned} \tag{4.6.4}$$

The expressions for the coefficients $A_{\alpha\beta}^{mnk}$ and $B_{\alpha\beta}^{mnk}$ relating them to the bracket integrals of the Sonine polynomials and tensorial spherical harmonics are given in Appendix 1.

It is to be pointed out that at $m \neq 1, 2$ the left-hand side of (4.6.4) is equal to zero, then due to the homogeneity of the equations all the coefficients b_α^{mn} at $m \neq 1, 2$ vanish. As the result, the system (4.6.4) is split into two independent systems of equations for the coefficients $\mathbf{b}_\beta^{1n}$ and $b_{\beta rs}^{2n}$ for the vector ($m = 1$) and tensor ($m = 2$) values, respectively.

The equations for $\mathbf{b}_{1\beta}^{1n}$ (at $T_\alpha = T$) are written as

$$\begin{aligned} & \sum_\beta \sum_{k=0}^{\xi-1} C_{\alpha\beta}^{10k} \mathbf{b}_\beta^{1k} = -p\mathbf{d}_\alpha - \frac{\rho_\alpha}{\rho}\mathbf{j} \times \mathbf{B} + n_\alpha \mathbf{b}_\alpha^{10} \times \boldsymbol{\omega}_\alpha, \\ & \sum_\beta \sum_{k=0}^{\xi-1} C_{\alpha\beta}^{11k} \mathbf{b}_\beta^{1k} = -\frac{5}{2}\frac{k}{m_\alpha} p_\alpha \nabla T + n_\alpha \mathbf{b}_\alpha^{11} \times \boldsymbol{\omega}_\alpha, \\ & \sum_\beta \sum_{k=0}^{\xi-1} C_{\alpha\beta}^{1nk} \mathbf{b}_\beta^{1k} = n_\alpha \mathbf{b}_\alpha^{1n} \times \boldsymbol{\omega}_\alpha, \qquad 2 \le n \le \xi - 1, \end{aligned} \tag{4.6.5}$$

and for the coefficients $b^{20}_{\beta rs}$

$$\sum_{\beta}\sum_{k=0}^{\xi-1} C^{20k}_{\alpha\beta} b^{20}_{\beta rs} = -2p_\alpha \varepsilon_{rs} + 2n_\alpha \left\{ b^{20}_{\alpha lr} \varepsilon_{slt} \omega_{\alpha t} \right\},$$

$$\sum_{\beta}\sum_{k=0}^{\xi-1} C^{2nk}_{\alpha\beta} b^{2k}_{\beta rs} = 2n_\alpha \left\{ b^{2n}_{\alpha lr} e_{slt} \omega_{\alpha t} \right\}, \quad 1 \le n \le \xi - 1. \tag{4.6.6}$$

The coefficients $C^{mnk}_{\alpha\beta}$ are given by

$$C^{mnk}_{\alpha\beta} = \delta_{\alpha\beta} \sum_{\gamma} A^{mnk}_{\alpha\gamma} + B^{mnk}_{\alpha\beta},$$

$$A^{1nk}_{\alpha\beta} = n_\alpha n_\beta Q_{1nk} \gamma_\alpha^{k-n} \left[S^n_{3/2}\left(W^2\right) \mathbf{W}, S^k_{3/2}\left(W^2\right) \mathbf{W} \right]'_{\alpha\beta},$$

$$A^{2nk}_{\alpha\beta} = n_\alpha n_\beta Q_{2nk} \gamma_\alpha^{k-n} \left[S^n_{5/2}\left(W^2\right) \left(W_r W_s - \frac{1}{3}\delta_{rs} W^2\right), S^k_{5/2}\left(W^2\right) \left(W_r W_s - \frac{1}{3}\delta_{rs} W^2\right) \right]'_{\alpha\beta},$$

$$B^{1nk}_{\alpha\beta} = n_\alpha n_\beta Q_{1nk} \frac{\gamma_\beta^{k-\frac{1}{2}}}{\gamma_\alpha^{n-\frac{1}{2}}} \left[S^n_{3/2}\left(W^2\right) \mathbf{W}, S^k_{3/2}\left(W^2\right) \mathbf{W} \right]''_{\alpha\beta}, \tag{4.6.7}$$

$$B^{2nk}_{\alpha\beta} = n_\alpha n_\beta Q_{2nk} \frac{\gamma_\beta^k}{\gamma_\alpha^n} \left[S^n_{5/2}\left(W^2\right) \left(W_r W_s - \frac{1}{3}\delta_{rs} W^2\right), S^n_{5/2}\left(W^2\right) \left(W_r W_s - \frac{1}{3}\delta_{rs} W^2\right) \right]''_{\alpha\beta},$$

$$\mathbf{W} = \left(\frac{m_\alpha}{2kT} \right)^{1/2} \mathbf{c},$$

where $[F,G]'_{\alpha\beta}$ and $[F,G]''_{\alpha\beta}$ are the bracket integrals whose expressions in terms of linear combinations of integrals $\Omega^{lr}_{\alpha\beta}$ can be found in [5,6]. A great part of the expressions required for practical calculations is given in Appendix 2.

The coefficients Q_{1nk} and Q_{2nk} in the expressions (4.6.7) are in the form

$$Q_{1nk} = \frac{2!n!\,(k+1)!}{(2k+3)!} 2^{2k+1} (-1)^{n+k} = (-1)^n n! \frac{2}{3} \left\{ 1, -\frac{2}{5}, \frac{4}{35}, \ldots \right\},$$

$$Q_{2nk} = \frac{4!n!\,(k+2)!}{(2k+5)!}2^{2k}\,(-1)^{n+k} = (-1)^n\,n!\frac{2}{5}\left\{1, -\frac{2}{7}, \frac{4}{63}, \ldots\right\},$$

$$n, k = 0, 1, \ldots .$$

The system of vector and tensorial linear algebraic equations (4.6.5) and (4.6.6) can be resolved with respect to $\mathbf{b}_\gamma^{1k}$ and $b_{\gamma rs}^{2k}$ for any finite value of ξ. Finally, the expressions of interest for $\mathbf{b}_\alpha^{10}$, $b_{\alpha rs}^{20}$ and $\mathbf{b}_\alpha^{11}$ or the values $\mathbf{w}_\alpha$, $\pi_{\alpha rs}$, and $\mathbf{h}_\alpha$ in any order of approximation with respect to ξ can be found.

Note that the $13N$-moment approximation corresponds to the use of the first two Eqs. (4.6.5) for $\mathbf{b}_\beta^{10}$ and $\mathbf{b}_\beta^{11}$ ($\xi = 2$) and one Eq. (4.6.6) for $b_{\beta rs}^{20}$ ($\xi = 1$). Comparison of these equations and (4.4.1)–(4.4.3) for the case of a one-temperature plasma proves that the coefficients $A_{\alpha\beta}^{mnk}$ and $B_{\alpha\beta}^{mnk}$ are related to $B_{\alpha\beta}^{(p)}$ as follows:

$$B_{\alpha\beta}^{(1)} = -m_\alpha A_{\alpha\beta}^{100} = m_\beta B_{\alpha\beta}^{100},$$

$$B_{\alpha\beta}^{(2)} = kT\left(\frac{m_\alpha}{\mu_{\alpha\beta}}\right)A_{\alpha\beta}^{101} = -kT\left(\frac{m_\beta}{\mu_{\alpha\beta}}\right)B_{\alpha\beta}^{101},$$

$$B_{\alpha\beta}^{(3)} = -\left(m_\alpha + m_\beta\right)A_{\alpha\beta}^{200}, \quad B_{\alpha\beta}^{(4)} = -\left(m_\alpha + m_\beta\right)B_{\alpha\beta}^{200},$$

$$B_{\alpha\beta}^{(5)} = -m_\alpha A_{\alpha\beta}^{111}, \qquad B_{\alpha\beta}^{(6)} = -m_\beta B_{\alpha\beta}^{111}.$$

As it follows from the structure of Eqs. (4.6.5) and (4.6.6) the $13N$-moment approximation is adequate to obtain all the linear relationships between the moments $\rho_\alpha \mathbf{w}_\alpha, \pi_{\alpha rs}$, and $\mathbf{h}_\alpha$ and gradients of the hydrodynamic values as well as the fields $\langle \mathbf{F}_\alpha^* \rangle$. Taking into account further linearized moment equations, which do not contain inhomogeuous terms proportional to the gradients, adds complexity to the coefficient matrices used to express the transport coefficients.

4.7 Relation to the results of the Chapman–Enskog method

We show now that the results in the moment method approximation linear in ε closely correspond to the results which can be obtained on the basis of the Chapman–Enskog method widely used in the kinetic theory of gases. The solution of the kinetic equation in it is known to be sought in the formal expansion in powers ε [5]

$$f_\alpha = f_\alpha^{(0)}\left[1+\varepsilon\Phi_\alpha^{(1)}+\varepsilon^2\Phi_\alpha^{(2)}+...\right], \tag{4.7.1}$$

where the perturbation function $\Phi_\alpha^{(1)}$, $\Phi_\alpha^{(2)}$ etc. correspond to the first, second (Burnett) etc. approximations. It is seen that the linearized kinetic equation (4.6.3) for Φ_α is in fact equivalent to the equation for the first order correction $\Phi_\alpha^{(1)}$ to the distribution function in the Chapman–Enskog approach [5,6]. To avoid complexity, we shall confine ourselves to the case of one-temperature plasma ($T_\alpha = T$) in the absence of the magnetic field. Then Eq. (4.6.3) can be represented as

$$\begin{aligned} D\left(f_\alpha^{(0)}\right) &= f_\alpha^{(0)}\left[\frac{n}{n_\alpha}\mathbf{c}_\alpha\cdot\mathbf{d}_\alpha+\left(\frac{m_\alpha c_\alpha^2}{2kT}-\frac{5}{2}\right)\mathbf{c}_\alpha\cdot\nabla\ln T\right.\\ &\left.+\frac{m_\alpha}{kT}\left(c_{\alpha r}c_{\alpha s}-\frac{1}{3}\delta_{rs}c_\alpha^2\right)\varepsilon_{rs}\right] = -\sum_{\beta=1}^{N}n_\alpha n_\beta I_{\alpha\beta}\left(\Phi_\alpha^{(1)}\right), \end{aligned} \tag{4.7.2}$$

where $\mathbf{d}_\alpha$ is defined by (4.4.4) and the definition of the integral operator is introduced [6]

$$\begin{aligned} n_\alpha n_\beta I_{\alpha\beta}\left(\Phi_\alpha^{(1)}\right) &= -\left[J_{\alpha\beta}\left(f_\alpha^{(0)}\Phi_\alpha^{(1)}, f_\beta^{(0)}\right)+J_{\alpha\beta}\left(f_\alpha^{(0)}, f_\beta^{(0)}\Phi_\beta^{(1)}\right)\right]\\ &= \iint f_\alpha^{(0)}f_\beta^{(0)}\left(\Phi_\alpha^{(1)}+\Phi_\beta^{(1)}-\Phi_\alpha^{(1)\prime}-\Phi_\beta^{(1)\prime}\right)\\ &\times g\sigma_{\alpha\beta}d\Omega d\mathbf{v}_\beta. \end{aligned} \tag{4.7.3}$$

The form of the expression for $D\left(f_\alpha^{(0)}\right)$ prompts the search for the solution of $\Phi_\alpha^{(1)}$ as

$$\Phi_\alpha^{(1)} = -\frac{1}{n}\sum_\beta \mathbf{D}_\alpha^\beta\cdot\mathbf{d}_\beta-\frac{1}{n}\mathbf{A}_\alpha\cdot\nabla\ln T-\frac{1}{n}B_{\alpha rs}\varepsilon_{rs}. \tag{4.7.4}$$

The substitution of (4.7.4) into (4.7.2) brings in separate integral equations for the coefficients

$$\begin{aligned} &\sum_\beta\frac{n_\alpha n_\beta}{n^2}I_{\alpha\beta}\left(\mathbf{D}^\gamma\right) = \frac{1}{n_\alpha}f_\alpha^{(0)}\left(\delta_{\alpha\gamma}-\frac{\rho_\alpha}{\rho}\right)\mathbf{c}_\alpha,\\ &\sum_\beta\frac{n_\alpha n_\beta}{n^2}I_{\alpha\beta}\left(\mathbf{A}\right) = \frac{1}{n}f_\alpha^{(0)}\left(\frac{m_\alpha c_\alpha^2}{2kT}-\frac{5}{2}\right)\mathbf{c}_\alpha, \end{aligned} \tag{4.7.5}$$

$$\sum_{\beta} \frac{n_\alpha n_\beta}{n^2} I_{\alpha\beta}\left(B_{rs}\right) = \frac{m_\alpha}{nkT} f_\alpha^{(0)} \left(c_{\alpha r} c_{\alpha s} - \frac{1}{3}\delta_{rs} c_\alpha^2\right),$$

$$\alpha, \gamma = 1, \ldots N.$$

Then the coefficients are expanded in infinite series with the Sonine polynomials [5,6]

$$\mathbf{A}_\alpha = -\left(\frac{m_\alpha}{2kT}\right)^{1/2} \sum_{k=0}^{\infty} a_{\alpha,k} S_{3/2}^k \left(W_\alpha^2\right) \mathbf{W}_\alpha,$$

$$B_{\alpha rs} = -\sum_{k=0}^{\infty} b_{\alpha,k} S_{5/2}^k \left(W_\alpha^2\right) \left(W_{\alpha r} W_{\alpha s} - \frac{1}{3}\delta_{rs} W_\alpha^2\right), \qquad (4.7.6)$$

$$\mathbf{D}_\alpha^\beta = \left(\frac{m_\alpha}{2kT}\right)^{1/2} \sum_{k=0}^{\infty} d_{\alpha,k}^\beta S_{3/2}^k \left(W_\alpha^2\right) \mathbf{W}_\alpha,$$

where

$$\mathbf{W}_\alpha = \left(\frac{m_\alpha}{2kT}\right)^{1/2} \mathbf{c}.$$

It is easily seen that the expansion of the non-equilibrium correction of the first approximation $\Phi_\alpha^{(1)}$ to the distribution function in the Chapman–Enskog approach is carried out in fact on the same system of basic functions (polynomials $\mathbf{G}_\alpha^{1k}$ and $G_{\alpha rs}^{2k}$) as in the linear approximation of the Grad's moment method analyzed earlier. The choice of such a system of functions is closely associated with the form of $D\left(f_\alpha^{(0)}\right)$ in the original linearized equation (4.7.2) which includes the linear dependence on the polynomials $\mathbf{G}_\alpha^{10}$, $\mathbf{G}_\alpha^{11}$ and G_{rs}^{20}. Owing to the orthogonality of irreducible Hermite polynomials, it results in vanishing of all the expansion coefficients in the distribution function except for $\mathbf{a}_\alpha^{1k}$ and $a_{\alpha rs}^{2k}$.

From the viewpoint of the procedure applied the difference between the two methods under consideration is as follows.

In the Chapman–Enskog approach, the macroscopic parameters of interest $\mathbf{J}_\alpha = \rho_\alpha \mathbf{w}_\alpha, \pi_{rs} = \sum_\alpha \pi_{\alpha rs}, \mathbf{q} = \sum_\alpha \mathbf{q}_\alpha$ in terms of their definition (1.1.3), (1.1.8) and (1.1.10) and with a substitute of the first approximation (4.7.4) can be represented as linear combinations of the driving forces $\mathbf{d}_\alpha, \varepsilon_{rs}$ and $\nabla \ln T$. The appropriate expressions are written as

$$\mathbf{J}_\alpha = -\rho_\alpha \sum_\beta D_{\alpha\beta} \mathbf{d}_\beta - \rho_\alpha D_\alpha^T \nabla \ln T, \qquad (4.7.7)$$

$$\pi_{rs} = -2\eta\varepsilon_{rs}, \tag{4.7.8}$$

$$\mathbf{q} = -\lambda' \nabla T - p \sum_{\beta} D_{\beta}^{T} \mathbf{d}_{\beta} + \frac{5}{2} \sum_{\alpha} p_{\alpha} \mathbf{w}_{\alpha}, \tag{4.7.9}$$

where λ' and η are the thermal conductivity and viscosity of a gas mixture, $D_{\alpha\beta}$ and D_{β}^{T} are the generalized coefficients of multicomponent diffusion and thermodiffusion, respectively. The transport coefficients are expressed with the help of the bracket integrals of the coefficients $D_{\alpha}^{\beta}, A_{\alpha}$ and $B_{\alpha rs}$ or, due to the orthogonality of the polynomials in (4.7.6), with the help of the coefficients at the Sonine polynomials of lower orders ($k = 0$, or $k = 1$)

$$[D_{\alpha\beta}]_{\xi} = \frac{1}{2n} d_{\alpha,0}^{\beta(\xi)}, \qquad \left[D_{\beta}^{T}\right]_{\xi} = -\frac{1}{2n} a_{\alpha,0}^{(\xi)},$$
$$\left[\lambda'\right]_{\xi} = \frac{5}{4} k \sum_{\alpha=1}^{N} \frac{n_{\alpha}}{n} a_{\alpha,1}^{(\xi)}, \quad [\eta]_{\xi} = \frac{1}{2} kT \sum_{\alpha=1}^{N} \frac{n_{\alpha}}{n} b_{\alpha,0}^{(\xi)}. \tag{4.7.10}$$

The full algebraic sets of equations for the coefficients $d_{\alpha,k}^{\beta}, a_{\alpha k}$ and $b_{\alpha,k}$ from which the coefficients $d_{\alpha,0}^{\beta}, a_{\alpha 0}, a_{\alpha 1}$ and $b_{\alpha,0}$ can be determined in the ξ-approximation, are obtained by substituting the expansions (4.7.6) into the integral equations (4.7.5) to be multiplied by the polynomials G_{α}^{1k} and $G_{\alpha rs}^{2k}$ and integrated over velocities.

In the Grad's moment method (its linearized form based on Eq. (4.6.4), to be precise) the system of linear vector and tensorial equations is obtained for the quantities of interest – $\mathbf{w}_{\alpha}, \pi_{\alpha rs}$, and $\mathbf{h}_{\alpha} = \mathbf{q}_{\alpha} - (5/2)\, p_{\alpha} \mathbf{w}_{\alpha}$ – and the finite number of other expansion coefficients depending on the chosen order of approximation of ξ. The linear transport relations, analogous to those derived in the Chapman–Enskog approach, follow from the solution of separate systems of vector and tensorial quantities. Note that the diffusion velocities $\mathbf{w}_{\alpha}$ and reduced heat fluxes $\mathbf{h}_{\alpha}$ are defined in the framework of the same system of equations. The resulting transport coefficients can be written as the ratio of pertinent determinants whose elements, similar to the analogous coefficients in the Chapman–Enskog method, are linearly expressed with the help of the bracket integrals of the Sonine polynomials.

From the formal viewpoint at the level of ε-linear approximation the results obtained by the Grad's technique are totally equivalent to the results of the Chapman–Enskog approach. However, the equations written directly for the macroscopic fluxes of interest turn out to be more convenient for use in many cases. In particular, the analysis of the multicomponent diffusion in plasma and the derivation of the generalized Ohm's law becomes simplified and the expressions for the coefficients of thermal conductivity and thermal diffusion of the mixture can be obtained in a more straightforward way [11] (see the next chapter). The particles with differing masses (e.g. electrons or light ions in the plasma) available in the mixture permit us to make simplifications (using the relevant small parameters) starting from the original transport equations and to separate the equations for the light particles from the general system of equations. Comparison of the results derived in this way with the exact solutions following from the complete system allows one to justify the procedure of separating equations for electrons from those for heavy particles more accurately than in the case of doing so starting from the system of kinetic equations for plasma [17,20].

New possibilities arise when going beyond the limits of the ε-linear approximation in the method, i.e. when using the nonlinear system of equations for the coefficients $n_\alpha b_\alpha^{mn}$. In practice, even from the moment equations obtained in terms of the linearized kinetic equation (4.6.2) keeping the term $\mathbf{c}_\alpha \cdot \nabla\Phi_\alpha$ in them, a few new effects occur. They are related to accounting for the terms with the derivatives of the fluxes with respect to the transverse coordinate of the order of a in the transport equations. It is seen, in particular, when considering non-one-dimensional problems with different scales of inhomogeneity $a \ll L$ when these terms can be of the first order in the small parameter $\varepsilon = \lambda/L$ as the basic terms. Some of the examples pertinent to the slow viscous gas flows [11] are considered in the next chapter. Another example refers to the theory of gradient instabilities in an inhomogeneous plasma [20] when the transverse velocity of species α of the plasma $\mathbf{u}_{\alpha\perp}$ is taken to be of the same order of value as $\mathbf{q}_{\alpha\perp}$. Therefore, in the expression of the viscous-stress tensor one must retain the derivatives $\{\partial q_{\alpha r}/\partial x_s\}$ along with the derivatives $\{\partial u_r/\partial x_s\}$. A number of other cases where account of similar terms is important can also be mentioned; however, their analysis is beyond the scope of the book.

Chapter 5

Transport phenomena in a multicomponent gas mixture

5.1 Original system of equations

We are postponing the general analysis of the transport phenomena in a nonisothermal multispecies plasma in the presence of magnetic fields to another chapter. Here we consider a simpler case of an isothermal (one-temperature) plasma in the absence of the magnetic field. The case in point is a description of transport phenomena in a common multicomponent gas mixture under the influence of gradients of local thermodynamic parameters and external fields where the electric field $\mathbf{E}$ is an inherent explicit part. Let us limit ourselves to the $13N$-moment approximation[1] in which case the original equations are taken to be Eqs. (4.4.1)–(4.4.3) with $\mathbf{B}=0$. It is convenient to present these equations as [1]

$$\sum_{\beta\neq\alpha}\frac{n_\alpha n_\beta kT}{n\left[\mathfrak{D}_{\alpha\beta}\right]_1}(\mathbf{w}_\alpha-\mathbf{w}_\beta)=-p\mathbf{d}_\alpha+\sum_{\beta\neq\alpha}\xi_{\alpha\beta}\left(\frac{\mathbf{h}_\beta}{m_\beta n_\beta}-\frac{\mathbf{h}_\alpha}{m_\alpha n_\alpha}\right), \tag{5.1.1}$$

$$p^2\sum_{\beta=1}^{N}H_{\alpha\beta}\frac{\pi_{\beta rs}}{p_\beta}=-2p_\alpha\varepsilon_{rs}, \tag{5.1.2}$$

[1] To be precise, it is a $12N+1$-approximation as all the temperatures T_α are equal to T.

$$\frac{p^2}{T}\sum_{\beta=1}^{N}\Lambda_{\alpha\beta}\frac{\mathbf{h}_\beta}{p_\beta}=-\frac{p_\alpha}{T}\nabla T-\sum_{\beta\neq\alpha}\frac{kT}{m_\alpha}\xi_{\alpha\beta}\left(\mathbf{w}_\alpha-\mathbf{w}_\beta\right). \tag{5.1.3}$$

It will be recalled that

$$\begin{aligned}\mathbf{d}_\alpha &= \nabla y_\alpha+\left(y_\alpha-\frac{\rho_\alpha}{\rho}\right)\nabla\ln p\\ &\quad -p^{-1}\left(n_\alpha\mathbf{X}_\alpha-\frac{\rho_\alpha}{\rho}\sum_\alpha n_\alpha\mathbf{X}_\alpha\right)-p^{-1}n_\alpha e_\alpha\mathbf{E}',\end{aligned} \tag{5.1.4}$$

where $y_\alpha=n_\alpha/n$ is the molar concentration of the species α of the mixture.

Coefficients $H_{\alpha\beta}$, $\Lambda_{\alpha\beta}$ and $\xi_{\alpha\beta}$, if the expressions for $B^{(p)}_{\alpha\beta}$ (4.2.18) are taken into account, look like

$$\begin{aligned}H_{\alpha\alpha}&=\frac{y_\alpha^2}{[\eta_{\alpha\alpha}]_1}+\sum_{\beta\neq\alpha}\frac{2y_\alpha y_\beta}{(m_\alpha+m_\beta)\,n\,[\mathfrak{D}_{\alpha\beta}]_1}\left(1+\frac{3}{5}\frac{m_\beta}{m_\alpha}A^*_{\alpha\beta}\right),\\ H_{\alpha\beta}&=-\frac{2y_\alpha y_\beta}{(m_\alpha+m_\beta)\,n\,[\mathfrak{D}_{\alpha\beta}]_1}\left(1-\frac{3}{5}A^*_{\alpha\beta}\right),\quad \beta\neq\alpha,\end{aligned} \tag{5.1.5}$$

$$\begin{aligned}\Lambda_{\alpha\alpha}&=\frac{y_\alpha^2}{[\lambda_{\alpha\alpha}]_1}+\frac{4}{25}\frac{T}{p}\sum_{\beta\neq\alpha}\frac{y_\alpha y_\beta}{(m_\alpha+m_\beta)^2\,[\mathfrak{D}_{\alpha\beta}]_1}\\ &\quad\times\left(\frac{15}{2}m_\alpha^2+\frac{25}{4}m_\beta^2-3m_\beta^2B^*_{\alpha\beta}+4m_\alpha m_\beta A^*_{\alpha\beta}\right),\\ \Lambda_{\alpha\beta}&=-\frac{4}{25}\frac{T}{p}\frac{m_\alpha m_\beta}{(m_\alpha+m_\beta)^2}\frac{y_\alpha y_\beta}{[\mathfrak{D}_{\alpha\beta}]_1}\\ &\quad\times\left(\frac{55}{4}-3B^*_{\alpha\beta}-4A^*_{\alpha\beta}\right),\qquad \beta\neq\alpha,\end{aligned} \tag{5.1.6}$$

$$\xi_{\alpha\beta}=\frac{n_\alpha n_\beta}{n\,[\mathfrak{D}_{\alpha\beta}]_1}\mu_{\alpha\beta}\left(\frac{6}{5}C^*_{\alpha\beta}-1\right). \tag{5.1.7}$$

The values of

$$\begin{aligned}&[\mathfrak{D}_{\alpha\beta}]_1=\frac{3}{16}\frac{kT}{n\mu_{\alpha\beta}\Omega^{11}_{\alpha\beta}},\quad [\eta_{\alpha\alpha}]_1=\frac{5}{8}\frac{kT}{\Omega^{22}_{\alpha\alpha}},\\ &[\lambda_{\alpha\alpha}]_1=\frac{15}{4}\frac{k}{m_\alpha}[\eta_{\alpha\alpha}]_1\end{aligned} \tag{5.1.8}$$

are consistent with the binary diffusion coefficients of species α and β as well as the viscosity and thermal conductivity of the pure gas of species α calculated in the first Chapman–Cowling approximation (with respect to the number of the Sonine polynomials in the expansion) [2]. Then the expressions for Ω-integrals (4.2.20) are valid if $\gamma_{\alpha\beta} = \mu_{\alpha\beta}/kT$ or

$$\Omega^{lr}_{\alpha\beta} = \left(\frac{kT}{2\pi\mu_{\alpha\beta}}\right)^{1/2} \int_0^\infty \zeta^{2r+3} \exp\left(-\zeta^2\right) Q^{(l)}_{\alpha\beta}(g)\, d\zeta, \qquad (5.1.9)$$

where $Q^{(l)}_{\alpha\beta}(g)$ is defined by the relation (1.3.2). The values of $A^*_{\alpha\beta}$, $B^*_{\alpha\beta}$ and $C^*_{\alpha\beta}$ represent the ratios of Ω-integrals of the form

$$A^*_{\alpha\beta} = \frac{\Omega^{22}_{\alpha\beta}}{2\Omega^{11}_{\alpha\beta}}, \quad B^*_{\alpha\beta} = \frac{5\Omega^{12}_{\alpha\beta} - \Omega^{13}_{\alpha\beta}}{3\Omega^{11}_{\alpha\beta}}, \quad C^*_{\alpha\beta} = \frac{\Omega^{12}_{\alpha\beta}}{3\Omega^{11}_{\alpha\beta}}. \qquad (5.1.10)$$

It is worth recalling that the Ω-integrals are easy to calculate for the hard elastic sphere model. In that case using the definition of $Q^{(l)}_{\alpha\beta}$ (1.3.7), we have

$$\left[\Omega^{lr}_{\alpha\beta}\right]_{h,s} = \left(\frac{kT}{2\pi\mu_{\alpha\beta}}\right)^{1/2} \frac{1}{2}(r+1)! \left[1 - \frac{1}{2}\frac{1+(-1)^l}{l+1}\right] \pi d^2_{\alpha\beta}.$$

Earlier, we have used the hard elastic sphere model in estimating the electrical conductivity in a weakly ionized gas. In what follows the same model can help us to evaluate other transport coefficients and also to test the convergence of approximations to the kinetic coefficients in expanding over the Sonine polynomials. In the kinetic theory of gas mixtures, the so called reduced Ω^*-integrals are also introduced representing the ratio of common Ω-integrals to $[\Omega]_{h,s}$-integrals calculated for the hard elastic sphere model [3]

$$\Omega^{lr*}_{\alpha\beta} = \frac{\Omega^{lr}_{\alpha\beta}}{\left[\Omega^{lr}_{\alpha\beta}\right]_{h,s}}.$$

For a real potential of the interactions of neutral particles these integrals weakly depend on temperature. With their help, we can define the values of A^*, B^* and C^* as

$$A^*_{\alpha\beta} = \frac{\Omega^{22*}_{\alpha\beta}}{\Omega^{11*}_{\alpha\beta}}, \quad B^*_{\alpha\beta} = \frac{5\Omega^{12*}_{\alpha\beta} - 4\Omega^{13*}_{\alpha\beta}}{\Omega^{11*}_{\alpha\beta}}, \quad C^*_{\alpha\beta} = \frac{\Omega^{12*}_{\alpha\beta}}{\Omega^{11*}_{\alpha\beta}}. \qquad (5.1.11)$$

It is clear that for the hard elastic sphere model

$$A^*_{\alpha\beta} = B^*_{\alpha\beta} = C^*_{\alpha\beta} = 1.$$

For the Coulomb interaction of particles

$$A^*_{\alpha\beta} = B^*_{\alpha\beta} = 1, \qquad C^*_{\alpha\beta} = \frac{1}{3}.$$

5.2 Viscous-stress tensor and heat flux

Solving Eqs. (5.1.2) with respect to $\pi_{\gamma rs}/p_\gamma$ we come to the expression of the partial viscous-stress tensor

$$\pi_{\alpha rs} = -2\eta_\alpha \varepsilon_{rs}, \tag{5.2.1}$$

where

$$\eta_\alpha = y_\alpha \sum_{\beta=1}^{N} y_\beta \frac{|H|_{\beta\alpha}}{|H|} \tag{5.2.2}$$

is the partial viscosity of species α. The notation $|A|$ corresponds to the determinant with elements $A_{\alpha\beta}$ and $|A|_{\alpha\beta}$ is the cofactor of the element $\alpha\beta$ of the determinant.

Summation (5.2.1) over α gives the relationship for the total viscous-stress tensor

$$\pi_{rs} = -2\eta\varepsilon_{rs}, \tag{5.2.3}$$

where the mixture viscosity η is given by

$$\eta = \sum_\alpha \eta_\alpha = \sum_\alpha \sum_\beta y_\alpha y_\beta \frac{|H|_{\beta\alpha}}{|H|} \tag{5.2.4}$$

which can be written as the ratio of the two determinants of the $(N+1)$th and Nth orders [3,5]

$$\eta = -\frac{\begin{vmatrix} H_{11} & \cdots & H_{1N} & y_1 \\ \cdots & \cdots & \cdots & \cdots \\ H_{N1} & \cdots & H_{NN} & y_N \\ y_1 & \cdots & y_N & 0 \end{vmatrix}}{\begin{vmatrix} H_{11} & \cdots & H_{1N} \\ \cdots & \cdots & \cdots \\ H_{N1} & \cdots & H_{NN} \end{vmatrix}}. \tag{5.2.4$'$}$$

The expression is analogous to the result of the first Chapman–Cowling approximation. Viscosity of the pure gas ($N = 1$) in the same approximation is defined as

$$[\eta]_1 = \frac{1}{H_{11}} = \frac{5}{16}\frac{(\pi mkT)^{1/2}}{\pi d^2 \Omega^{22*}}. \tag{5.2.5}$$

Here it as appropriate to give the expression of viscosity of the binary gas mixture [3]

$$[\eta]_1 = \frac{1+Z_\eta}{X_\eta + Y_\eta},$$

$$X_\eta = \frac{y_1^2}{[\eta_{11}]_1} + \frac{2y_1y_2}{[\eta_{12}]_1} + \frac{y_2^2}{[\eta_{22}]_1},$$

$$Y_\eta = \frac{3}{5}A_{12}^*\left\{\frac{y_1^2}{[\eta_{11}]_1}\frac{m_1}{m_2} + \frac{2y_1y_2}{[\eta_{12}]_1}\frac{(m_1+m_2)^2}{4m_1m_2}\frac{[\eta_{12}]_1^2}{[\eta_{11}]_1[\eta_{22}]_1} + \frac{y_1^2}{[\eta_{22}]_1}\frac{m_2}{m_1}\right\},$$

$$Z_\eta = \frac{3}{5}A_{12}^*\left\{y_1^2\frac{m_1}{m_2} + 2y_1y_2\left[\frac{(m_1+m_2)^2}{4m_1m_2}\left(\frac{[\eta_{12}]_1}{[\eta_{11}]_1} + \frac{[\eta_{12}]_1}{[\eta_{22}]_1} - 1\right)\right] + y_2^2\frac{m_2}{m_1}\right\}. \tag{5.2.6}$$

Here $[\eta_{11}]_1$ and $[\eta_{22}]_1$ are the viscosities of the pure gases in the first approximation defined by Eq. (5.2.5) and

$$[\eta_{12}]_1 = \frac{5}{3}\frac{m_1m_2}{m_1+m_2}\frac{n[\mathfrak{D}_{12}]_1}{A_{12}^*}.$$

As the further analysis proves (see Chapter 7), the contribution of the electron component to the viscosity of the partially ionized gas can be neglected. Consequently, expression (5.2.6) may be used to approximately calculate the viscosity of the three-component partially ionized plasma in the absence of a magnetic field, treated as a two-component mixture of ions and atoms of equal masses.

At the moment we shall consider expressions of heat fluxes. Solving Eqs. (5.1.3) with respect to $\mathbf{h}_\gamma/p_\gamma$, we obtain

$$\mathbf{h}_\alpha = -\lambda_\alpha \nabla T - \frac{kT^2}{p}\sum_{\gamma=1}^{N}\sum_{\delta=1}^{N}\frac{y_\alpha}{m_\gamma}\xi_{\gamma\delta}\frac{|\Lambda|_{\gamma\alpha}}{|\Lambda|}(\mathbf{w}_\gamma - \mathbf{w}_\delta) \tag{5.2.7}$$

with

$$\lambda_\alpha = y_\alpha \sum_{\beta=1}^{N} y_\beta \frac{|\Lambda|_{\beta\alpha}}{|\Lambda|} \tag{5.2.8}$$

as the partial thermal conductivity of species α.

For the thermal conductivity of a pure gas ($N = 1$), we have

$$[\lambda]_1 = \frac{1}{\Lambda_{11}} = \frac{75}{64}\left(\frac{\pi kT}{m}\right)^{1/2} \frac{1}{\pi d^2 \Omega^{22*}}. \tag{5.2.9}$$

Summing (5.2.7) over α results in the following expression of the total heat flux in the gas mixture

$$\mathbf{q} = -\lambda \nabla T + \sum_{\alpha}\sum_{\beta} \frac{n_\alpha n_\beta kT}{n_\alpha m_\alpha} \frac{D^T_{\alpha\beta}}{n\,[D_{\alpha\beta}]_1} (\mathbf{w}_\alpha - \mathbf{w}_\beta) + \frac{5}{2}\sum_{\alpha} p_\alpha \mathbf{w}_\alpha, \tag{5.2.10}$$

where

$$\lambda = \sum_{\alpha=1}^{N} \lambda_\alpha = \sum_{\alpha=1}^{N}\sum_{\beta=1}^{N} y_\alpha y_\beta \frac{|\Lambda|_{\beta\alpha}}{|\Lambda|} \tag{5.2.11}$$

is the thermal conductivity of the gas mixture and

$$D^T_{\alpha\beta} = -\frac{m_\alpha m_\beta}{m_\alpha + m_\beta}\left(\frac{6}{5}C^*_{\alpha\beta} - 1\right)\frac{\lambda_\alpha}{k} \tag{5.2.12}$$

is the generalized thermal diffusion coefficient whose meaning is clarified by the discussion to follow.

Note that the thermal conductivity λ (5.2.11) is the ratio of the determinants of the $(N+1)$th and Nth order

$$\lambda = -\frac{\begin{vmatrix} \Lambda_{11} & \cdots & \Lambda_{1N} & y_1 \\ \cdots & \cdots & \cdots & \cdots \\ \Lambda_{N1} & \cdots & \Lambda_{NN} & y_N \\ y_1 & \cdots & y_N & 0 \end{vmatrix}}{\begin{vmatrix} \Lambda_{11} & \cdots & \Lambda_{1N} \\ \cdots & \cdots & \cdots \\ \Lambda_{N1} & \cdots & \Lambda_{NN} \end{vmatrix}}. \tag{5.2.11'}$$

The expressions for λ and $D^T_{\alpha\beta}$ correspond to the results of the complete second Chapman–Cowling approximation (in the expansion with respect to the Sonine polynomials). They are different from the expressions derived elsewhere [3] with a simpler form since the kinetic coefficients are written in terms of the ratios of the determinants of the $(N+1)$th and Nth orders rather than $(2N+1)$th and $2N$th. The latter is due to the fact that in calculation of kinetic coefficients [3] only those expansion terms were used which produce the first non-vanishing contribution to the proper kinetic coefficients. In particular, to get a non-zero result for the binary diffusion coefficient, only one term would do which in our calculation corresponds to the coefficient a^{10}_α.

Yet to calculate λ and $D^T_{\alpha\beta}$ would demand two expansion coefficients (a^{10}_α and a^{11}_α in our case). As was shown in [4], successive application of the complete second approximation in the framework of the Chapman–Enskog method results in the expression of the thermal conductivity of the mixture which agrees with a simpler result (5.2.11′). We shall see below that the 13N-moment approximation also adds some terms to the equations of the multicomponent diffusion of the species.

Let us consider the case of a binary gas mixture.

The expression of the total heat flux may be written as

$$\mathbf{q} = -\lambda \nabla T + p y_1 y_2 \alpha_T (\mathbf{w}_1 - \mathbf{w}_2) + \frac{5}{2} kT (n_1 \mathbf{w}_1 + n_2 \mathbf{w}_2). \quad (5.2.13)$$

The structure of the coefficient of the thermal conductivity of a binary mixture λ is the same as viscosity [Eq. (5.2.6)] and can be represented as [3]

$$[\lambda]_1 = \frac{1 + Z_\lambda}{X_\lambda + Y_\lambda},$$

where

$$X_\lambda = \frac{y_1^2}{[\lambda_{11}]_1} + \frac{2y_1y_2}{[\lambda_{12}]_1} + \frac{y_2^2}{[\lambda_{22}]_1},$$

$$Y_\lambda = \frac{y_1^2}{[\lambda_{11}]_1} U^{(1)} + \frac{2y_1y_2}{[\lambda_{12}]_1} U^{(Y)} + \frac{y_2^2}{[\lambda_{22}]_1} U^{(2)}, \quad (5.2.14)$$

$$Z_\lambda = y_1^2 U^{(1)} + 2y_1y_2 U^{(Z)} + y_2^2 U^{(2)},$$

and

$$U^{(1)} = \frac{4}{15} A^*_{12} - \frac{1}{12}\left(\frac{12}{5} B^*_{12} + 1\right)\frac{m_1}{m_2} + \frac{1}{2}\frac{(m_1 - m_2)^2}{m_1 m_2},$$

$$U^{(2)} = \frac{4}{15} A^*_{12} - \frac{1}{12}\left(\frac{12}{5} B^*_{12} + 1\right)\frac{m_2}{m_1} + \frac{1}{2}\frac{(m_2 - m_1)^2}{m_1 m_2},$$

$$U^{(Y)} = \frac{4}{15} A^*_{12} \frac{(m_1 + m_2)^2}{4 m_1 m_2} \frac{[\lambda_{12}]_1^2}{[\lambda_{11}]_1 [\lambda_{22}]_1} - \frac{1}{12}\left(\frac{12}{5} B^*_{12} + 1\right) - \frac{5}{32 A^*_{12}}\left(\frac{12}{5} B^*_{12} - 5\right)\frac{(m_1 - m_2)^2}{m_1 m_2},$$

$$U^{(Z)} = \frac{4}{15}A_{12}^{*}\left[\frac{(m_1+m_2)^2}{4m_1m_2}\left(\frac{[\lambda_{12}]_1}{[\lambda_{11}]_1}+\frac{[\lambda_{12}]_1}{[\lambda_{22}]_1}\right)-1\right] - \frac{1}{12}\left(\frac{12}{5}B_{12}^{*}+1\right).$$

Here $[\lambda_{11}]_1$ and $[\lambda_{22}]_1$ are the thermal conductivity of pure gases in the first Chapman–Cowling approximation defined by (5.2.9) and

$$[\lambda_{12}]_1 = \frac{15}{4}k\frac{m_1+m_2}{2m_1m_2}[\eta_{12}]_1\,. \tag{5.2.15}$$

The thermal diffusion factor (or thermal diffusion constant) α_T in the expression (5.2.13) is given by the relation

$$\alpha_T\,[D_{12}]_1 = \frac{D_{12}^T}{m_1n_1} - \frac{D_{21}^T}{m_2n_2}. \tag{5.2.16}$$

We shall discuss the value α_T for a binary mixture in detail in the next section. Note that the third term on the right-hand side of Eq. (5.2.13) cancels if the total heat flux is defined in the reference frame moving with the molar-average velocity $\mathbf{u}_m = n^{-1}\sum_\alpha n_\alpha \mathbf{u}_\alpha$. This is accounted for by the fact that the term $(5/2)\,kT\,(n_1\mathbf{w}_1+n_2\mathbf{w}_2)$ is the density enthalpy flux in the reference frame where the mass-average velocity $\mathbf{u}$ cancels. The total heat flux in the reference frame where $\mathbf{u}_m = 0$ is defined as [5]

$$\mathbf{q}_m = -\lambda\nabla T + kT\alpha_T\mathbf{J}_{1m}, \tag{5.2.17}$$

where

$$\mathbf{J}_{1m} = n_1(\mathbf{u}_1-\mathbf{u}_m) = ny_1y_2(\mathbf{u}_1-\mathbf{u}_2) \tag{5.2.18}$$

is the diffusion flux of the particles of species 1 in the reference frame moving with the velocity $\mathbf{u}_m$. It means that the quantity of the energy which is transferred by the mutual diffusion of the species per particle equals $kT\alpha_T$.

Now consider the contribution of the terms proportional to the spatial derivatives of the fluxes to the expressions for π_{rs} and $\mathbf{q}$. In this case, we retain the terms $\sim\{\partial q_{\alpha r}/\partial x_s\}$ and $\sim\{\partial \pi_{\alpha rs}/\partial x_s\}$ in Eqs. (4.3.3) and (4.3.4). Adding the corresponding term to the right-hand side of Eqs. (5.1.2) and solving them with respect to $\pi_{\gamma rs}/p_\gamma$, we come to the expression after summing over α and changing the indices $\beta \rightleftharpoons \alpha$ under the sign of the sum

$$\pi_{rs} = -2\eta\varepsilon_{rs} - \frac{4}{5p}\sum_{\alpha}\sum_{\beta} y_\alpha \frac{|H|_{\beta\alpha}}{|H|}\left\{\frac{\partial q_{\beta r}}{\partial x_s}\right\}$$
$$= -2\eta\varepsilon_{rs} - \frac{4}{5p}\sum_{\alpha}\frac{\eta_\alpha}{y_\alpha}\left\{\frac{\partial q_{\alpha r}}{\partial x_s}\right\}. \tag{5.2.19}$$

The same procedure for Eqs. (5.1.3) brings the result

$$\mathbf{h} = \mathbf{q} - \frac{5}{2}\sum_{\alpha} p_\alpha \mathbf{w}_\alpha$$
$$= -\lambda\nabla T + \sum_{\alpha}\sum_{\beta}\frac{n_\beta kT}{m_\alpha n\,[\mathfrak{D}_{\alpha\beta}]_1} D^T_{\alpha\beta}(\mathbf{w}_\alpha - \mathbf{w}_\beta) - \frac{2}{5}\frac{T}{p}\sum_{\alpha}\frac{\lambda_\alpha}{y_\alpha}\nabla\,\hat{\pi}\,. \tag{5.2.20}$$

For the viscous-stress tensor and heat flux of a pure gas, we have

$$\pi_{rs} = -2\eta\left(\varepsilon_{rs} + \frac{2}{5}\frac{1}{p}\left\{\frac{\partial q_r}{\partial x_s}\right\}\right),$$
$$\mathbf{q} = -\lambda\left(\nabla T + \frac{2}{5}\frac{T}{p}\nabla\,\hat{\pi}\right). \tag{5.2.21}$$

The relations go beyond the limits of the standard linear transport relations derived in the conventional (non-Burnett) approximation by the Chapman–Enskog method and used in the classical nonequilibrium thermodynamics [6]. Assuming as usual that these terms are small in the second order in the Knudsen number we may use the iteration procedure. Substitute the values of $\mathbf{q}$ and π_{rs} obtained in the first approximation, when the second order terms are dropped, into the terms containing the derivatives of fluxes in Eqs. (5.2.21), then

$$\pi = -2\eta\left(\varepsilon_{rs} - \frac{2}{5}\frac{\lambda}{p}\{\nabla\nabla\}T\right),$$
$$\mathbf{q} = -\lambda\left(\nabla T - \frac{2}{5}\frac{T}{p}\eta\nabla^2\mathbf{u}\right). \tag{5.2.21'}$$

The representation of Eqs. (5.2.21′) is known to correspond to the results of the linearized Burnett approximation of the Chapman–Enskog method [2].

It should be emphasized that expressions (5.2.21) are more general than Eqs. (5.2.21′). This can be readily seen in the case of the steady viscous gas flow, when from the motion equation (2.2.17) it follows that $\nabla\,\hat{\pi} = -\nabla p$. Then the expression for $\mathbf{q}$ in Eq. (5.2.21) takes the form

$$\mathbf{q} = -\lambda\left(\nabla T - \frac{2}{5}\frac{T}{p}\nabla p\right) \tag{5.2.22}$$

whence it follows that the additional terms, i.e. heat transfer due to the pressure gradient, which occurs, for example, in narrow capillary channels, is of the same (first) order of magnitude as the conventional term accounting for the heat transfer because of the temperature gradient. Section 5.5 shows that the presence of additional terms in the expressions of the heat flux and the viscous-stress tensor does not result in any conflict with the relations of the generalized nonequilibrium thermodynamics [7].

5.3 Diffusion in a multicomponent mixture

Let us turn back to Eqs. (5.1.1). Substituting the expressions of the partial reduced heat fluxes (5.2.7) into them gives the following system of diffusion equations in a multicomponent gas mixture [1]:

$$\sum_\beta \frac{n_\alpha n_\beta kT}{n\left[\mathfrak{D}_{\alpha\beta}\right]_1}(\mathbf{w}_\alpha - \mathbf{w}_\beta) + \frac{k^2T^3}{p^2}\sum_\beta\sum_\gamma\sum_\delta \frac{\omega_{\alpha\beta\gamma}}{m_\gamma}\xi_{\gamma\delta}\xi_{\alpha\beta}$$
$$\times(\mathbf{w}_\gamma - \mathbf{w}_\delta) = -p\mathbf{d}_\alpha - \sum_\beta \frac{n_\alpha n_\beta kT}{n\left[\mathfrak{D}_{\alpha\beta}\right]_1}\left(\frac{D^T_{\alpha\beta}}{m_\alpha n_\alpha} - \frac{D^T_{\beta\alpha}}{m_\beta n_\beta}\right)\nabla \ln T, \tag{5.3.1}$$

where

$$\omega_{\alpha\beta\gamma} = \frac{1}{m_\beta}\frac{|\Lambda|_{\gamma\beta}}{|\Lambda|} - \frac{1}{m_\alpha}\frac{|\Lambda|_{\gamma\alpha}}{|\Lambda|}.$$

It is shown in [8] that instead of the first and second terms on the left-hand side of Eq. (5.3.1) we can write a single term as

$$\sum \frac{n_\alpha n_\beta kT}{n\left[\mathfrak{D}_{\alpha\beta}\right]_1}(1 - \Delta_{\alpha\beta})(\mathbf{w}_\alpha - \mathbf{w}_\beta), \tag{5.3.2}$$

where $\Delta_{\alpha\beta}$ is the second-order correction to the binary diffusion coefficient. Such a representation is also valid for the higher n-approximation [9,10] if the parameter $\Delta^{(n)}_{\alpha\beta}$ is introduced instead of $\Delta_{\alpha\beta} = \Delta^{(2)}_{\alpha\beta}$. The expressions for $\Delta^{(n)}_{\alpha\beta}$ are cumbersome and are not given here.

Note that the coefficients $D^T_{\alpha\beta}$ and $D^T_{\beta\alpha}$ in the diffusion equations (5.3.1) are not, generally speaking, the coefficients of multicomponent thermal diffusion in the sense they are used in the Chapman–Enskog theory [3]. To compare the results, let us determine the thermal diffusion factor in a multicomponent mixture with the help of the expression

$$\alpha^T_{\alpha\beta}\left[\mathfrak{D}_{\alpha\beta}\right]_1 = \frac{D^T_{\alpha\beta}}{m_\alpha n_\alpha} - \frac{D^T_{\beta\alpha}}{m_\beta n_\beta} \tag{5.3.3}$$

or

$$\alpha^T_{\alpha\beta}\left[\mathfrak{D}_{\alpha\beta}\right]_1 = \frac{1}{k}\left(\frac{6}{5}C^*_{\alpha\beta} - 1\right)\mu_{\alpha\beta}\left(\frac{\lambda_\beta}{m_\beta n_\beta} - \frac{\lambda_\alpha}{m_\alpha n_\alpha}\right). \tag{5.3.3$'$}$$

Taking into consideration the relations (5.3.2) and (5.3.3′), we can represent the diffusion equations (5.3.1) as

$$\sum_\beta \frac{y_\alpha y_\beta}{\left[\mathfrak{D}_{\alpha\beta}\right]_1}\left(1 - \Delta_{\alpha\beta}\right)\left(\mathbf{w}_\alpha - \mathbf{w}_\beta\right) = -\,\mathbf{d}_{\alpha\beta} - \sum_\beta y_\alpha y_\beta \alpha^T_{\alpha\beta} \nabla \ln T. \tag{5.3.4}$$

Equations (5.3.4) together with the additional requirement of $\sum\limits_\alpha \rho_\alpha \mathbf{w}_\alpha = 0$ can be solved with respect to the species diffusion velocity. A similar formal solution was considered in Section 3.3. The result agrees with the expression (4.7.7) which can be obtained within the framework of the Chapman–Enskog method [5]

$$\mathbf{J}_\alpha = -\rho_\alpha \sum_\beta D_{\alpha\beta}\mathbf{d}_\beta - \rho_\alpha D^T_\alpha \nabla \ln T, \tag{5.3.5}$$

where $D_{\alpha\beta}$ and D^T_α are the multicomponent diffusion and thermal diffusion coefficients. The coefficients $D_{\alpha\beta}$ were shown in [12] to be symmetric and the diagonal coefficients $D_{\alpha\alpha}$ to be positive. The values of $D_{\alpha\beta}$ can be calculated through the solution of the set of equations [12,9]

$$\begin{gathered} \sum_\alpha \rho_\alpha D_{\alpha\beta} = 0, \\ \sum_\beta \frac{y_\alpha y_\beta}{\left[\mathfrak{D}_{\alpha\beta}\right]_1}\left(1 - \Delta_{\alpha\beta}\right)\left(D_{\alpha\gamma} - D_{\beta\gamma}\right) = \delta_{\alpha\gamma} - \frac{\rho_\alpha}{\rho}. \end{gathered} \tag{5.3.6}$$

Two sets of equations (5.3.6) may be combined to obtain the set

$$\sum_{\beta\neq\alpha}\left[\frac{y_\alpha y_\beta}{\left[\mathfrak{D}_{\alpha\beta}\right]_1}\left(1 - \Delta_{\alpha\beta}\right) + \sum_{\delta\neq\alpha}\frac{y_\delta y_\beta m_\beta}{m_\alpha\left[\mathfrak{D}_{\alpha\delta}\right]_1}\left(1 - \Delta_{\alpha\delta}\right)\right]D_{\beta\gamma} = \frac{\rho_\alpha}{\rho} - \delta_{\alpha\gamma}. \tag{5.3.6$'$}$$

The set of equations to calculate the coefficients D^T_α can be obtained in a similar way.

Note that various definitions of the multicomponent diffusion and thermal diffusion coefficients are met elsewhere [3,5,11,12]. The present definition coincides with the Onsager symmetry relations in the nonequilibrium thermodynamics [6] and looks more preferable than the definition given in [3].

Let us return back to the expression (5.2.10) for the total heat flux $\mathbf{q}$. The third term on the right-hand side of Eq. (5.2.10) can be transformed with the help of Eq. (5.3.5). The result is [5]

$$\mathbf{q} = -\lambda' \nabla T - p \sum_\alpha D_\alpha^T \mathbf{d}_\alpha + \frac{5}{2} kT \sum_\alpha n_\alpha \mathbf{w}_\alpha, \tag{5.3.7}$$

where

$$\lambda' = \lambda + nk \sum_\alpha y_\alpha y_\beta \alpha_\alpha^T D_\alpha^T. \tag{5.3.8}$$

The expression (5.3.7) agrees with the result (4.7.7) obtained by the Chapman–Enskog method and with the linear relations of nonequilibrium thermodynamics (2.5.21). In practice, the form (5.2.10) is preferable as in experiments the thermal conductivity of a gas mixture is usually defined under stationary conditions when there is no diffusion transport of the mixture species, therefore the coefficient λ defines the proper (true) thermal conductivity of the mixture.

Now let us consider the diffusion in a binary gas mixture. Define the molar diffusion flux of the component 1 in the reference frame, where $\mathbf{u}_m = 0$. Using the definition (5.2.18), we find from Eq. (5.3.4)

$$\mathbf{J}_{1m} = -n \left[D_{12}\right]_2 \left(\nabla y_1 + \alpha_p y_1 y_2 \nabla \ln p + \left[\alpha_T\right]_1 y_1 y_2 \nabla \ln T\right). \tag{5.3.9}$$

Here $[D_{12}]_2 = [D_{12}]_1 / (1 - \Delta_{12})$ corresponds to the second approximation to the binary diffusion coefficient, α_p and $[\alpha_T]_1$ are the barodiffusion factor and thermal diffusion factor, respectively[2]. The coefficients α_p, $[\alpha_T]_1$ and Δ_{12} are defined as

$$\alpha_p = \frac{m_2 - m_1}{m_1 y_1 + m_2 y_2}, \tag{5.3.10}$$

$$\left[\alpha_T\right]_1 = \frac{5}{2} \left(a_2 y_2 - a_1 y_1\right), \tag{5.3.11}$$

$$\Delta_{12} = \frac{5}{2} \mu_{12} \left(\frac{6}{5} C_{12}^* - 1\right) \left(\frac{a_2}{m_2} + \frac{a_1}{m_1}\right), \tag{5.3.12}$$

[2]Subscript index 1 with the thermal diffusion factor means in fact that the first non-vanishing approximation when calculating the coefficient as the contribution of a_α^{10} to thermal diffusion is zero. Conversely, when calculating the binary diffusion coefficient both a_α^{10} and a_α^{11} are used.

where

$$a_1 = \frac{2}{5}\frac{T}{pn}\xi_{12}\left(\frac{\Lambda_{22}}{m_1} + \frac{\Lambda_{21}}{m_2}\right)\frac{1}{|\Lambda|},$$

$$a_2 = \frac{2}{5}\frac{T}{pn}\xi_{12}\left(\frac{\Lambda_{12}}{m_1} + \frac{\Lambda_{11}}{m_2}\right)\frac{1}{|\Lambda|}, \tag{5.3.13}$$

$$|\Lambda| = \Lambda_{11}\Lambda_{22} - \Lambda_{12}^2.$$

Substituting $\Lambda_{\alpha\alpha}$ and $\Lambda_{\alpha\beta}$ into a_1 and a_2 we can reduce the expressions for $[\alpha_T]_1$ and Δ_{12} to the well-known form, which was obtained by the Chapman–Enskog method [3,5]

$$[\alpha_T]_1 = \frac{1}{6}\frac{1}{[\lambda_{12}]_1}\frac{S^{(1)}y_1 - S^{(2)}y_2}{X_\lambda + Y_\lambda}\left(6C_{12}^* - 5\right), \tag{5.3.14}$$

$$\Delta_{12} = \frac{W}{X_\lambda + Y_\lambda}\frac{\left(6C_{12}^* - 5\right)^2}{60}, \tag{5.3.15}$$

where

$$S^{(1)} = \frac{m_1 + m_2}{2m_2}\frac{[\lambda_{12}]_1}{[\lambda_{11}]_1} - \frac{15}{4A_{12}^*}\frac{m_2 - m_1}{2m_1} - 1,$$

$$S^{(2)} = \frac{m_1 + m_2}{2m_1}\frac{[\lambda_{12}]_1}{[\lambda_{22}]_1} - \frac{15}{4A_{12}^*}\frac{m_1 - m_2}{2m_2} - 1,$$

$$W = \frac{y_1^2}{[\lambda_{11}]_1}\frac{m_1}{m_2} + \frac{2y_1y_2}{[\lambda_{12}]_1}\left[1 + \frac{15}{8A_{12}^*}\frac{(m_1 - m_2)^2}{m_1m_2}\right] + \frac{y_2^2}{[\lambda_{22}]_1}\frac{m_2}{m_1} \tag{5.3.16}$$

and X_λ, Y_λ are described by the expressions (5.2.14).

5.4 Barodiffusion in a viscous flow

So far when analyzing the transport phenomena in a mixture we have referred to Eqs. (5.1.1)–(5.1.3). As was pointed out earlier, in some cases the account of terms proportional to the space derivatives of fluxes which were kept in Eqs. (4.3.2)–(4.3.4) becomes essential. For instance, let us consider the stationary viscous flow of a gas mixture in a channel. In this case, $d\mathbf{u}/dt = 0$ and it follows from the motion equation (2.2.17) that

$$\frac{\partial p}{\partial x_r} = -\frac{\partial \pi_{rs}}{\partial x_s} + \sum_\alpha n_\alpha X_{\alpha r}. \tag{5.4.1}$$

For this case, Eq. (4.3.2) is written as

$$\frac{\partial p_\alpha}{\partial x_r} + \frac{\partial \pi_{\alpha rs}}{\partial x_s} - n_\alpha \left(X_{\alpha r} + e_\alpha E_r\right) = R_{\alpha r}. \tag{5.4.2}$$

Since

$$\frac{\partial \pi_{\alpha rs}}{\partial x_s} = -2\eta_\alpha \frac{\partial \varepsilon_{rs}}{\partial x_s}, \tag{5.4.3}$$

at first sight this value proportional to the second derivative of the velocity is of the $(\lambda/L)^2$ order and may be omitted. Yet, from (5.4.1) it follows that

$$2\eta \frac{\partial \varepsilon_{rs}}{\partial x_s} = \frac{\partial p}{\partial x_r} - \sum_\alpha n_\alpha X_{\alpha r}, \tag{5.4.1$'$}$$

i.e. for the stationary viscous flows this value can be of the (λ/L) order and be defined in essence by the value of the pressure gradient. Then, eliminating $\partial \varepsilon_{rs}/\partial x_s$ in the expression for $\partial \pi_{\alpha rs}/\partial x_s$ with the help of (5.4.1$'$), we can represent Eq. (5.4.2) as

$$\left(\nabla p_\alpha - \frac{\eta_\alpha}{\eta}\nabla p\right) - \left(n_\alpha \mathbf{X}_\alpha - \frac{\eta_\alpha}{\eta}\sum_\alpha n_\alpha \mathbf{X}_\alpha\right) - n_\alpha e_\alpha \mathbf{E} = \mathbf{R}_\alpha. \tag{5.4.2$'$}$$

If the contribution of $\partial \pi_{\alpha rs}/\partial x_s$ is taken into account in the equations for $h_{\gamma r}$ following Eq. (4.3.4) and the results obtained for $h_{\alpha r}$ and $h_{\beta r}$ are substituted into $R_{\alpha r}$ (the right-hand side of Eq. (5.1.1)), we arrive at the final system of diffusion equations in the viscous flow of the mixture [1]

$$\begin{aligned}
&\sum_\beta \frac{n_\alpha n_\beta kT}{n\left[\mathfrak{D}_{\alpha\beta}\right]_1}\left(\mathbf{w}_\alpha - \mathbf{w}_\beta\right) + \frac{k^2T^3}{p^2}\sum_\beta\sum_\gamma\sum_\delta \frac{\omega_{\alpha\beta\gamma}}{m_\gamma}\xi_{\alpha\beta}\xi_{\gamma\delta}\left(\mathbf{w}_\gamma - \mathbf{w}_\delta\right) \\
&= -p\nabla y_\alpha - \left(y_\alpha - \frac{\eta_\alpha + \omega_\alpha}{\eta}\right)\nabla p + \left(n_\alpha \mathbf{X}_\alpha - \frac{\eta_\alpha + \omega_\alpha}{\eta}\sum_\alpha n_\alpha \mathbf{X}_\alpha\right) \\
&\quad + n_\alpha e_\alpha \mathbf{E} - \sum_\beta \frac{n_\alpha n_\beta kT}{n}\alpha^T_{\alpha\beta}\nabla \ln T,
\end{aligned} \tag{5.4.4}$$

where

$$\omega_\alpha = \frac{2}{5}k\left(\frac{T}{p}\right)^2 \sum_\beta\sum_\gamma \xi_{\alpha\beta}\eta_\gamma\omega_{\alpha\beta\gamma}.$$

In the case of the binary mixture, it follows from Eqs. (5.4.4) that

$$\mathbf{J}_{1m} = -n\left[D_{12}\right]_2\left(\nabla y_1 + \left[\alpha_p\right]_2 y_1y_2\nabla\ln p + \left[\alpha_T\right]_1 y_1y_2\nabla\ln T\right). \tag{5.4.5}$$

The barodiffusion factor in the viscous flow calculated in the second approximation turns out to be substantially a kinetic parameter and is defined by [1]

$$\left[\alpha_p\right]_2 = \left[\alpha_p\right]_1\left(1-\Delta_p\right) - \frac{2}{5}\left[\alpha_T\right]_1, \tag{5.4.6}$$

where

$$\left[\alpha_p\right]_1 = \frac{1}{\left[\eta\right]_1}\left(\frac{\left[\eta_2\right]_1}{y_2} - \frac{\left[\eta_1\right]_1}{y_1}\right) \tag{5.4.7}$$

is the barodiffusion factor calculated in the first approximation. Using the expressions for $[\eta]_1$ (5.2.2) and the values of $H_{\alpha\alpha}$ and $H_{\alpha\beta}$ we can represent $[\alpha_p]_1$ as

$$\left[\alpha_p\right]_1 = \frac{3}{5}A_{12}^*\frac{(m_1+m_2)}{2m_1m_2}\frac{K^{(1)}y_1 - K^{(2)}y_2}{1+Z_\eta}, \tag{5.4.7'}$$

where

$$K^{(1)} = \frac{\left[\eta_{12}\right]_1}{\left[\eta_{11}\right]_1} - \frac{2m_1}{m_1+m_2}, \quad K^{(2)} = \frac{\left[\eta_{12}\right]_1}{\left[\eta_{22}\right]_1} - \frac{2m_2}{m_1+m_2}$$

and Z_η is defined by (5.2.6). The correction of the second approximation Δ_p is written as

$$\Delta_p = a_1 + a_2 = \frac{2}{5}\left(\frac{m_2-m_1}{(m)_y}y_1y_2\left[\alpha_T\right]_1 + \frac{m_2+m_1}{(m)_y}\frac{\Delta_{12}}{\frac{6}{5}C_{12}^*-1}\right), \tag{5.4.8}$$

where $(m)_y = m_1y_1 + m_2y_2$.

The barodiffusion factor (5.4.6) markedly differs from the value of α_p for non-viscous flows (5.3.10). The difference between (5.4.6) and (5.3.10) is explained by the fact that the barodiffusion factor in the viscous flow depends on the interaction of the particles and, similar to the thermal diffusion factor, can generally be of any sign. This is clearly exhibited in the case of a small relative difference of particle masses and average cross sections. In particular, for the model of hard elastic spheres of diameters d_1 and d_2, expansion of the relevant expressions in series over small values brings [1]

$$[\alpha_p]_1 = 1.13\frac{m_2 - m_1}{m_2 + m_1} - 1.50\frac{d_2 - d_1}{d_2 + d_1}. \tag{5.4.9}$$

When the corrections of the second approximation are taken into account, the coefficients in this expression are respectively changed for 1.41 and 1.26.

5.5 Entropy production. Generalized nonequilibrium thermodynamics of a gas mixture

In Section 2.5 the derivation of expressions of entropy flux density and entropy production was considered using the Gibbs thermodynamic identity and conservation equations. Analogous results can be directly obtained in terms of kinetic expressions of these values using the linearized Boltzmann equation and expansion of the distribution function in series over the system of orthogonal tensorial polynomials. This enables one to generalize the common results of the nonequilibrium thermodynamics allowing for additional terms in the expression of the entropy production proportional to the products of the fluxes and by the spatial derivatives of the fluxes (or moments) of another tensor rank. It is important that despite the formal complexity of generalized thermodynamic forces because of the added terms the expression of the entropy production still preserves the form of the bilinear combination of generalized thermodynamic forces and fluxes. The linear phenomenological relationships of generalized nonequilibrium thermodynamics are in perfect agreement with those following from the linearized moment equations. Because of that, the effects considered earlier and related to the account of spatial derivatives of the heat flux and viscous-stress tensor in the linearized moment equations inherently constitute part of the scheme of generalized nonequilibrium thermodynamics.

Let the distribution function f_α be represented as

$$f_\alpha = f_\alpha^{(0)}\left(1 + \Phi_\alpha\right), \qquad |\Phi_\alpha| \ll 1.$$

The correction Φ_α, as we have already seen, satisfies the linearized kinetic equation (4.6.2). For simplicity, consider the stationary state of a gas mixture and for the time being omit the term of $f_\alpha^{(0)}\left(\mathbf{c}_\alpha \cdot \boldsymbol{\nabla}\right)\Phi_\alpha$

and the term following it which depends on the external forces on the left-hand side of Eq. (4.6.2). Then, this equation can be rewritten as

$$f_\alpha^{(0)}\left[\frac{n}{n_\alpha}\mathbf{c}_\alpha\cdot\mathbf{d}_\alpha+\left(\frac{m_\alpha c_\alpha^2}{2kT}-\frac{5}{2}\right)\mathbf{c}_\alpha\cdot\nabla\ln T\right.$$
$$\left.+\frac{m_\alpha}{kT}\left(c_{\alpha r}c_{\alpha s}-\frac{1}{3}\delta_{rs}c_\alpha^2\right)\varepsilon_{rs}\right] = -\sum_{\beta=1}^{N} n_\alpha n_\beta I_{\alpha\beta}(\Phi_\alpha). \qquad (5.5.1)$$

This equation is coincident with Eq. (4.7.2) for the correction of the first approximation by the Chapman–Enskog method. The integral operator $n_\alpha n_\beta I_{\alpha\beta}(\Phi_\alpha)$ is defined in this case by (4.7.3).

It can be seen that in the linear approximation over Φ_α the expression of the local entropy production is written as

$$\sigma = k\sum_{\alpha=1}^{N}\sum_{\beta=1}^{N} n_\alpha n_\beta\int\Phi_\alpha I_{\alpha\beta}(\Phi_\alpha)\,d\mathbf{v}_\alpha. \qquad (5.5.2)$$

Substituting the left-hand side of the linearized kinetic equation (5.5.1) for the integral collisional operator, integrating over velocities and using the definitions of the fluxes $\mathbf{h}, \pi_{rs}$ and diffusion velocities $\mathbf{w}_\alpha$, we find

$$\sigma = -\frac{1}{T}\left[\mathbf{h}\cdot\nabla\ln T+\pi_{rs}\varepsilon_{rs}+p\sum_{\alpha=1}^{N-1}(\mathbf{w}_\alpha-\mathbf{w}_N)\cdot\mathbf{d}_\alpha\right]. \qquad (5.5.3)$$

When putting down (5.5.3), the condition $\sum_{\alpha=1}^{N}\mathbf{d}_\alpha = 0$ is used. The result obtained fully agrees with the expression of entropy production (2.5.17) derived earlier in the framework of nonequilibrium thermodynamics.

Let us now consider the problem as the allowance for the term $f_\alpha^{(0)}(\mathbf{c}_\alpha\cdot\nabla)\Phi_\alpha$ in the linearized kinetic equation (4.6.2) is compatible with the nonequilibrium thermodynamics results. It has been previously shown that the terms with the spatial derivatives of fluxes (moments) of a different tensor rank occur in the moment equations. The terms are in some cases of the same order with respect to the Knudsen number as the main linear terms. As can be seen from the structure of the generalized linearized moment equations, which are given in Appendix 1, there are only the tensor moments $(m-1)$th and $(m+1)$th ranks under the sign of the spatial derivatives. Similar results were also obtained in [14]. The direct calculation of entropy production [7,13] shows that if the integral collision operator in the expression (5.5.2) is replaced by the left-hand side of the kinetic equation (4.6.2) with the additional term, we get a bilinear combination of tensor moments of the rank m and spatial derivatives of moments of $(m-1)$th and $(m+1)$th ranks in this expression. We shall not write here the general expression of σ (see e.g. [13]) but limit ourselves by the $13N$-moment approximation. The total expression of the entropy production in this approximation has the form

$$\sigma = -\frac{1}{T}\sum_{\alpha}\left[h_{\alpha r}\left(\frac{\partial \ln T}{\partial x_r} + \frac{2}{5}\frac{1}{p_\alpha}\frac{\partial \pi_{\alpha rs}}{\partial x_s}\right)\right]$$
$$-\frac{1}{T}\sum_{\alpha}\left[\pi_{\alpha rs}\left(\varepsilon_{rs} + \frac{2}{5}\frac{1}{p_\alpha}\left\{\frac{\partial q_{\alpha r}}{\partial x_s}\right\}\right)\right]$$
$$-\frac{1}{T}\sum_{\alpha=1}^{N-1}\left(w_{\alpha r} - w_{Nr}\right) p d^*_{\alpha r}, \tag{5.5.4}$$

where

$$p d^*_{\alpha r} = p d_{\alpha r} + \left(\frac{\partial \pi_{\alpha rs}}{\partial x_s} - \frac{\rho_\alpha}{\rho}\frac{\partial \pi_{rs}}{\partial x_s}\right). \tag{5.5.5}$$

The set of linear phenomenological relations corresponding to the representation (5.5.4) may be written for the difference of diffusion velocities $w_{\alpha r} - w_{Nr}$, partial fluxes $\mathbf{h}_\alpha$ and $\pi_{\alpha rs}$ if the generalized thermodynamic forces $d^*_{\alpha r}$, $\nabla \ln T + (2/5)\,p_\alpha^{-1}\nabla\,\hat{\pi}_\alpha$ and $\varepsilon_{rs} + (2/5)\,p_\alpha^{-1}$ $\{\partial q_{\alpha r}/\partial x_s\}$ are used. A number of new Onsager symmetry relations between the coefficients in the expressions of the total and partial fluxes arises in this case in addition to the usual relations in the expressions for $w_{\alpha r} - w_{Nr}$ and $\mathbf{h} = \sum_\alpha \mathbf{h}_\alpha$ (see (2.5.21). In particular, we have for π_{rs} and $\pi_{\alpha rs}$

$$\pi_{rs} = -\Lambda_{00}\varepsilon_{rs} + \sum_{\beta=1}^{N}\Lambda_{0\beta}\frac{2}{5}\frac{1}{p_\beta}\left\{\frac{\partial q_\beta}{\partial x_s}\right\},$$
$$\pi_{\alpha rs} = -\Lambda_{\alpha 0}\varepsilon_{rs} + \sum_{\beta=1}^{N}\Lambda_{\alpha\beta}\frac{2}{5}\frac{1}{p_\beta}\left\{\frac{\partial q_\beta}{\partial x_s}\right\}, \tag{5.5.6}$$

where $\Lambda_{\alpha 0} = \Lambda_{0\alpha}$. The similar relations can be written for the crossed coefficients in the expressions for $\mathbf{h}$ and $\mathbf{h}_\alpha$.

Let us consider the case of a pure gas $(N = 1)$. The relations of $\mathbf{q}$ and π_{rs} can be written as

$$q_r = -L_{00}\left[\frac{\partial \ln T}{\partial x_r} + \frac{2}{5}\frac{1}{p}\frac{\partial \pi_{rs}}{\partial x_s}\right],$$

$$\pi_{rs} = -\Lambda_{00}\left[\varepsilon_{rs} + \frac{2}{5}\frac{1}{p}\left\{\frac{\partial q_r}{\partial x_s}\right\}\right], \tag{5.5.7}$$

The relations obtained are completely consistent with the results (5.2.21) following from the direct solution of the moment equations of 13 moment approximation with allowance for spatial derivatives of fluxes on their left-hand sides. In this case, $L_{00} = [\lambda]_1 T$ and $\Lambda_{00} = 2[\eta]_1$.

We note that the complication of the form of the thermodynamic forces because of the account of the derivatives of the fluxes in Eqs. (5.5.7) does not demand any changes in the values of the coefficients L_{00} and Λ_{00} which are determined by the ordinary viscosity and thermal conductivity. Moreover, just as in the standard classical scheme [6,11], the entropy production is found to be a linear combination of terms which are quadratic in the fluxes

$$\sigma = \frac{1}{T}\left(\frac{1}{L_{00}}q_r q_r + \frac{1}{\Lambda_{00}}\pi_{rs}\pi_{rs}\right). \tag{5.5.8}$$

The condition $\sigma \geq 0$, i.e. the positive entropy production, is ensured by the positive values of the coefficients L_{00} and Λ_{00} or the obvious conditions $\lambda > 0$ and $\eta > 0$. The appropriate analysis brings us to the conclusion that the similar conditions are valid for a multicomponent gas mixture.

Strictly speaking, the systems of equations (5.5.6) and (5.5.7) must be solved self-consistently since the thermodynamic forces contain derivatives of the fluxes. This situation is somewhat different from the conventional representation of the phenomenological equations of nonequilibrium thermodynamics, where the fluxes appear on the left-hand side and the gradient of the standard hydrodynamic variables appear on the right-hand side.

We can also represent our expressions in the canonical form with the help of the iteration procedure using a small parameter such as the Knudsen number or decompose the fluxes into constituents corresponding to the contributions of the first and successive orders of magnitude. Similar results can be obtained in the linearized Burnett approximation of the Chapman–Enskog method. The linear phenomenological relations of the generalized nonequilibrium thermodynamics may be given [15,16] as

$$\mathbf{h} = -L_{00}\nabla \ln T - \sum_{\beta=1}^{N-1} L_{0\beta} p \mathbf{d}_\beta - \widetilde{L}_{0N} \nabla^2 \mathbf{u},$$

$$\mathbf{w}_\alpha - \mathbf{w}_N = -L_{\alpha 0}\nabla \ln T - \sum_{\beta=1}^{N-1} L_{\alpha\beta} p \mathbf{d}_\beta - \widetilde{L}_{\alpha N} \nabla^2 \mathbf{u}, \tag{5.5.9}$$

$$\pi_{rs} = -\Lambda_{00}\varepsilon_{rs} - \widetilde{\Lambda}_{0N}\frac{1}{T}\{\nabla\nabla\}T - \sum_{\beta=1}^{N-1} \widetilde{\Lambda}_{0\beta}\{\nabla\nabla\}y_\alpha.$$

A consistent development of the Burnett approximation [15,16] brings in additional 'nonphysical' fluxes which are of the same tensor rank as the main fluxes. Allowing for the additional moments, the system of phenomenological equations for the fluxes and forces greatly expands, while the cross coefficients in the expressions for the fluxes with the same tensor rank satisfy the Onsager symmetry relation. The same result can also be obtained by the moment method if one uses the highest moments of the distribution function and the moment equations corresponding to them. The appropriate investigation of this case is given elsewhere [7,13].

Chapter 6

Transport phenomena in a multispecies plasma

6.1 Multispecies nonisothermal plasma in the magnetic field

We shall turn back to the general system of Eqs. (4.3.5)–(4.3.7) for the nonisothermal multispecies plasma. Let us recall that $G^{(1)}_{\alpha\beta} = -n_\alpha\mu_{\alpha\beta}\tau^{-1}_{\alpha\beta}$ and introduce the notation $A^{(n)}_{\alpha\beta} = G^{(n)}_{\alpha\beta}/G^{(1)}_{\alpha\beta}$. Then represent this system as [1,2][1]

$$n_\alpha \sum_\beta \mu_{\alpha\beta}\tau^{-1}_{\alpha\beta}\left[\mathbf{w}_\alpha - \mathbf{w}_\beta + \frac{\gamma_\beta}{\gamma_\alpha+\gamma_\beta}A^{(2)}_{\alpha\beta}\left(\frac{\mathbf{h}_\alpha}{p_\alpha} - \frac{\gamma_\alpha}{\gamma_\beta}\frac{\mathbf{h}_\beta}{p_\beta}\right)\right]$$

$$= -\left(\nabla p_\alpha - \frac{\rho_\alpha}{\rho}\nabla p\right) + n_\alpha e_\alpha \mathbf{E}' + \rho_\alpha \mathbf{w}_\alpha \times \omega_\alpha - \frac{\rho_\alpha}{\rho}\mathbf{j}\times\mathbf{B}, \tag{6.1.1}$$

$$\sum_\beta a_{\alpha\beta}\pi_{\beta rs} = -\overline{\eta}_\alpha W_{rs} + \frac{1}{2}\left(\pi_{\alpha lr}e_{slm}\omega_{\alpha m} + \pi_{\alpha ls}e_{rlm}\omega_{\alpha m}\right)\tau_\alpha, \tag{6.1.2}$$

$$\sum_\beta b_{\alpha\beta}\mathbf{h}_\beta = -\overline{\lambda}_\alpha \mathbf{N}_\alpha + \mathbf{h}_\alpha \times \omega_\alpha \tau^*_\alpha, \tag{6.1.3}$$

where

[1]The forces of non-electromagnetic character are omitted for simplicity.

$$W_{rs} = 2\varepsilon_{rs} = \frac{\partial u_r}{\partial x_s} + \frac{\partial u_s}{\partial x_r} - \frac{2}{3}\delta_{rs}\frac{\partial u_l}{\partial x_l},$$

$$\mathbf{N}_\alpha = \nabla T_\alpha + \frac{1}{k}\sum_\beta \mu_{\alpha\beta}\tau_{\alpha\beta}^{-1}\left[\frac{\gamma_\beta}{\gamma_\alpha+\gamma_\beta}A_{\alpha\beta}^{(7)}(\mathbf{w}_\alpha - \mathbf{w}_\beta) + 2\theta_{\alpha\beta}\mathbf{w}_\alpha\right]. \tag{6.1.4}$$

In addition

$$\overline{\eta}_\alpha = \frac{1}{2}p_\alpha\tau_\alpha, \qquad \overline{\lambda}_\alpha = \frac{5}{2}\frac{k}{m_\alpha}p_\alpha\tau_\alpha^*,$$

$$\tau_\alpha^{-1} = \frac{1}{8}\left(A_{\alpha\alpha}^{(3)} + A_{\alpha\alpha}^{(4)}\right)\tau_{\alpha\alpha}^{-1} + \frac{1}{2}\sum_{\beta\neq\alpha}\frac{m_\beta}{m_\alpha+m_\beta} \times\left(1+\frac{\gamma_\beta}{\gamma_\alpha}\right)^{-1}A_{\alpha\beta}^{(3)}\tau_{\alpha\beta}^{-1},$$

$$(\tau_\alpha^*)^{-1} = \frac{1}{2}\left(A_{\alpha\alpha}^{(5)} + A_{\alpha\alpha}^{(6)}\right)\tau_{\alpha\alpha}^{-1} + \sum_{\beta\neq\alpha}\frac{m_\beta}{m_\alpha+m_\beta}A_{\alpha\beta}^{(5)}\tau_{\alpha\beta}^{-1}, \tag{6.1.5}$$

$$a_{\alpha\beta} = \frac{2}{3}\frac{m_\beta}{m_\alpha+m_\beta}\left(1+\frac{\gamma_\beta}{\gamma_\alpha}\right)^{-1}\frac{p_\alpha}{p_\beta}\tau_\alpha\tau_{\alpha\beta}^{-1}A_{\alpha\beta}^{(4)},$$

$$b_{\alpha\beta} = \frac{m_\beta}{m_\alpha+m_\beta}\frac{p_\alpha}{p_\beta}\tau_\alpha^*\tau_{\alpha\beta}^{-1}A_{\alpha\beta}^{(6)},$$

$$a_{\alpha\alpha} = b_{\alpha\alpha} = 1.$$

Note that in the case of the isothermal plasma ($T_\alpha = T$)

$$a_{\alpha\beta} = \frac{y_\alpha H_{\alpha\beta}}{y_\beta H_{\alpha\alpha}}, \qquad b_{\alpha\beta} = \frac{y_\alpha \Lambda_{\alpha\beta}}{y_\beta \Lambda_{\alpha\alpha}},$$
$$\tau_\alpha^{-1} = \frac{p}{2y_\alpha}H_{\alpha\alpha}, \quad (\tau_\alpha^*)^{-1} = \frac{5}{2}\frac{kT}{m_\alpha}\frac{p}{y_\alpha}\Lambda_{\alpha\alpha}, \tag{6.1.6}$$

where $H_{\alpha\beta}$, $H_{\alpha\alpha}$, $\Lambda_{\alpha\beta}$ and $\Lambda_{\alpha\alpha}$ are given by expressions (5.1.5), (5.1.6) of the previous chapter.

In the absence of the magnetic field ($|\mathbf{B}| = 0$) the general solutions of Eqs. (6.1.2), (6.1.3) can be written in the explicit form

$$\pi_{\alpha rs} = -\sum_\beta \frac{|a|_{\beta\alpha}}{|a|}\overline{\eta}_\beta W_{rs}, \quad \mathbf{h}_\alpha = -\sum_\beta \frac{|b|_{\beta\alpha}}{|b|}\overline{\lambda}_\beta \mathbf{N}_\beta. \tag{6.1.7}$$

At $T_\alpha = T$ and if Eq. (6.1.6) is taken into account, the expressions (6.1.7) agree with the results of Eqs. (5.2.1) and (5.2.7) obtained in the previous chapter.

To solve the sets of Eqs. (6.1.2) and (6.1.3) in an arbitrarily oriented magnetic field, we shall write a sequence of equations to make contractions of the tensor $\pi_{\alpha rs}$ with the tensor $e_{lrk}\omega_{\alpha s}$, vector $\omega_{\alpha s}$ and tensors $e_{lrk}\omega_{\alpha k}\omega_{\alpha s}$ and $\omega_{\alpha k}\omega_{\alpha s}$, on the basis of Eqs. (6.1.2) and equations for $h_{\alpha r}e_{lrk}\omega_{\alpha k}$ and $h_{\alpha r}\omega_{\alpha r}$, on the basis of Eq. (6.1.3). Using the properties of the antisymmetric unit tensor e_{rsl} after cumbersome but simple transformations, we can arrive at the expression for $\pi_{\alpha rs}$ written with the help of five coefficients of viscosity [3]

$$\pi_{\alpha rs} = -\eta_\alpha^{(0)} W_{rs}^{(0)} - \eta_\alpha^{(1)} W_{rs}^{(1)} - \eta_\alpha^{(2)} W_{rs}^{(2)} + \eta_\alpha^{(3)} W_{rs}^{(3)} + \eta_\alpha^{(4)} W_{rs}^{(4)}, \tag{6.1.8}$$

where [2]

$$\begin{aligned} \eta_\alpha^{(0)} &= \sum_\beta \frac{|a|_{\beta\alpha}}{|a|} \bar{\eta}_\beta, \quad \eta_\alpha^{(1)} = \sum_\beta \frac{|a^*|_{\beta\alpha}}{|a^*|} \bar{\eta}_\beta, \\ \eta_\alpha^{(2)} &= \sum_\beta \frac{|a^{**}|_{\beta\alpha}}{|a^{**}|} \bar{\eta}_\beta, \quad \eta_\alpha^{(3)} = \sum_\beta \omega_\beta \tau_\beta \frac{|a^*|_{\beta\alpha}}{|a^*|} \eta_\beta^{(0)}, \\ \eta_\alpha^{(4)} &= \frac{1}{2} \sum_\beta \omega_\beta \tau_\beta \frac{|a^{**}|_{\beta\alpha}}{|a^{**}|} \eta_\beta^{(0)}. \end{aligned} \tag{6.1.9}$$

By analogy, the expressions for $\mathbf{h}_\alpha$ can be found and conveniently written by introducing the components of $\mathbf{N}_\alpha$ parallel and perpendicular to the magnetic field [2]

$$\mathbf{h}_\alpha = -\sum_\beta \frac{|b|_{\beta\alpha}}{|b|} \bar{\lambda}_\beta \mathbf{N}_{\beta\|} - \sum_\beta \frac{|b^*|_{\beta\alpha}}{|b^*|} \times \left(\bar{\lambda}_\beta \mathbf{N}_{\beta\perp} + \omega_\beta \tau_\beta^* \sum_\beta \frac{|b|_{\gamma\beta}}{|b|} \bar{\lambda}_\gamma \mathbf{N}_\gamma \times \mathbf{k} \right), \tag{6.1.10}$$

where

$$\mathbf{N}_{\beta\|} = \mathbf{k} \cdot (\mathbf{k} \cdot \mathbf{N}_\beta), \quad \mathbf{N}_{\beta\perp} = \mathbf{k} \times (\mathbf{N}_\beta \times \mathbf{k}).$$

The elements of the determinants with one or two asterisks are related to the coefficients $a_{\alpha\beta}$ and $b_{\alpha\beta}$ via the expressions

$$a^{*}_{\alpha\beta} = a_{\alpha\beta} + \frac{|a|_{\beta\alpha}}{|a|}\omega_\alpha\tau_\alpha\omega_\beta\tau_\beta,$$

$$a^{**}_{\alpha\beta} = a_{\alpha\beta} + \frac{1}{4}\frac{|a|_{\beta\alpha}}{|a|}\omega_\alpha\tau_\alpha\omega_\beta\tau_\beta, \tag{6.1.11}$$

$$b^{*}_{\alpha\beta} = b_{\alpha\beta} + \frac{|b|_{\beta\alpha}}{|b|}\omega_\alpha\tau^{*}_\alpha\omega_\beta\tau^{*}_\beta.$$

Expressions for the tensor $W^{(p)}_{rs}$ which are various contractions of the tensor W_{kl} and the tensorial values of the type

$$k_r k_s k_k k_l,\ \delta_{rs} k_k k_l,\ e_{rmk} k_s k_m k_l \text{ and } e_{rmk}\delta_{ls} k_m,$$

can be referred to [3]. The tensors $W^{(p)}_{rs}$ are clearly simplified in a specially chosen Cartesian system of coordinates with the x axis directed along the magnetic field.

For the components of the tensor $\pi_{\alpha rs}$ in such a system, we have

$$\begin{aligned}
\pi_{\alpha xx} &= -\eta^{(0)}_\alpha W_{xx},\\
\pi_{\alpha yy} &= -\frac{1}{2}\eta^{(0)}_\alpha (W_{yy}+W_{zz}) - \frac{1}{2}\eta^{(1)}_\alpha (W_{yy}-W_{zz}) - \eta^{(3)}_\alpha W_{yz},\\
\pi_{\alpha zz} &= -\frac{1}{2}\eta^{(0)}_\alpha (W_{zz}+W_{yy}) - \frac{1}{2}\eta^{(1)}_\alpha (W_{zz}-W_{yy}) + \eta^{(3)}_\alpha W_{yz},\\
\pi_{\alpha yz} = \pi_{\alpha zy} &= -\eta^{(1)}_\alpha W_{yz} + \frac{1}{2}\eta^{(3)}_\alpha (W_{yy}-W_{zz}),\\
\pi_{\alpha xy} = \pi_{\alpha yx} &= -\eta^{(2)}_\alpha W_{xy} - \eta^{(4)}_\alpha W_{xz},\\
\pi_{\alpha xz} = \pi_{\alpha zx} &= -\eta^{(2)}_\alpha W_{xz} + \eta^{(4)}_\alpha W_{xy}.
\end{aligned} \tag{6.1.12}$$

The expressions (6.1.8)–(6.1.11) permit one to calculate the viscous-stress tensor and the reduced heat flux for any of the species of the nonisothermal multispecies plasma. The total viscous-stress tensor π_{rs} and heat flux $\mathbf{q}$ are obtained by simply adding the respective partial values

$$\pi_{rs} = \sum_\alpha \pi_{\alpha rs}, \qquad \mathbf{q} = \sum_\alpha \mathbf{h}_\alpha + \frac{5}{2}\sum_\alpha p_\alpha \mathbf{w}_\alpha. \tag{6.1.13}$$

Now let us consider the diffusion equations (6.1.1). They can be resolved with respect to the species diffusion velocities after the expressions (6.1.10) for $\mathbf{h}_\alpha$ and $\mathbf{h}_\beta$ are substituted on their left-hand side. Notice that if the contribution of the reduced heat fluxes in the diffusion equations are neglected, the latter are in full agreement with the equations of the quasihydrodynamic approach (3.1.21) whose solution was considered in Chapter 3.

Consideration of $\mathbf{h}_\alpha$ and $\mathbf{h}_\beta$ complicates the coefficients $c_{\alpha\beta}$ and $b_{\alpha\beta}$ in Eqs. (3.3.1) due to the terms proportional to the diffusion velocities of the components in $\mathbf{N}_\alpha$ (6.1.4). Besides, new terms appear which are related to the species temperature gradients describing the thermal diffusion. The formal solution of the system of equations derived in this manner was presented elsewhere [4,5]. Because the expressions obtained are cumbersome, we do not intend to write these solutions in the general form but limit ourselves to the analysis of special cases of interest.

6.2 Separation of equations for electrons and heavy particles. Electron transport properties

The general set of Eqs. (6.1.1)–(6.1.3) can be simplified to a certain extent because of the small parameter, vis. the ratio of the electron mass m_e to the mass of heavy particles m_β ($\beta \neq e$). If the requirement $\gamma_e \ll \gamma_\beta$ (or $m_e T_\beta / m_\beta T_e \ll 1$) is met, then in Eq. (6.1.1) for electrons ($\alpha = e$) the terms proportional to $\mathbf{h}_\beta$ can be neglected and the terms $\sim (\rho_e/\rho)$ can be omitted. As a result, it can be written as

$$\sum_{\beta \neq e} \tau_{e\beta}^{-1} \left(\mathbf{w}_e - \mathbf{w}_\beta + \zeta_{e\beta} \frac{\mathbf{h}_e}{p_e} \right) = -\frac{e}{m_e} \mathbf{E}^* - \omega_e \mathbf{w}_e \times \mathbf{k}, \tag{6.2.1}$$

where

$$\mathbf{E}^* = \mathbf{E}' + \frac{\nabla p_e}{n_e e}, \quad \omega_e = e\frac{|\mathbf{B}|}{m_e}, \quad \zeta_{e\beta} = \frac{6}{5} C_{e\beta}^* - 1.$$

Estimation of the order of the coefficients $a_{e\beta}$ and $b_{e\beta}$ ($\beta \neq e$) in Eqs. (6.1.2) and (6.1.3) under the same conditions yields

$$a_{e\beta} \sim \frac{n_e m_e \tau_e}{n_\beta m_\beta \tau_{e\beta}} \ll 1,$$

$$b_{e\beta} \sim \frac{n_e m_e \tau_e^*}{n_\beta m_\beta \tau_{e\beta}} \ll 1.$$

Then if in Eqs. (6.1.2) and (6.1.3) the cross terms are omitted, the viscous-stress tensor and reduced heat flux of the electrons can be determined irrespective of the other species from the equations

$$\pi_{ers} = -\bar{\eta}_e W_{rs} - \frac{1}{2}(\pi_{elr} e_{slm} k_m + \pi_{els} e_{rlm} k_m)\omega_e \tau_e, \tag{6.2.2}$$

$$\mathbf{h}_e = -\bar{\lambda}_e \mathbf{N}_e - \mathbf{h}_e \times \mathbf{k}\omega_e \tau_e^*, \tag{6.2.3}$$

where

$$\mathbf{N}_e = \nabla T_e + \frac{m_e}{k}\sum_\beta \tau_{e\beta}^{-1}\zeta_{e\beta}(\mathbf{w}_e - \mathbf{w}_\beta). \tag{6.2.4}$$

The solution of Eq. (6.2.2) can be written in the form of (6.1.8) at $\alpha = e$, where

$$\begin{gathered}\eta_e^{(0)} = \frac{1}{2}p_e\tau_e, \quad \eta_e^{(1)} = \frac{\eta_e^{(0)}}{1+\beta_e^2}, \quad \eta_e^{(2)} = \eta_e^{(1)}\left[\frac{\beta_e}{2}\right], \\ \eta_e^{(3)} = \beta_e\eta_e^{(1)}, \quad \eta_e^{(4)} = \eta_e^{(3)}\left[\frac{\beta_e}{2}\right], \qquad \beta_e = \omega_e\tau_e.\end{gathered} \tag{6.2.5}$$

For the electron heat flux, solving Eq. (6.2.3) with respect to $\mathbf{h}_e$ gives

$$\mathbf{h}_e = -\lambda_{e\|}\mathbf{N}_{e\|} - \lambda_{e\perp}\mathbf{N}_{e\perp} - \lambda_{e\wedge}\mathbf{k}\times\mathbf{N}_e \tag{6.2.6}$$

with the thermal conductivity coefficients

$$\begin{gathered}\lambda_{e\|} = \frac{5}{2}\frac{k}{m_e}p_e\tau_e^*, \quad \lambda_{e\perp} = \frac{\lambda_{e\|}}{1+\beta_e^{*2}}, \\ \lambda_{e\wedge} = \beta_e^*\lambda_{e\perp}, \qquad \beta_e^* = \omega_e\tau_e^*.\end{gathered} \tag{6.2.7}$$

The characteristic times τ_e and τ_e^* defined by

$$\begin{gathered}\tau_e^{-1} = 0.3\tau_{ee}^{-1} + 0.6\sum_{\beta\neq e} A_{e\beta}^*\tau_{e\beta}^{-1}, \\ (\tau_e^*)^{-1} = 0.4\tau_{ee}^{-1} + \sum_{\beta\neq e}\left(2.5 - 1.2B_{e\beta}^*\right)\tau_{e\beta}^{-1},\end{gathered} \tag{6.2.8}$$

where the coefficients $A_{e\beta}^*$, $B_{e\beta}^*$ (as well as $C_{e\beta}^*$ in the expression for $\zeta_{\alpha\beta}$) are found from Eq. (5.1.11) where $\Omega_{e\beta}^{lr}$ is defined by (4.2.20) if $\gamma_{e\beta} = \gamma_e = m_e/kT_e$. Note that for the Coulomb interactions (with the accuracy up to the terms $(\ln \Lambda)^{-1}$ as compared with unity) $A^*_{e\beta} = B^*_{e\beta} = 1$, $C^*_{e\beta} = 1/3$, and $\zeta_{e\beta} = -0.6$. For electron–atom interactions at $Q_{e\beta}(v) = \text{const}$ (the model of hard elastic spheres) $A^*_{e\beta} = B^*_{e\beta} = C^*_{e\beta} = 1$ and $\zeta_{e\beta} = 0.2$.

The expressions of (6.2.5) and (6.2.7) for the electron transport coefficients can be also derived directly from the general expressions (6.1.9)–(6.1.10) if the small values $a_{e\beta}$ and $b_{e\beta}$ are used and expansion over the small parameter in the determinants is made.

We shall use this procedure in the next chapter when defining the electron transport properties in the case of nonisothermal three-component plasma ($T_i = T_a = T$, $T_e \neq T$). The analysis proves that the only exception when the separation of equations for electrons from the equations of heavy particles used above appears to be incorrect refers to the definition of coefficients $\eta_e^{(1)}$ and $\eta_e^{(2)}$ at $\omega_i \tau_i \geq 1$. In Chapter 8, this case will be closely investigated for a two-component fully ionized plasma.

Now let us dwell on the equations for heavy particles. The analysis for the coefficients $a_{\alpha e}$ and $b_{\beta e}$ $(\alpha, \beta \neq e)$ shows that they are of the order of $(m_e/m_\alpha)^{1/2}$, therefore the terms with π_{ers} and $\mathbf{h}_e$ in the equations can be omitted. The resulting viscous-stress tensors and heat fluxes of the heavy particles are defined from the system of equations independent of the equations for electrons. This implies that the viscosities and thermal conductivities of ions and atoms are determined by (6.1.9) and (6.1.10) in which the terms related to the electron component should be eliminated in summation. If the temperature of the heavy components is assumed to be equal, then the values $a_{\alpha\beta}$, $b_{\alpha\beta}$, τ_α and τ^*_α involved in these equations are related to (6.1.6) with the coefficients $H_{\alpha\beta}$ and $\Lambda_{\alpha\beta}$ used in Chapter 5. For the longitudinal transport coefficients in this case the results obtained in the absence of a magnetic field for an isothermal multicomponent mixture are valid (Eqs. (5.2.2) and (5.2.8)).

Let us now consider the equations for the diffusion velocity of the species. Strictly speaking, the equation for electrons (6.2.1) cannot be separated from the equations for heavy particles since the diffusion velocities $\mathbf{w}_\beta$ $(\beta \neq e)$ participate there as well as in $\mathbf{h}_e$ by definition of $\mathbf{N}_e$. The analysis of the solutions of the diffusion equations written in the quasihydrodynamic approximation shows that in defining $\mathbf{w}_e$ the introduction of $\mathbf{w}_\beta$ into Eq. (6.2.1) becomes necessary only when calculating the transverse components of the electron diffusion velocity at $\beta_i \geq 1$, where $\beta_i = \omega_i \tau_{i\beta}$. If $\beta_i^2 \ll 1$, the equation for electrons becomes entirely independent. When one is interested in deriving the expression for $\mathbf{j} = -\sum_\beta n_\beta e_\beta (\mathbf{w}_e - \mathbf{w}_\beta)$, then instead of $\mathbf{w}_e$ and $\mathbf{w}_\beta$

the variables $\mathbf{j}$ and $\mathbf{x}_\beta = \mathbf{w}_\beta - \mathbf{w}_k$ should be used with $\beta \neq e$ and k as the index of an arbitrarily chosen species the equation for which is eliminated from the system as Eq. (6.1.1) are linearly dependent. When doing so, the term of $(\rho_\alpha/\rho)\mathbf{j} \times \mathbf{B}$ in the equations for the heavy species should be regarded as an independent parameter until the final result is obtained. Let us use a simple model of quasi two-species medium. It is valid as we have shown in Chapter 3 when the condition $\beta_e\beta_i \ll 1$ is fulfilled. For this case, $\mathbf{j} = -n_e e\mathbf{w}_e$, and we can put $\mathbf{w}_\beta = 0$ in Eq. (6.2.1) as well as in the expressions for $\mathbf{N}_e$ as part of $\mathbf{h}_e$. Then equations for $\mathbf{j}$ and $\mathbf{h}_e$ can be written as

$$\begin{aligned} \mathbf{j} + \omega_e\tau_0\mathbf{j} \times \mathbf{k} &= \sigma_0\mathbf{E}^* + \nu_0\tau_0\frac{e}{kT_e}\mathbf{h}_e, \\ \mathbf{h}_e + \omega_e\tau_e^*\mathbf{h}_e \times \mathbf{k} &= -\overline{\lambda}_e\nabla T_e + \frac{kT_e}{e}\alpha_T\mathbf{j}, \end{aligned} \tag{6.2.9}$$

where σ_0 and τ_0 are defined by Eqs. (3.4.6), (3.4.7) and for α_T and ν_0 we have

$$\alpha_T = \frac{5}{2}\nu_0\tau_e^*, \qquad \nu_0 = \sum_{\beta\neq e}\zeta_{e\beta}\tau_{e\beta}^{-1}. \tag{6.2.10}$$

Solving the second equation of (6.2.9) with respect to $\mathbf{h}_e$, we obtain

$$\begin{aligned} \mathbf{h}_e \;=\; & -\lambda_{e\parallel}\nabla_\parallel T_e - \lambda_{e\perp}\nabla_\perp T_e - \lambda_{e\wedge}\mathbf{k} \times \nabla T_e \\ & + \chi_{e\parallel}\mathbf{j}_\parallel + \chi_{e\perp}\mathbf{j}_\perp + \chi_{e\wedge}\mathbf{k} \times \mathbf{j}, \end{aligned} \tag{6.2.11}$$

where the components λ_e are defined by Eqs. (6.2.7); and with respect to χ_e

$$\chi_{e\parallel} = \frac{kT_e}{e}\alpha_T, \quad \chi_{e\perp} = \frac{\chi_{e\parallel}}{1+\beta_e^{*2}}, \quad \chi_{e\wedge} = \beta_e^*\chi_{e\perp}. \tag{6.2.12}$$

The expression

$$\mathbf{j} = \sigma_\parallel\mathbf{E}_\parallel^* + \sigma_\perp\mathbf{E}_\perp^* + \sigma_\wedge\mathbf{k} \times \mathbf{E}^* + \varphi_\parallel\nabla_\parallel T_e + \varphi_\perp\nabla_\perp T_e + \varphi_\wedge\mathbf{k} \times \nabla T_e \tag{6.2.13}$$

follows for the current density $\mathbf{j}$ from (6.2.9) if (6.2.11) is taken into account, with

$$\sigma_\parallel = \frac{\sigma_0}{1-\Delta_0}, \qquad \sigma_\perp = \frac{\sigma_0 A}{A^2 + B^2\omega_e^2\tau_0^2},$$

$$\sigma_\wedge = \frac{\sigma_0 B\omega_e\tau_0}{A^2 + B^2\omega_e^2\tau_0^2},$$

$$\varphi_{\|} = -\alpha_T \frac{k}{e}\sigma_{\|}, \qquad \varphi_{\perp} = -\alpha_T \frac{k}{e}\frac{\sigma_{\perp} - \beta_e^* \sigma_{\wedge}}{1+\beta_e^{*2}},$$

$$\varphi_{\wedge} = -\alpha_T \frac{k}{e}\frac{\beta_e^* \sigma_{\perp} + \sigma_{\wedge}}{1+\beta_e^{*2}}, \tag{6.2.14}$$

where

$$A = 1 - \frac{\Delta_0}{1+\beta_e^{*2}}, \qquad B = 1 + \frac{\gamma \Delta_0}{1+\beta_e^{*2}},$$

$$\Delta_0 = \nu_0 \tau_0 \alpha_T, \qquad \gamma = \tau_0^{-1}\tau_e^*. \tag{6.2.15}$$

It is to be noted that $\mathbf{h}_e$ as well as $\mathbf{j}$ is easily represented by a linear relationship connecting it with the components of the field $\mathbf{E}^*$ and temperature gradients for which the components of $\mathbf{j}$ from (6.2.13) should be substituted into (6.2.11). The result will be

$$\begin{aligned} \mathbf{h}_e \;\; = \;\; & -\lambda'_{e\|}\nabla_{\|}T_e - \lambda'_{e\perp}\nabla_{\perp}T_e - \lambda'_{e\wedge}\mathbf{k}\times\nabla T_e \\ & -\psi_{\|}\mathbf{E}^* - \psi_{\perp}\mathbf{E}^* - \psi_{\wedge}\mathbf{k}\times\mathbf{E}^*, \end{aligned} \tag{6.2.16}$$

where

$$\lambda'_{e\|} = \lambda_{e\|} - \varphi_{\|}\chi_{e\|}, \qquad \lambda'_{e\perp} = \lambda_{e\perp} - (\chi_{e\perp}\varphi_{\perp} - \chi_{e\wedge}\varphi_{\wedge}),$$

$$\lambda'_{e\wedge} = \lambda_{e\wedge} + \varphi_{\wedge}\chi_{e\perp} + \varphi_{\perp}\chi_{e\wedge}, \tag{6.2.17}$$

$$\psi_{\|} = \varphi_{\|}T_e, \qquad \psi_{\perp} = \varphi_{\perp}T_e, \qquad \psi_{\wedge} = \varphi_{\wedge}T_e,$$

The relationship between ψ and φ ensures that the Onsager reciprocity relations be fulfilled. When $\mathbf{h}_e$ is known, the expression for the total heat flux of electrons can be readily written as

$$\mathbf{q}_e = \mathbf{h}_e - \frac{5}{2}\frac{kT_e}{e}\mathbf{j}, \tag{6.2.18}$$

where, depending on the way of representing $\mathbf{h}_e$ either in the form of (6.2.11) or (6.2.16), $\mathbf{j} = \mathbf{j}_{\|} + \mathbf{j}_{\perp}$ is conserved either explicitly or with the help of Eq. (6.2.13).

It should be stressed that it is the coefficient λ_e rather than λ'_e that corresponds to the true thermal conductivity of electrons in the plasma, since in experiments the thermal conductivity is usually measured under stationary conditions when there is no diffusion transport, i.e. at $\mathbf{j} = 0$. Respectively, the expression for λ_e looks simpler than λ'_e, with this peculiarity kept valid even in higher approximations of the

Grad's moment method. Note that in calculating the thermal conductivity with the Chapman–Enskog method the expression of the heat flux is commonly presented in the form of (6.2.16) and afterwards it is reduced to (6.2.11), which in the general case demands rather time consuming transformations and those in turn bring about the simplified determinants discussed in Chapter 5.

In terms of accuracy, the expressions obtained here for the electron transport coefficients correspond to the second Chapman–Cowling approximation in the expansion over the Sonine polynomials. The contribution of the second approximation to the electrical conductivity is defined here with the value of Δ_0. For the magnetized plasma ($\beta_e^* \gg 1$) the effect of corrections of the second approximation on the transverse components of electrical conductivity sharply drops. Let us estimate the contribution of Δ_0 to the longitudinal electrical conductivity of the plasma in some cases of practical interest. Thus, at the limit of a weakly ionized gas, when the interaction of the charged particles is negligible, the expression for τ_0^{-1} can be written as

$$\tau_0^{-1} = \frac{16}{3}\left(\frac{kT_e}{2\pi m_e}\right)^{1/2} \sum_{\beta=1}^{K} n_\beta \overline{Q}_{e\beta}, \tag{6.2.19}$$

where $\overline{Q}_{e\beta}$ is the averaged cross section of electron scattering on the neutrals of species β (K is the number of neutral spesies). In this case

$$\alpha_T = \frac{\sum\limits_\beta \zeta_{e\beta} n_\beta \overline{Q}_{e\beta}}{\sum\limits_\beta \left(1 - 0.48 B_{e\beta}^*\right) n_\beta \overline{Q}_{e\beta}}, \quad \Delta_0 = \frac{\sum\limits_\beta \zeta_{e\beta} n_\beta \overline{Q}_{e\beta}}{\sum\limits_\beta n_\beta \overline{Q}_{e\beta}} \alpha_T. \tag{6.2.20}$$

If $Q_{e\beta}^{(1)}(v) = \text{const}$, which corresponds to the hard elastic sphere model, then $\zeta_{e\beta} = 0.2$, $B_{e\beta}^* = 1$, $\alpha_T = 0.385$, $\Delta_0 = 0.077$. The expression of electrical conductivity takes the form

$$[\sigma]_2 = 1.08\,[\sigma]_1 = 0.510 \frac{n_e e^2}{(m_e kT_e)^{1/2}} \frac{1}{\sum\limits_{\beta=1}^{K} n_\beta \overline{Q}_{e\beta}}, \tag{6.2.21}$$

where $[\sigma]_1 = \sigma_0$ corresponds to the first Chapman–Cowling approximation (3.4.11).

This result is just 4% different from the precise Lorentz value of $\sigma_{||}$ for this model of interaction (0.510 instead of 0.532 [6]).

This is a more essential contribution of Δ_0 in another limiting case when the gas is fully ionized. For this case at $Z = 1$, τ_{ei} is defined by

(3.2.19). Besides,

$$\tau_{ee}^{-1} = \sqrt{2}\tau_{ei}^{-1}, \qquad \zeta_{ei} = -0.6, \qquad B_{ei}^{*} = 1,$$
$$\alpha_T = -0.802, \qquad \Delta_0 = 0.481. \tag{6.2.22}$$

The expression of $\sigma_{||}$ is written as

$$[\sigma]_2 = 1.93\,[\sigma]_1 = 0.578\,(4\pi\varepsilon_0)^2 \frac{(kT_e)^{3/2}}{e^2 m_e^{1/2}} \frac{1}{\ln\Lambda}, \tag{6.2.23}$$

which closely approaches the exact result of Spitzer–Harm (0.578 instead of 0.591) [7].

It would be appropriate to mention expressions for other electron transport coefficients for the cases of weak and full ionizations. Thus, for a weakly ionized gas ($Q_{e\beta}^{(1)} = \text{const}$)

$$[\eta_e]_1 = \frac{5}{6} p_e \tau_0 = 0.392 n_e m_e^{1/2} (kT_e)^{1/2} \left(\sum_\beta n_\beta \overline{Q}_{e\beta}\right)^{-1},$$
$$[\lambda_e]_2 = \frac{25}{13}\frac{k}{m_e} p_e \tau_0 = 0.905 n_e k m_e^{-1/2} (kT_e)^{1/2} \left(\sum_\beta n_\beta \overline{Q}_{e\beta}\right)^{-1},$$
$$[\varphi]_2 = -0.385 \frac{k}{e} [\sigma]_2 = -0.196 e n_e k m_e^{-1/2} (kT_e)^{-1/2} \left(\sum_\beta n_\beta \overline{Q}_{e\beta}\right)^{-1}. \tag{6.2.24}$$

And for a fully ionized gas ($Z = 1$)

$$[\eta_e]_1 = 0.488 p_e \tau_{ei} = 0.146\,(4\pi\varepsilon_0)^2 (kT_e)^{5/2} m_e^{1/2} \frac{1}{e^4 \ln\Lambda},$$
$$[\lambda_e]_2 = 1.34 \frac{k}{m_e} p_e \tau_{ei} = 0.399\,(4\pi\varepsilon_0)^2 k \frac{(kT_e)^{5/2}}{m_e^{1/2}} \frac{1}{e^4 \ln\Lambda}, \tag{6.2.25}$$
$$[\varphi]_2 = 0.802 \frac{k}{e} [\sigma]_2 = 0.463\,(4\pi\varepsilon_0)^2 k \frac{(kT_e)^{3/2}}{m_e^{1/2}} \frac{1}{e^3 \ln\Lambda}.$$

The indices 1 and 2 with the kinetic coefficients correspond to the level of approximation or the number of polynomials used in the expansion when calculating these values. Note that for λ_e and φ the second approximation is in fact the first non-vanishing approximation in the expansion over Sonine polynomials.

To sum up, a few remarks should be made concerning the diffusion equations for the heavy species. They can be basically simplified due to, for example, the small value of the force of the ion–electron "diffusion friction" $\mathbf{R}_{ie} \sim m_e n_\beta \tau_{\beta e}^{-1}(\mathbf{w}_\beta - \mathbf{w}_e)$ compared with the "diffusion friction" force of heavy particles. The same reasoning allows us to omit the terms of the heat fluxes of electrons $\mathbf{h}_e$. The analysis proves that such neglect is almost always justified if the concentration of neutral particles is big enough for a partially ionized gas or of ions of another species for a multispecies fully ionized plasma since this guaranties an intense exchange of momenta between the heavy particles. It should be pointed out that at $|\mathbf{w}_\beta| \ll |\mathbf{w}_e|$ the values $\mathbf{R}_{ie}$ and $\mathbf{h}_e$ could be found from the independent equations for electrons. When they must be taken into account, it is possible to formally include the appropriate terms into the external forces affecting the heavy particles [8,12]. As the expression of $\mathbf{j}$ is also found in this case, it can be substituted into the term $\sim (\rho_\alpha/\rho)\mathbf{j} \times \mathbf{B}$ with the resulting diffusion equation for ions and atoms which get absolutely independent of the electron equations. The general solution of these equations can be obtained with the procedure used in Chapter 3.

6.3 Higher approximations

So far we have confined ourselves to the analysis of the transport equations written in $13N$-moment approximation. Calculation of the kinetic coefficients shows that this approximation sometimes does not ensure the accuracy required. This is true for the case of a fully ionized Coulomb plasma in which, to get satisfactory results for the thermal conductivity, thermal diffusion and viscosity, we would require using the additional polynomials for vector and tensor values in expansion of the distribution function. When considering electron–atom and ion–atom interactions in the cases when the effective cross sections do not depend on velocity (as in the model of hard elastic spheres) or do not diminish abruptly with growing velocities, the $13N$-moment approximation proves, as a rule, to be sufficient. The same reasoning works in calculating the common transport properties in gas mixtures in terms of the Lennard–Jones potential for atom–atom collisions, though in some cases such as the calculation of thermal diffusion, the consideration of higher approximation may turn out to be useful [6,9]. However, if the effective cross sections of electron scattering by atoms change nonmonotonously with energy (e.g. in the range of the Ramsauer minimum) and particularly if the cross sections increase with growing velocities, then it is necessary to consider a greater number of polynomials in the expansion of f_e [10] to obtain a satisfactory result.

The transport phenomena in a fully ionized plasma with several species of ions accounting for higher approximations than the 13N-moment approximation will be discussed in Chapter 8. Below we intend to improve the results of the electron transport properties in an arbitrary multicomponent plasma.

Estimates of the order of values of cross coefficients in Eqs. (4.6.5)–(4.6.6) show that the separation of the electron equation from the equations of heavy particles can be carried out in all the succeeding approximations of the method. As we have earlier shown, taking into account the cross terms when defining the electron transport properties becomes crucial only at $\beta_i \geq 1$. As $\beta_e \gg 1$ in this case, the contribution of all the consequent approximations to the transverse electron kinetic coefficient becomes negligibly small and, therefore, the cross terms in the equations of higher orders may be omitted. Thus, the equations for the electrons in all approximations turn out to be completely independent and, respectively, the systems of equations for the coefficients b_{er}^{1k} and b_{ers}^{2k} acquire the form

$$\sum_{k=0}^{\xi-1} C_{ee}^{10k} \mathbf{b}_e^{1k} = -n_e e \mathbf{E}^* - n_e \mathbf{b}_e^{10} \times \mathbf{k}\omega_e, \tag{6.3.1}$$

$$\begin{aligned} \sum_{k=0}^{\xi-1} C_{ee}^{11k} \mathbf{b}_e^{1k} &= -\frac{5}{2}\frac{k}{m_e} p_e \nabla T_e - n_e \mathbf{b}_e^{11} \times \mathbf{k}\omega_e, \\ \sum_{k=0}^{\xi-1} C_{ee}^{1nk} \mathbf{b}_e^{1k} &= -n_e \mathbf{b}_e^{1n} \times \mathbf{k}\omega_e, \qquad 1 < n \leq \xi - 1, \end{aligned} \tag{6.3.2}$$

$$\begin{aligned} \sum_{k=0}^{\xi-1} C_{ee}^{20k} b_{ers}^{2k} &= -p_e W_{rs} - 2n_e b_{ers}^{20} e_{slm} k_m \omega_e, \\ \sum_{k=0}^{\xi-1} C_{ee}^{2nk} b_{ers}^{2k} &= -2n_e b_{ers}^{2n} e_{slm} k_m \omega_e, \qquad 0 < n \leq \xi - 1. \end{aligned} \tag{6.3.3}$$

It is preferable to use the coefficients q^{nk} and p^{nk} which were introduced by Devoto [11] rather than the coefficients C_{ee}^{1nk} and C_{ee}^{2nk}. Unlike the coefficients C_{ee}^{mnk}, those coefficients are symmetric about a reversal of indices n and k and are defined as

$$\begin{aligned} q^{nk} &= \left(\frac{2\pi m_e}{kT_e}\right)^{1/2} \left\{ n_e^2 \left[S_{3/2}^n (W^2)\, \mathbf{W},\ S_{3/2}^k (W^2)\, \mathbf{W} \right]_e \right. \\ &\quad \left. + \sum_{\beta \neq e} n_e n_\beta \left[S_{3/2}^n (W^2)\, \mathbf{W},\ S_{3/2}^k (W^2)\, \mathbf{W} \right]_{e\beta}' \right\}, \end{aligned}$$

$$
\begin{aligned}
p^{nk} = \left(\frac{2\pi m_e}{kT_e}\right)^{1/2} \Bigg\{ & n_e^2 \Bigg[S_{5/2}^n (W^2) \left(W_r W_s - \frac{1}{3}\delta_{rs} W^2 \right), \\
& S_{5/2}^k (W^2) \left(W_r W_s - \frac{1}{3}\delta_{rs} W^2 \right) \Bigg]_e \\
& + \sum_{\beta \neq e} n_e n_\beta \Bigg[S_{5/2}^n (W^2) \left(W_r W_s - \frac{1}{3}\delta_{rs} W^2 \right), \\
& S_{5/2}^k (W^2) \left(W_r W_s - \frac{1}{3}\delta_{rs} W^2 \right) \Bigg]_{e\beta} \Bigg\}. \qquad (6.3.4)
\end{aligned}
$$

where $\mathbf{W} = \left(\frac{m_e}{2kT_e}\right)^{1/2} \mathbf{c}$ and

$$
[F, G]_e = [F, G]'_{ee} + [F, G]''_{ee}.
$$

The coefficients C_{ee}^{1nk} and C_{ee}^{2nk} are related to q^{nk} and p^{nk} as

$$
C_{ee}^{1nk} = (2\pi)^{-1/2} Q_{1nk} \left(\frac{m_e}{kT_e}\right)^{k-n-1/2} q^{nk},
$$

$$
C_{ee}^{2nk} = (2\pi)^{-1/2} Q_{2nk} \left(\frac{m_e}{kT_e}\right)^{k-n-1/2} p^{nk}.
$$

First, let us consider expressions for the electron transport properties in the absence of a magnetic field. Assuming $\omega_e = 0$ in Eqs. (6.3.1)–(6.3.3), allowing for $n_e \mathbf{b}_e^{10} = \rho_e \mathbf{w}_e$, $n_e \mathbf{b}_e^{11} = \mathbf{h}_e$, $n_e b_{ers}^{20} = \pi_{ers}$, and solving the appropriate equations with respect to $\mathbf{b}_e^{10}$, $\mathbf{b}_e^{11}$, b_{ers}^{20}, we find

$$
\begin{aligned}
&\mathbf{j} = -n_e e \mathbf{w}_e = \sigma \mathbf{E}^* + \varphi \nabla T_e, \\
&\mathbf{h}_e = -\lambda'_e \nabla T_e - \psi \mathbf{E}^*, \quad \pi_{ers} = -\eta_e W_{rs},
\end{aligned} \qquad (6.3.5)
$$

Going to the coefficients q^{nk} and p^{nk}, we obtain

$$
\sigma = \frac{3}{2} n_e^2 e^2 \left(\frac{2\pi}{m_e kT_e}\right)^{1/2} |q|^{-1} \begin{vmatrix} q^{11} & q^{12} & \dots & q^{1n} \\ q^{21} & q^{22} & \dots & q^{2n} \\ \vdots & \vdots & & \vdots \\ q^{n1} & q^{n2} & \dots & q^{nn} \end{vmatrix},
$$

$$\lambda_e' = \frac{75}{8} n_e^2 \left(\frac{2\pi}{m_e kT_e}\right)^{1/2} k^2 T_e \, |q|^{-1} \begin{vmatrix} q^{00} & q^{02} & \dots & q^{0n} \\ q^{20} & q^{22} & \dots & q^{2n} \\ \vdots & \vdots & & \vdots \\ q^{n0} & q^{n2} & \dots & q^{nn} \end{vmatrix},$$

$$\varphi = \frac{\psi}{T_e} = \frac{15}{4} n_e^2 e \left(\frac{2\pi}{m_e kT_e}\right)^{1/2} k \, |q|^{-1} \begin{vmatrix} q^{10} & q^{12} & \dots & q^{1n} \\ q^{20} & q^{22} & \dots & q^{2n} \\ \vdots & \vdots & & \vdots \\ q^{n0} & q^{n2} & \dots & q^{nn} \end{vmatrix},$$

$$\eta_e = \frac{5}{2} n_e^2 (2\pi m_e kT_e)^{1/2} \, |p|^{-1} \begin{vmatrix} p^{11} & p^{12} & \dots & p^{1n} \\ p^{20} & p^{22} & \dots & p^{2n} \\ \vdots & \vdots & & \vdots \\ p^{n1} & p^{n2} & \dots & p^{nn} \end{vmatrix}, \tag{6.3.6}$$

where $|q|$ and $|p|$ are the determinants of the matrices of elements q^{nk} and p^{nk} $(n, k = 0, ..., \xi - 1)$.

If $\mathbf{E}^*$ is expressed in terms of $\mathbf{j}$ of the first Eq. (6.3.5) and Eq. (6.2.18) is also used, then the total electron heat flux can be written as

$$\mathbf{q}_e = -\lambda_e \nabla T_e - \alpha \mathbf{j},$$

$$\lambda_e = \lambda_e' - \frac{\varphi^2 T_e}{\sigma}, \qquad \alpha = \left(\frac{5}{2}\frac{k}{e} + \frac{\varphi}{\sigma}\right) T_e. \tag{6.3.7}$$

It is easily seen that the coefficient λ_e can be found directly from the set of Eqs. (6.3.2), if $\mathbf{b}_e^{10} = 0$ is formally taken and they are solved with respect to $\mathbf{b}_e^{11}$. In this case

$$\lambda_e = \frac{75}{8} n_e^2 \left(\frac{2\pi}{m_e kT_e}\right)^{1/2} k^2 T_e \frac{1}{|q|'} \begin{vmatrix} q^{22} & q^{23} & \dots & q^{2n} \\ q^{32} & q^{33} & \dots & q^{3n} \\ \vdots & \vdots & & \vdots \\ q^{n2} & q^{n3} & \dots & q^{nn} \end{vmatrix}, \tag{6.3.8}$$

where $|q|'$ is the determinant $|q|$ without the q^{nk} with n or $k = 0$. It is this definition of λ_e that results in the agreement with $\lambda_{e\|}$ in the lowest approximation which was obtained in the $13N$-moment approximation [see Eqs. (6.2.7)]. In the presence of a magnetic field ($\omega_e \neq 0$), the expressions (6.3.6) and (6.3.8) are also valid for the longitudinal transport coefficients.

To define the transverse components of the current density and electron heat flux, it is convenient to solve Eqs. (6.3.1) and (6.3.2) replacing the vector products of the values with their complex representation. In that case, the expressions (6.3.6) and (6.3.8) will be correct if we make the substitution

$$q^{nk} \to q^{nk} - in_e\omega_e \left(\frac{2\pi m_e}{kT_e}\right)^{1/2} \frac{2}{\pi^{1/2}} \frac{(k+\frac{3}{2})!}{k!} \delta_{nk}.$$

For example, the components $\lambda_{e\perp}$ and $\lambda_{e\wedge}$ of electron thermal conductivity are given by the real and imaginary parts, respectively, of the resulting expression for λ_e.

The expressions of the vector electron transport coefficients derived above agree with the results [11] when they were defined on the basis of the Chapman–Enskog method and disconnected systems of the linearized kinetic equations for electrons and heavy particles. The expressions of the elements q^{nk} for $n, k = 0, 1, 2, 3$ were obtained by Devoto [11] and are given in Appendix 3. These values allow us to calculate the coefficients of electrical conductivity, thermal conductivity and thermal diffusion up to and including the fourth approximation ($\xi = 4$) to these coefficients. The coefficients p^{nk} for $n, k = 0, 1$ which can be used to define the electron viscosity coefficient up to and including the second approximation ($\xi = 2$) are also given in Appendix 3.

Convergence of the successive approximations in calculating the longitudinal electron transport coefficients is illustrated by the data of Table 6.1 and 6.2 for a weakly ionized gas ($\sigma_{e\beta}$ = const) and for a two-component fully ionized plasma ($Z = 1$).

The transport coefficients in the ξth approximation can be found using the formulas of the type

$$[\sigma]_\xi = [\sigma]_1 \, f_\sigma^{(\xi)}, \qquad [\eta_e]_\xi = [\eta_e]_1 \, f_\eta^{(\xi)},$$

$$[\lambda_e]_\xi = [\lambda_e]_2 \, f_\lambda^{(\xi)}, \qquad [\varphi]_\xi = [\varphi]_2 \, f_\varphi^{(\xi)},$$

Table 6.1. Corrections for the electron transport coefficients for a weakly ionized gas

f	$\xi = 2$	$\xi = 3$	Accurate solution
f_σ	1.080	1.11	1.130
f_λ	1.000	1.10	1.180
f_φ	1.000	1.19	1.360
f_η	1.058	-	1.083

Table 6.2. Corrections for the electron transport coefficients for a fully ionized plasma

f	$\xi = 2$	$\xi = 3$	Accurate solution
f_σ	1.93	1.950	1.975
f_λ	1.00	2.360	2.400
f_φ	1.00	0.895	0.900
f_η	1.50	–	–

where $[\sigma]_1 = \sigma_0$, and $[\eta_e]_1$, $[\lambda_e]_2$, $[\varphi]_2$ are defined by expressions (3.4.11), (6.2.24) for a weakly ionized gas and (3.4.14), (6.2.25) for a fully ionized plasma.

Comparison with the exact results obtained in the Lorentz approximation for a weakly ionized gas [6,12] and with a numerical method for a fully ionized plasma [7] shows that the third approximation ($\xi = 3$) ensures a satisfactory accuracy. The electrical conductivity of plasma can be calculated in fact in the second approximation to a quite sufficient accuracy. The analysis [12] proves that the requirement of higher approximations for a weakly ionized gas arises mainly when the dependence of electron transport cross section $Q^{(1)}_{e\beta}$ upon velocity has the form of $Q^{(1)}_{e\beta}(v) \sim v^{m-1}$ at $m > 1$. Poor convergence is also observed when the dependence of cross section on velocity has a pronounced Ramsauer minimum (e.g. for the plasma whose component is argon [10]). The rate of convergence of the approximations to the electron transport coefficients up to and including the twelfth approximation in the case of a partially ionized argon at 1 atm pressure and for a maximum temperature of 20000 K was studied in detail in [12].

Chapter 7

Two-temperature partially ionized gas

7.1 Viscous-stress tensor

We shall use the results mentioned in the previous chapter for a more detailed analysis of the transport coefficients for the case of a three-component plasma or a partially ionized gas of electrons, ions and atoms ($m_i = m_a = m$). The temperature of ions and atoms is taken to be the same ($T_i = T_a = T$) and the electron temperature to be different from the temperature of heavy particles in the general case ($T_e \neq T$). When deriving the relative expressions, we shall not rely on disconnecting the original equations for electrons and heavy particles but use general solutions written in the form of the ratio of the determinants. Such an approach permits us to verify the limits of applicability of the technique to disconnect the equations used earlier. To simplify our analysis, we shall confine ourselves to the expressions used in the $13N$-moment approximation ($N = 3$).

First consider the problem of the viscous-stress tensor in a two-temperature partially ionized gas [1,2]. When simplifying the determinants to be used in putting down the kinetic coefficients of interest, the following small parameters will be essentially used:

$$\delta = \frac{m_e}{m} \ll 1, \quad \delta\theta \ll 1, \quad \delta\theta^{-1} \ll 1 \quad \left(\theta = \frac{T}{T_e}\right). \tag{7.1.1}$$

All told, for the coefficients which are constituents of the equations used to define the partial viscous-stress tensor (6.1.2), we have

$$a_{ee} = a_{ii} = a_{aa} = 1,$$

$$a_{ei} = -0.4\delta\tau_e\tau_{ie}^{-1}, \quad a_{ie} = -0.2\delta(3\theta - 1)\tau_i\tau_{ei}^{-1},$$

$$a_{ea} = -4\delta f_{ea}\tau_e\tau_{ae}^{-1}, \quad a_{ae} = -4\delta\left[f_{ea} + \frac{1}{4}\zeta_{ea}(1-\theta)\right]\tau_a\tau_{ea}^{-1},$$

$$a_{ia} = -f_{ia}\tau_i\tau_{ai}^{-1}, \quad a_{ai} = -f_{ia}\tau_a\tau_{ia}^{-1},$$

$$f_{\alpha\beta} = \frac{1}{4}\left(1 - 0.6A^*_{\alpha\beta}\right), \quad \zeta_{\alpha\beta} = \frac{6}{5}C^*_{\alpha\beta} - 1. \tag{7.1.2}$$

To an accuracy up to the values of the order of $(\delta\theta)^{3/2}$ with respect to the terms left after the expansion, the determinants $|a|$, $|a^*|$ and $|a^{**}|$ are equal to

$$\begin{aligned} |a| &= 1 - a_{ia}a_{ai} = \Delta_\eta, \\ |a^*| &= \Delta_\eta\left(1 + \beta_e^2\right)\left(1 + \Delta_\eta^{-2}\beta_i^2\right), \\ |a^{**}| &= \Delta_\eta\left(1 + \frac{1}{4}\beta_e^2\right)\left(1 + \frac{1}{4}\Delta_\eta^{-2}\beta_i^2\right), \end{aligned} \tag{7.1.3}$$

where

$$\beta_e = \omega_e\tau_e, \quad \beta_i = \omega_i\tau_i, \quad \omega_e = \frac{e}{m_e}|\mathbf{B}|, \quad \omega_i = \frac{Ze}{m_i}|\mathbf{B}|,$$

$$\begin{aligned} \tau_e^{-1} &= 0.3\tau_{ee}^{-1} + 0.6\tau_{ei}^{-1} + 0.6A^*_{ea}\tau_{ea}^{-1}, \\ \tau_i^{-1} &= 0.3\tau_{ii}^{-1} + f'_{ia}\tau_{ia}^{-1} + \delta\tau_{ie}^{-1}, \\ \tau_a^{-1} &= 0.3A^*_{aa}\tau_{aa}^{-1} + f'_{ia}\tau_{ai}^{-1} + \delta\tau_{ae}^{-1}, \\ f'_{\alpha\beta} &= \frac{1}{4}\left(1 + 0.6A^*_{\alpha\beta}\right). \end{aligned} \tag{7.1.4}$$

Using Eqs. (7.1.2) and (7.1.3) in the expressions (6.1.9), it is easy to find that at $\beta_i^2 \ll 1$ to an accuracy up to the terms $\delta^{1/2}\theta^{5/2}$ with respect to the terms left the electron viscosity coefficients are defined by the expressions (6.2.5) obtained earlier from Eqs. (6.2.2) written for the electron species neglecting the cross terms, i.e.

$$\eta_e^{(0)} = \frac{1}{2} p_e \tau_e, \qquad \eta_e^{(1)} = \frac{\eta_e^{(0)}}{1+\beta_e^2},$$

$$\eta_e^{(2)} = \frac{\eta_e^{(0)}}{1+\frac{1}{4}\beta_e^2}, \qquad \eta_e^{(3)} = \frac{\beta_e \eta_e^{(0)}}{1+\beta_e^2}, \tag{7.1.5}$$

$$\eta_e^{(4)} = \frac{1}{2} \frac{\beta_e \eta_e^{(0)}}{1+\frac{1}{4}\beta_e^2}.$$

The analysis of the general solutions shows that the consideration of the cross terms may turn out to be essential in defining $\eta_e^{(1)}$ and $\eta_e^{(2)}$, if $\beta_i \geq 1$. Let us investigate this case in detail, using a two-component fully ionized plasma as an example. Using Eqs. (6.1.9) and (7.1.3), we find that at $\beta_e^2 \gg 1$ to an accuracy up to the terms of $\delta^{1/2}\theta^{-1/2}$ as compared with unity

$$\eta_e^{(1)} = \frac{\eta_e^{(0)}}{\beta_e^2}\left(1 - \frac{\beta_i^2}{1+\beta_i^2} a_{ei} \frac{\bar{\eta}_i}{\bar{\eta}_e} \frac{\beta_e}{\beta_i}\right) = \frac{p_e}{2\omega_e^2 \tau_e}\left[1 + q(Z)\,\theta \frac{\beta_i^2}{1+\beta_i^2}\right],$$

$$\eta_e^{(2)} = \frac{2p_e}{\omega_e^2 \tau_e}\left[1 + q(Z)\,\theta \frac{(\beta_i/2)^2}{1+(\beta_i/2)^2}\right], \tag{7.1.6}$$

where

$$q(Z) = 0.4Z^{-1}\tau_e \tau_{ei}^{-1} \cong (1.5Z + 0.11)^{-1}.$$

For the case when $\beta_i \gg 1$ and $Z = 1$, we have

$$\eta_e^{(1)} = \frac{p_e}{2\omega_e^2 \tau_e}(1 + 0.39\theta), \qquad \eta_e^{(2)} = 4\eta_e^{(1)}. \tag{7.1.7}$$

These expressions differ from the well-known equations for a magnetized fully ionized plasma [3,4] by an additional term in brackets[1] which is related to the cross terms in the equations for π_{ers}.

The coefficients $\eta_i^{(p)}$ and $\eta_a^{(q)}$ for ions and atoms following from the general expressions (6.1.9) can be written as

[1] The discrepancy with [3] is explained by the fact that the transport coefficients for each species were found there from the solution of the system of "disconnected" kinetic equations for electrons and ions. In [4] the cross terms were eliminated in the proper transport equations as we did in the previous chapter.

$$\eta_i^{(0)} = \frac{1}{2} p_i \tau_i \xi_i \Delta_\eta^{-1}, \quad \eta_i^{(1)} = \frac{\eta_i^{(0)}}{1 + \Delta_\eta^{-2} \beta_i^2},$$

$$\eta_i^{(3)} = \beta_i \Delta_\eta^{-1} \eta_i^{(1)}, \quad \eta_i^{(2)} = \eta_i^{(1)} \left[\frac{\beta_i}{2}\right], \quad \eta_i^{(4)} = \eta_i^{(3)} \left[\frac{\beta_i}{2}\right],$$

$$\eta_a^{(0)} = \frac{1}{2} p_a \tau_a \xi_a \Delta_\eta^{-1},$$

$$\eta_a^{(1)} = \eta_a^{(0)} \frac{1 + \xi_a^{-1} \Delta_\eta^{-1} \beta_i^2}{1 + \Delta_\eta^{-2} \beta_i^2}, \tag{7.1.8}$$

$$\eta_a^{(3)} = \Delta_\eta^{-1} \beta_i \eta_a^{(0)} \frac{1 - \xi_a^{-1} \Delta_\eta}{1 + \Delta_\eta^{-2} \beta_i^2},$$

$$\eta_a^{(2)} = \eta_a^{(1)} \left[\frac{\beta_i}{2}\right], \quad \eta_a^{(4)} = \eta_a^{(3)} \left[\frac{\beta_i}{2}\right],$$

$$\xi_i = 1 + f_{ia} \tau_a \tau_{ia}^{-1}, \quad \xi_a = 1 + f_{ia} \tau_i \tau_{ai}^{-1}.$$

In (7.1.8), we omit the terms the maximum order of which is $\delta\theta^{-1}$ relative to the terms left. Thus, the expressions obtained are valid when the condition $T_e m_e / T m \ll 1$ is met.

To assess the relative contribution of each plasma species to the full viscosity $\eta = \sum_\alpha \eta_\alpha$ depending on the degree of ionization $\alpha = n_i / (n_i + n_a)$, the parameter θ and Hall's parameters β_e, β_i, we need the ratios of various collision frequencies $\tau_{\alpha\beta}^{-1} / \tau_{\delta\gamma}^{-1}$. We shall use the representation of $\tau_{\alpha\beta}^{-1}$ as (3.1.12). In that case for collisions of heavy particles ($T_\alpha = T_\beta = T$), we shall have

$$\tau_{\alpha\beta}^{-1} = \frac{16}{3} n_\beta \left(\frac{kT}{\pi m}\right)^{1/2} \overline{Q}_{\alpha\beta}(T),$$

and for electron–electron and electron–heavy particle interaction

$$\tau_{ee}^{-1} = \frac{16}{3} n_e \left(\frac{kT_e}{\pi m_e}\right)_{ee}^{1/2} \overline{Q}(T_e),$$

$$\tau_{e\beta}^{-1} = \frac{16}{3} n_\beta \left(\frac{kT_e}{2\pi m_e}\right)^{1/2} \overline{Q}_{e\beta}(T_e).$$

For the averaged cross sections of atom–atom, electron–atom and ion–atom collisions, the theoretical and experimental data [5–7] can be useful. In terms of rough estimations, it can be taken as

$$\frac{\overline{Q}_{ea}}{\overline{Q}_{aa}} \approx 1 - 10, \qquad \frac{\overline{Q}_{ia}}{\overline{Q}_{aa}} \approx 1 - 10.$$

For Coulomb interactions, we have

$$\frac{\overline{Q}_{ee}}{\overline{Q}_{ii}} \sim Z^{-4}\theta^2, \qquad \frac{\overline{Q}_{ei}}{\overline{Q}_{ii}} \sim Z^{-2}\theta^2,$$

$$\overline{Q}_{ii} = \frac{\pi}{2}\frac{Z^4 e^4}{(4\pi\varepsilon_0 kT)^2}\ln\Lambda.$$

When estimating $\overline{Q}_{ii}$, one should remember that within the characteristic temperature range for a partially ionized plasma ($T = 2\cdot 10^3$–10^4) and for the densities of charged particles $10^{15} \le n_i \le 10^{20}$ cm^{-3} the Coulomb logarithm $\ln\Lambda$ is from 5 to 13 (see Table 1.2). Hence, when the results of [5] for $\overline{Q}_{aa}$ are used, the estimate can be made

$$\beta = \frac{\overline{Q}_{aa}}{\overline{Q}_{ii}} \approx 10^{-3} - 10^{-5}.$$

The relative contribution of the partial viscosities to the full viscosity of the plasma can be easily calculated by introducing the dimensionless viscosity coefficients which are defined as

$$\eta_\alpha^{(p)*} = \frac{\eta_\alpha^{(p)}\overline{Q}_{aa}}{(mkT)^{1/2}}.$$

Taking into account that ξ_i, ξ_a and Δ_η are close to unity and using the estimates made above in (7.1.5) and (7.1.8), we have

$$\eta_e^{(0)*} \sim \frac{\delta^{1/2}\theta^{-5/2}\alpha\beta}{[\alpha + (1-\alpha)\beta\theta^{-2}]},$$
$$\eta_i^{(0)*} \sim \frac{\alpha\beta}{[\alpha + (1-\alpha)\beta]}, \qquad \eta_a^{(0)*} \sim \frac{1-\alpha}{1+\alpha}. \tag{7.1.9}$$

As is seen, the relative contribution of the electron viscosity to the full viscosity is negligibly small at $\theta = 1$, but it can become comparable to the ion contribution at $\theta \sim \delta^{1/5}$, i.e. at $T_e/T \approx 5 - 10$. Note that the part of $\eta_i^{(0)}$ is essential only at higher degrees of ionization $\alpha \gtrsim 1 - \beta$. At $\alpha \ll 1 - \beta$, the main contribution is due to the atom viscosity $\eta_a^{(0)}$. When the contribution of the electron viscosity is negligible, the viscosity of a partially ionized gas in the absence of a magnetic field is described by

$$\eta = \frac{1}{2}\left(p_a\tau_a\xi_a + p_i\tau_i\xi_i\right)\Delta_\eta^{-1} \tag{7.1.10}$$

which includes the parameters of atom–atom, ion–atom and ion–ion interactions. The application of this expression gives the result similar to the one calculated by using the formulas for the viscosity of a two-species mixture under the assumption of $m_i = m_a$. At higher degrees of ionization, though, when the viscosity becomes dominated by the ion–ion interaction, the accuracy of Eq. (7.1.10) corresponding to the first approximation in the expansion over Sonine polynomials decreases. Comparison with the result for a fully ionized gas calculated in the approximation of two polynomials (cf. Chapter 8) shows that the approximations of the first and second orders differ by about 15%.

To analyse the relative contribution of the transverse constituents of viscosity coefficients $\eta_a^{(1)}$ and $\eta_a^{(3)}$, some additional estimations would be necessary, besides Eqs. (7.1.9):

$$\frac{\beta_i}{\beta_e} \sim \delta^{1/2}\theta^{-1/2}\frac{\alpha\theta^2 + (1-\alpha)\beta}{\alpha + (1-\alpha)\beta},$$

$$\xi_a - \Delta_\eta \sim \frac{\alpha\beta}{\alpha + (1-\alpha)\beta}.$$

Comparison of the expressions for the relative viscosity coefficients discussed above suggests that the contributions of various species to $\eta^{(1)}$ are similar in their behaviour to those to $\eta^{(0)}$ with the only difference that at $\beta_i \gg 1$ the effect of $\eta_i^{(1)}$ becomes essential at higher degrees of ionization $\alpha \sim 1 - \beta/\omega_i^2\tau_i^2$.

The contributions of $\eta_e^{(3)}$, $\eta_i^{(3)}$ and $\eta_a^{(3)}$ to $\eta^{(3)}$ prove to be the same in weak magnetic fields ($\beta_e \leq 1$) at $\alpha \ll 1$. As the degree of ionization increases, the contribution of $\eta_a^{(3)}$ becomes smaller, whereas the magnetic field rises; the role of $\eta_e^{(3)}$ diminishes and at $\beta_i \gg 1$ we have $\eta_e^{(3)} \ll \eta_i^{(3)}, \eta_a^{(3)}$. For the coefficients $\eta^{(2)}$ and $\eta^{(4)}$, the conclusions made in the analysis of $\eta^{(1)}$ and $\eta^{(3)}$ are valid.

7.2 Heat fluxes

Now we shall analyse the expressions (6.1.10) for the partial heat fluxes $\mathbf{h}_\alpha$. Using the general expressions for the coefficients $b_{\alpha\beta}$ and considering (7.1.1), we have

$$b_{ee} = b_{ii} = b_{aa} = 1,$$

$$b_{ei} = -2.7\delta\tau_e^*\tau_{ie}^{-1}, \quad b_{ea} = -8\delta g_{ea}\tau_e^*\tau_{ae}^{-1},$$

$$b_{ia} = -g_{ia}\tau_i^*\tau_{ai}^{-1}, \quad b_{ai} = -g_{ia}\delta\tau_a^*\tau_{ia}^{-1},$$

where

$$(\tau_e^*)^{-1} = 0.4\tau_{ee}^{-1} + 1.3\tau_{ei}^{-1} + (2.5 - 1.2B_{ea}^*)\tau_{ea}^{-1},$$

$$(\tau_i^*)^{-1} = 0.4\tau_{ii}^{-1} + g_{ia}'\tau_{ia}^{-1} + 3\delta\tau_{ie}^{-1},$$

$$(\tau_a^*)^{-1} = 0.4A_{aa}^*\tau_{aa}^{-1} + g_{ia}'\tau_{ai}^{-1} + 3\delta\tau_{ae}^{-1}, \tag{7.2.1}$$

$$g_{ia} = \frac{11}{16} - 0.2A_{ia}^* - 0.15B_{ia}^*,$$

$$g_{ia}' = \frac{11}{16} + 0.2A_{ia}^* - 0.15B_{ia}^*.$$

The coefficients b_{ie} and b_{ae} are of the order of δ^2 and are not written here. Calculating the determinants $|b|$ and $|b^*|$ to an accuracy up to $\sim \delta^{5/2}\theta^{3/2}$, we find [2]

$$|b| = 1 - b_{ia}b_{ai} = \Delta_\lambda, \qquad |b^*| = \Delta_\lambda\left(1 + \beta_e^{*2}\right)\left(1 + \Delta_\lambda^{-2}\beta_i^{*2}\right),$$

$$\beta_e^* = \omega_e\tau_e^*, \qquad \beta_i^* = \omega_i\tau_i^*. \tag{7.2.2}$$

In accordance with the structure of $\mathbf{N}_\alpha$ the reduced heat flux of every species comprises several independent parts. Let us consider first the part defined by the gradient of the species temperature. The analysis shows that the terms depending on ∇T_e contribute greatly to $\mathbf{h}_e^T$ and, hence, the expressions for $\mathbf{h}_e^T$ can be written as

$$\mathbf{h}_e^T = -\lambda_{e\|}\nabla_\| T_e - \lambda_{e\perp}\nabla_\perp T_e - \lambda_{e\wedge}\mathbf{k}\times\nabla T_e. \tag{7.2.3}$$

The coefficients of thermal conductivity λ_e are defined as

$$\lambda_{e\|} = \frac{5}{2}\frac{k}{m_e}p_e\tau_e^*, \quad \lambda_{e\perp} = \frac{\lambda_{e\|}}{1+\beta_e^{*2}}, \quad \lambda_{e\wedge} = \frac{\lambda_{e\|}\beta_e^*}{1+\beta_e^{*2}}. \tag{7.2.4}$$

When writing down Eqs. (7.2.3), (7.2.4) we omitted the terms the order of which is

$$\delta^{3/2}\theta^{5/2}\frac{|\nabla_\| T|}{|\nabla_\| T_e|} \quad \text{and} \quad \delta\theta\frac{|\nabla_\perp T|}{\nabla_\perp T_e}$$

with respect to the other terms. Note that under the given conditions the expressions of (7.2.4) agree with those of (6.2.7) derived from the equations of the electron heat flux when the cross terms are neglected.

The analogous estimates in the expressions for $\mathbf{h}_i^T$ and $\mathbf{h}_a^T$ permit us to omit the terms proportional to ∇T_e. Then [1,2]

$$\mathbf{h}_{i,a}^T = -\lambda_{i,a\|}\nabla_{\|}T - \lambda_{i,a\perp}\nabla_{\perp}T - \lambda_{i,a\wedge}\nabla T \times \mathbf{k}, \tag{7.2.5}$$

where

$$\lambda_{i\|} = \frac{5}{2}\frac{k}{m}p_i\tau_i^*\xi_i^*\Delta_\lambda^{-1}, \quad \lambda_{i\perp} = \frac{\lambda_{i\|}}{1+\Delta_\lambda^{-2}\beta_i^{*2}}, \quad \lambda_{e\wedge} = \beta_i^*\Delta_\lambda^{-1}\lambda_{i\perp},$$

$$\lambda_{a\|} = \frac{5}{2}\frac{k}{m}p_a\tau_a^*\xi_a^*\Delta_\lambda^{-1},$$

$$\lambda_{a\perp} = \lambda_{a\|}\frac{1+\Delta_\lambda^{-1}\left(\xi_a^*\right)^{-1}\beta_i^{*2}}{1+\Delta_\lambda^{-2}\beta_i^{*2}},$$

$$\lambda_{a\wedge} = \beta_i^*\Delta_\lambda^{-1}\lambda_{a\|}\frac{1-\Delta_\lambda\left(\xi_a^*\right)^{-1}\beta_i^{*2}}{1+\Delta_\lambda^{-2}\beta_i^{*2}} \tag{7.2.6}$$

and

$$\xi_i^* = 1 + g_{ia}\tau_a^*\tau_{ia}^{*-1}, \qquad \xi_a^* = 1 + g_{ia}\tau_i^*\tau_{ai}^{*-1}.$$

In Eqs. (7.2.5), (7.2.6), the terms the maximum order of which is

$$\delta\frac{|\nabla T_e|}{|\nabla T|} \quad \text{and} \quad \delta\theta^{-1}\frac{|\nabla T_e|}{\nabla T}\left(5\ln\Lambda\right)^{-1}$$

with respect to the other terms, are omitted. At $|\nabla T_e| \sim |\nabla T|$, this leads to the condition $T_e/T \ll 5(m/m_e)\ln\Lambda$ which is less rigid than the condition $T_e/T \ll (m/m_e)$. If $|\nabla T_e|/|\nabla T| \sim T_e/T$, the limit for T_e/T appears to be more rigid, namely $(T_e/T)^2 \ll 5(m/m_e)\ln\Lambda$.

To analyse the contribution of every species to the total reduced heat flux $\mathbf{h}^T = \sum_\alpha \mathbf{h}_\alpha^T$ it is appropriate to recall that for τ_α^*, ξ_α^* and Δ_λ the same estimations as for τ_α, ξ_α, $\Delta\eta$ are correct.

Instead of λ_α, the dimensionless thermal conductivity is introduced

$$\lambda_\alpha^* = \lambda_\alpha\left(\frac{m}{kT}\right)^{1/2}\frac{Q_{aa}}{k}.$$

Then

$$\lambda^*_{e\|} \sim \delta^{-1/2}\theta^{-5/2}\frac{\alpha\beta}{\alpha + (1-\alpha)\,\beta\theta^{-2}},$$

$$\lambda^*_{i\|} \sim \frac{\alpha\beta}{\alpha + (1-\alpha)}, \qquad \lambda^*_{a\|} \sim \frac{1-\alpha}{1+\alpha}.$$

From comparison of the expressions for λ^*_α it follows that the total heat flux along the magnetic field is defined mainly by the values of $\mathbf{h}^T_{e\|}$ and $\mathbf{h}^T_{a\|}$ and the ion contribution is negligibly small. With the increase of the degree of ionization, the contribution of the electron heat flux $\mathbf{h}^T_{e\|}$ grows fast and at $\alpha \sim \delta^{1/2}\theta^{3/2}$ the fluxes $\mathbf{h}^T_{e\|}$ and $\mathbf{h}^T_{a\|}$ become comparable. The increase of the magnetic field markedly limits the electron heat flux across the field; therefore, at $\beta^*_e \gg 1$ the actual contribution in $\mathbf{h}^T_\perp$ is made by $\mathbf{h}^T_{a\perp}$ and $\mathbf{h}^T_{i\perp}$ with the ion contribution getting essential only at high degrees of ionization. The heat flux perpendicular both to the magnetic field and the temperature gradient at $\beta^*_e \leq 1$ is defined basically by $\mathbf{h}^T_{e\wedge}$. At the growth of the magnetic field, the contribution of $\mathbf{h}^T_{i\wedge}$ and $\mathbf{h}^T_{a\wedge}$ increases and is of the order of $\mathbf{h}^T_{e\wedge}$ at $\beta_i \gg 1$. The nonisothermal nature of plasma ($T_e > T$) makes the relative importance of $\mathbf{h}^T_{e\wedge}$ more prominent.

Now let us consider the part of the heat flux proportional to the relative velocity of the species. In the case of a three-component plasma, the part of the vector $\mathbf{N}_\alpha$ is conveniently expressed in terms of current density $\mathbf{j} = -\mathbf{n}_e e\,(\mathbf{w}_e - \mathbf{w}_i)$ and the ion-slip $\mathbf{s} = \mathbf{w}_i - \mathbf{w}_a$. Then in the expression for $\mathbf{h}_\alpha$ in addition to the contribution of $\mathbf{h}^T_\alpha$ discussed above, we can separate the terms $\mathbf{h}^j_\alpha$ and $\mathbf{h}^s_\alpha$.

The electron parts of the reduced heat flux are defined as

$$\begin{aligned}\mathbf{h}^j_e &= \chi_{e\|}\mathbf{j}_\| + \chi_{e\perp}\mathbf{j}_\perp + \chi_{e\wedge}\mathbf{k}\times\mathbf{j},\\ \mathbf{h}^s_e &= -\mu_{e\|}\mathbf{s}_\| - \mu_{e\perp}\mathbf{s}_\perp - \mu_{e\wedge}\mathbf{k}\times\mathbf{s}.\end{aligned} \tag{7.2.7}$$

The coefficients of χ_e to an accuracy up to the terms $\sim \delta^{5/2}\theta^{-1/2}$ are defined by Eqs. (6.2.12), i.e.

$$\chi_{e\|} = \frac{kT_e}{e}\alpha_T, \quad \chi_{e\perp} = \frac{\chi_{e\|}}{1+\beta^{*2}_e}, \quad \chi_{e\wedge} = \beta^*_e\chi_{e\perp},$$

$$\alpha_T = \frac{5}{2}\tau^*_e\left(\zeta_{ea}\tau^{-1}_{ea} + \zeta_{ei}\tau^{-1}_{ei}\right).$$

For coefficients μ_e, we have

$$\mu_{e\parallel} = d_e^s p_e, \quad \mu_{e\perp} = \frac{\mu_{e\parallel}}{1+\beta_e^{*2}}, \quad \mu_{e\wedge} = \beta_e^* \mu_{e\perp},$$
$$d_e^s = \frac{5}{2}\tau_e^* \left[\zeta_{ea}\tau_{ea}^{-1} + 2\delta(1-\theta)(1-\alpha)\tau_{ei}^{-1}\right]. \tag{7.2.8}$$

The expressions which are similar to Eqs. (7.2.7) can be written for $\mathbf{h}_\alpha^j$ and $\mathbf{h}_\alpha^s$ $(\alpha = i, a)$.

Estimates of χ_i and χ_a show that these coefficients are of the order of $\delta^{3/2}\theta^{5/2}$ and $\delta^{3/2}\theta^{-1/2}$ compared with χ_e, and their relative contribution to the total heat fluxes $\mathbf{h}^j = \sum_\alpha \mathbf{h}_\alpha^j$ can be neglected. As to the coefficients μ_α, their contribution to the full coefficients at the projections of the vector $\mathbf{s}$ in the expression of $\mathbf{h}^s$ turn out to be of the same order for all the species. Without mentioning specific expressions for each of these coefficients, we shall directly give the expressions for the total heat flux $\mathbf{q}$ in which an additional contribution is due to the term of $(5/2)\sum_\alpha p_\alpha \mathbf{w}_\alpha$ in (6.1.13), besides the contributions of $\mathbf{h}^j$ and $\mathbf{h}^s$. Then

$$\mathbf{q} = \mathbf{q}^T + \mathbf{q}^j + \mathbf{q}^s,$$
$$\mathbf{q}^T = \sum_\alpha \mathbf{h}_\alpha^T, \quad \mathbf{q}^j = \kappa_\parallel \mathbf{j}_\parallel + \kappa_\perp \mathbf{j}_\perp + \kappa_\wedge \mathbf{k}\times\mathbf{j}, \tag{7.2.9}$$
$$\mathbf{q}^s = -\mu_\parallel \mathbf{s}_\parallel - \mu_\perp \mathbf{s}_\perp - \mu_\wedge \mathbf{k}\times\mathbf{s},$$

where

$$\kappa_\parallel = \frac{kT_e}{e}\left(\alpha_T - \frac{5}{2}\right), \quad \kappa_\perp = \frac{kT_e}{e}\left[\frac{\alpha_T}{(1+\beta_e^{*2})} - \frac{5}{2}\right],$$

$$\kappa_\wedge = \frac{kT_e}{e}\frac{\beta_e^*}{1+\beta_e^{*2}}\alpha_T,$$

$$\mu_\parallel = \left[d_e^s - \frac{5}{2}(1-\alpha)\right]p_e + \Delta_\lambda^{-1}\left(d_i^s p_i \xi_i^* + d_a^s p_a \xi_a^*\right),$$

$$\mu_\perp = \left[\frac{d_e^s}{1+\beta_e^{*2}} - \frac{5}{2}(1-\alpha)\right]p_e + \frac{d_i^s p_i \xi_i^* + d_a^s p_a\left(\xi_a^* + \Delta_\lambda^{-1}\beta_i^{*2}\right)}{\Delta_\lambda\left(1+\Delta_\lambda^{-2}\beta_i^{*2}\right)},$$

$$\mu_\wedge = \frac{\beta_e^*}{1+\beta_e^{*2}} d_e^s p_e + \beta_i^* \frac{d_i^s p_i \xi_i^* + d_a^s p_a\left(\xi_a^* - \Delta_\lambda^{-1}\right)}{\Delta_\lambda\left(1+\Delta_\lambda^{-2}\beta_i^{*2}\right)}. \tag{7.2.10}$$

Here

$$d_i^s = \frac{5}{2}\tau_i^* \left[\frac{1}{4}\zeta_{ia}\tau_{ia}^{-1} - 2\delta\theta^{-1}(1-\theta)(1-\alpha)\tau_{ie}^{-1}\right],$$

$$d_a^s = -\frac{5}{2}\tau_a^* \left[\frac{1}{4}\zeta_{ia}\tau_{ai}^{-1} - 2\delta\theta^{-1}(1-\theta)\alpha\tau_{ae}^{-1}\right],$$

and the coefficients Δ_λ, ξ_i^* and ξ_a^* were defined earlier.

To sum up, when it is necessary to increase the accuracy of calculating the electron transport coefficients, the results following from the successive approximations of the expansion over Sonine polynomials can be useful. More accurate expressions for the thermal conductivity of ions in the limit of full ionization are discussed in Chapter 8.

7.3 Diffusion fluxes

Solution of the diffusion equations for a three-component partially ionized gas has been considered in Chapter 3 for the case when such equations are written in a quasihydrodynamic approximation. The diffusion equations in the $13N$-moment approximation differ in the additional terms of the reduced heat fluxes of the species. After the substitution of the expressions of the heat fluxes into the diffusion equations the methods of their solution do not differ from those used earlier.

Let us use the variables $\mathbf{w} = \mathbf{w}_e - \mathbf{w}_i$ and $\mathbf{s} = \mathbf{w}_i - \mathbf{w}_a$ and the relation

$$\mathbf{w}_e = \mathbf{w} + (1-\alpha)\mathbf{s} \tag{7.3.1}$$

following from the condition $\sum_\alpha n_\alpha m_\alpha \mathbf{w}_\alpha = 0$.

We represent Eq. (6.2.1) for the electrons as

$$\tau_0^{-1}\mathbf{w} + \tau_{ea}^{-1}\mathbf{s} + \nu_o\frac{\mathbf{h}_e}{p_e} = -\frac{e}{m_e}\mathbf{E}^* - \omega_e\mathbf{w}_e \times \mathbf{k}. \tag{7.3.2}$$

The expression for the electron heat flux has already been given, it can be subdivided into the terms with ∇T_e, $\mathbf{w}$ and $\mathbf{s}$.

The equation to define $\mathbf{s}$ results from the diffusion equation written for the neutral species. Using the procedure in Section 3.2, we find

$$\begin{aligned} \mathbf{s} \quad &= -\frac{\varepsilon}{1+\varepsilon}\left[\mathbf{w} + \xi_{ea}\frac{\mathbf{h}_e}{p_e} + (1-\alpha)\omega_e\tau_{ea}\mathbf{w}\times\mathbf{k}\right] \\ &+ \frac{2\tau_{ia}}{n_i m(1+\varepsilon)}\left[\alpha\nabla p - \nabla(p_e + p_i)\right] - \frac{1}{2(1+\varepsilon)}\xi_{ia}\left(\frac{\mathbf{h}_i}{p_i} - \frac{\mathbf{h}_a}{p_a}\right), \end{aligned} \tag{7.3.3}$$

where

$$\varepsilon = \frac{2n_e m_e \tau_{ea}^{-1}}{n_i m_i \tau_{ia}^{-1}} \sim \left(Z\frac{m_e T}{m T_e}\right)^{1/2} \ll 1.$$

Equations (7.3.1)–(7.3.3), after substituting the expressions for $\mathbf{h}_e$, $\mathbf{h}_i$ and $\mathbf{h}_a$, form a closed system of equations which help to define diffusion velocities (fluxes) of the plasma species. We do not intend to consider the expressions for these fluxes in the general form. Our aim is to study the changes introduced by the reduced heat fluxes into the generalized Ohm's law and the contribution of thermal diffusion to the diffusion fluxes in a weakly ionized plasma.

The analysis proves [1] that the additional contribution to the generalized Ohm's law is mainly associated with the reduced heat flux $\mathbf{h}_e$ in the equation for the electrons (7.3.2). Corrections to s due to $\mathbf{h}_i$, $\mathbf{h}_a$ in (7.3.3) are proportional to ξ_{ea}^2 and ξ_{ia}^2 and are not more than 2% (estimated for the model of hard elastic spheres). The terms proportional to the temperature gradient can contribute more extensively; however, in the final expression of the current density the corrections are of the order of a small value ε and, therefore, can be omitted.

Substituting s into (7.3.2) on account of (7.3.1) and applying the expression for $\mathbf{h}_e$ obtained in the previous section, we can see that in the relation derived the terms proportional to the small value ε can be neglected unless it is multiplied by ω_e. As a result, we come to the following expression for the generalized Ohm's law [1]

$$A\mathbf{j} + B\beta_0 \mathbf{j} \times \mathbf{k} - C\beta_0^2 \mathbf{j} \cdot (\mathbf{j} \cdot \mathbf{k}) = \sigma_0 \mathbf{E}_{eff}, \tag{7.3.4}$$

where

$$\begin{aligned}
A &= 1 - \frac{\Delta_0}{1+\beta_e^{*2}} + \delta_0 \beta_0^2, \\
B &= 1 + \gamma \frac{\Delta_0}{1+\beta_e^{*2}}, \qquad C = \delta_0 + \gamma^2 \frac{\Delta_0}{1+\beta_e^{*2}}, \\
\mathbf{E}_{eff} &= \mathbf{E}^* - \frac{k}{e}\alpha_T \left[\nabla_{\parallel} T_e + \frac{1}{1+\beta_e^{*2}} \left(\nabla_{\perp} T_e + \beta_e^* \mathbf{k} \times \nabla T_e\right)\right] \\
&\quad + \frac{\delta_0}{n_e e (1-\alpha)} \mathbf{P} \times \mathbf{k}, \\
\mathbf{P} &= \alpha \nabla p - \nabla (p_e + p_i),
\end{aligned} \tag{7.3.5}$$

and

$$\beta_0 = \omega_e \tau_0, \qquad \beta_e^* = \omega_e \tau_e^*,$$

$$\Delta_0 = \nu_0 \tau_0 \alpha_T = \frac{5}{2} \nu_0^2 \tau_0 \tau_e^*, \quad \gamma = \tau_0^{-1} \tau_e^*, \tag{7.3.6}$$

$$\delta_0 = (1-\alpha)^2 \varepsilon \tau_{ea} \tau_0^{-1} = 2(1-\alpha)^2 Z \left(\frac{m_e}{m}\right) \tau_0^{-1} \tau_{ia}.$$

The values α_T and ν_0 are defined by the expressions (6.2.10). When solving (7.3.4) with respect to $\mathbf{j}$, we can write the expression of the current density as

$$\begin{aligned} \mathbf{j} &= \sigma_{\|} \mathbf{E}_{\|}^* + \sigma_{\perp} \mathbf{E}_{\perp}^* + \sigma_{\wedge} \mathbf{k} \times \mathbf{E}^* + \varphi_{\|} \nabla_{\|} T_e \\ &+ \varphi_{\perp} \nabla_{\perp} T_e + \varphi_{\wedge} \mathbf{k} \times \nabla T_e + \beta_{\perp} \mathbf{P}_{\perp} + \beta_{\wedge} \mathbf{k} \times \mathbf{P}. \end{aligned} \tag{7.3.7}$$

The coefficients σ and φ are defined by (6.2.14). The coefficient B is kept in the previously set form (6.2.15), whereas an additional term $s = \delta_0 \beta_0^2$ arises in the coefficient A due to the effect of ion-slip. The coefficients $\beta_\perp$ and $\beta_\wedge$ are defined as

$$\beta_{\perp} = \sigma_{\perp} \frac{A}{B} \frac{\delta_0 \beta_e^2}{n_e e (1-\alpha)}, \quad \beta_{\wedge} = \sigma_{\wedge} \frac{\delta_0 \beta_e}{n_e e (1-\alpha)}.$$

Note that at $s \ll 1$ and when neglecting the terms dependent on $\mathbf{P}$ the results obtained agree with the expressions (6.2.13)–(6.2.15) considered in the previous chapter which also estimated the contribution of the second approximation to the longitudinal electrical conductivity of plasma. As to the transverse coefficients of electrical conductivity, as we mentioned earlier, with the growing magnetized state of plasma ($\beta_0 \gg 1$) the effect of corrections of the second approximation becomes inessential. Yet, the requirement of taking into account the ion-slip due to $s = \delta_0 \beta_0^2$ in the coefficient A may be of importance.

Now let us analyse the role of thermal diffusion in the case of a weakly ionized gas mixture. It is to be recalled that the diffusion equations (3.5.2) are valid if the right-hand side should include the term of the kind

$$-n_\alpha \sum_{\beta=1}^{N} \mu_{\alpha\beta}^2 \zeta_{\alpha\beta} \tau_{\alpha\beta}^{-1} \left(\frac{\mathbf{h}_\alpha}{m_\alpha p_\alpha} - \frac{\mathbf{h}_\beta}{m_\beta p_\beta} \right). \tag{7.3.8}$$

When the corrections of the second approximation to the diffusion coefficients are omitted (so as they are small as a rule in a weakly ionized mixture), the appropriate thermal diffusion terms in the diffusion equations are obtained by substituting in (7.3.8) those parts of heat fluxes which depend on the temperature gradient, i.e. $\mathbf{h}_\alpha^T$ and $\mathbf{h}_\beta^T$, calculated in the limit of a weakly ionized gas.

When writing the diffusion equation for electrons, this procedure would imply considering only $\mathbf{h}_e^T$ for which the solution (6.2.11) can be used. The ultimate expression for the diffusion flux $n_e\mathbf{w}_e$ in the absence of a magnetic field can be represented as

$$n_e\mathbf{w}_e = -D_e\nabla n_e - n_e b_e\mathbf{E} - n_e D_e\left(1-\alpha_T^e\right)\nabla\ln T_e, \tag{7.3.9}$$

where $\alpha_T^e = (5/2)\,\nu_0\tau_e^*$ is the electron thermal diffusion factor for a weakly ionized plasma.

For the ion species of plasma at $T_i = T$, the contribution of the thermal diffusion in the absence of a magnetic field is found by applying the expressions derived earlier for the case of an isothermal multicomponent gas mixture (see Section 5.3). In particular, if the mixture contains a single species of the neutrals, we have

$$n_i\mathbf{w}_i = -D_i\nabla n_i + n_i Z_i b_i\mathbf{E} - n_i D_i\left(1+\alpha_T^i\right)\nabla\ln T_i, \tag{7.3.10}$$

where $\alpha_T^i = [\alpha_T]_1$ is the thermal diffusion factor of a two-species mixture defined by (5.3.14).

In these expressions for the diffusion fluxes the coefficient at the temperature gradient is formed from the thermal diffusion contribution proper which is proportional to α_T and the contribution formed by separating the term with the temperature gradient from $\nabla\ln p_\alpha$ following from $\nabla\ln p_\alpha = \nabla\ln n_\alpha + \nabla\ln T_\alpha$. It is to be pointed out that in the literature on plasma physics the whole multiplier at $\nabla\ln T$ is often termed the thermal diffusion coefficient, whereas in terms of kinetic theory only the part proportional to α_T is implied. As a rule, α_T is considerable smaller than unity. In particular, for electrons $\alpha_T^e = 0$ if the effective momentum transfer frequency $\nu_{e\beta}(v)$ does not depend on velocity. For ions, this case corresponds to the polarization interaction of an ion and an atom ($U \sim r^{-4}$) which appears to be dominant at low energies. For more intricate models of the interaction potential which take into account both long range polarization attraction and short distance particle repulsion at the potential $U \sim r^{-n}$ $(n > 0)$, the thermal diffusion factor α_T^i can have any sign depending on the value kT/ε, where ε is the depth of the curve of the potential energy.

The expressions (7.3.9) and (7.3.10) permit us to consider the problem of ambipolar diffusion accounting for the temperature gradients. Using the condition $\sum\limits_\alpha n_\alpha e_\alpha\mathbf{w}_\alpha = 0$ and eliminating with its help the field $\mathbf{E}$ for the case of plasma with a single ion species with charge Ze provided that $b_e \gg b_i$, we have

$$\begin{aligned} n_e\mathbf{w}_e &= Zn_i\mathbf{w}_i = -D_A\nabla n_e - D_{TA}^e n_e\nabla\ln T_e \\ &\quad - D_{TA}^i n_e\nabla\ln T, \end{aligned} \tag{7.3.11}$$

where D_A is defined by the expressions (3.5.19), and for the ambipolar coefficients of thermal diffusion, we get

$$D_{TA}^{e} = Z\frac{T_e}{T}D_i\left(1-\alpha_T^{e}\right), \quad D_{TA}^{i} = D_i\left(1+\alpha_T^{i}\right). \tag{7.3.12}$$

Chapter 8

Multispecies fully ionized plasma

8.1 Original equations

This chapter treats the transport phenomena in a fully ionized multispecies plasma of electrons and several species of ions each of which can, generally speaking, be in various charge states. Accordingly, we shall use indices α and β for various particles ($m_\alpha \neq m_\beta$) and indices Z and ζ will refer to the charge states (for electrons, one can formally accept $Z = -1$ or use a single index of $\alpha = e$). Since, as we pointed out earlier, in the case of the Coulomb collisions of charged particles the $13N$-moment approximation does not ensure a required accuracy when calculating transport coefficients, we intend to use an approximation in which in expansion of the distribution function an additional polynomial is applied both for vector and tensor values. The appropriate expression for the distribution function of particles of species α in the Z charge state looks like

$$\begin{aligned} f_{\alpha Z} \; = \; & f_{\alpha Z}^{(0)}\Big[1+\gamma_{\alpha Z}\mathbf{w}_{\alpha Z}\cdot\mathbf{c}+\frac{1}{5}\gamma_{\alpha Z}^{2}p_{\alpha Z}^{-1}\mathbf{h}_{\alpha Z}\cdot\mathbf{c}\left(c^{2}-5\gamma_{\alpha Z}^{-1}\right) \\ & +\frac{1}{70}\gamma_{\alpha Z}^{4}p_{\alpha Z}^{-1}\mathbf{r}_{\alpha Z}\cdot\mathbf{c}\left(c^{4}-14\gamma_{\alpha Z}^{-1}c^{2}+35\gamma_{\alpha Z}^{-2}\right) \\ & +\frac{1}{2}\gamma_{\alpha Z}p_{\alpha Z}^{-1}\pi_{\alpha Zrs}\left(c_{r}c_{s}-\frac{1}{3}\delta_{rs}c^{2}\right) \\ & +\frac{1}{14}\gamma_{\alpha Z}^{3}p_{\alpha Z}^{-1}\sigma_{\alpha Zrs}\left(c_{r}c_{s}-\frac{1}{3}\delta_{rs}c^{2}\right)\left(c^{2}-7\gamma_{\alpha Z}^{-1}\right)\Big]. \end{aligned} \tag{8.1.1}$$

The new moments of the distribution function in (8.1.1) are defined by

$$\mathbf{r}_{\alpha Z} = \frac{m_\alpha}{4}\int\left(c^4 - 14\gamma_{\alpha Z}^{-1}c^2 + 35\gamma_{\alpha Z}^{-2}\right)\mathbf{c} f_{\alpha Z}d\mathbf{c},$$
$$\sigma_{\alpha Z rs} = \frac{m_\alpha}{2}\int\left(c^2 - 7\gamma_{\alpha Z}^{-1}\right)\left(c_r c_s - \frac{1}{3}\delta_{rs}c^2\right)f_{\alpha Z}d\mathbf{c}. \tag{8.1.2}$$

The total number of independent moments is equal to $21N$; therefore, the representation of $f_{\alpha Z}$ in the form (8.1.1) corresponds to the $21N$-moment approximation.

The moment equation systems used to calculate the diffusion velocities, heat fluxes and viscous-stress tensors of species will be written in the approximation when the time derivatives of the moments and all nonlinear terms are negligible (i.e. in the linear approximation with respect to ε of Grad's method). To simplify the right-hand sides of the equations (or the moments with respect to the collision integrals), it is assumed that the condition $|T_{\alpha Z} - T_{\beta\zeta}| \ll T_{\alpha Z}$ is fulfilled which in turn allows us to put down the coefficient at the moments on the right-hand sides of the expressions at the same temperature T. Such a limitation does not refer to light particles, electrons in particular, for which, as was already shown in the analysis of the $13N$-moment approximation, the same expressions are valid as for the isothermal plasma if T is simultaneously replaced by T_e (in which case the condition $T_e/T_\alpha \gg m_e/m_\alpha$ must be met). Note that the requirement of the approximate equality of plasma species temperatures does not impose any limits on the temperature gradients of the particles. Therefore, on the left-hand side of the equations both for electrons and ions of various species we shall retain different temperatures under the derivative signs.

With the above considerations made, the system of equations to define the diffusion velocities and heat fluxes in a multispecies fully ionized plasma is written as [1]

$$-\rho_{\alpha Z}\omega_{\alpha Z}\mathbf{w}_{\alpha Z}\times\mathbf{k} + \nabla p_{\alpha Z} - \rho_{\alpha Z}\left(\frac{\mathbf{F}_{\alpha Z}}{m_\alpha} - \frac{d\mathbf{u}}{dt}\right) = \mathbf{R}_{\alpha Z}$$
$$\equiv \sum_{\beta,\zeta}\left[G^{(1)}_{\alpha Z\beta\zeta}\left(\mathbf{w}_{\alpha Z} - \mathbf{w}_{\beta\zeta}\right) + \frac{\mu_{\alpha\beta}}{kT}G^{(2)}_{\alpha Z\beta\zeta}\left(\frac{\mathbf{h}_{\alpha Z}}{\rho_{\alpha Z}} - \frac{\mathbf{h}_{\beta\zeta}}{\rho_{\beta\zeta}}\right)\right.$$
$$\left. + \frac{\mu^2_{\alpha\beta}}{(kT)^2}G^{(8)}_{\alpha Z\beta\zeta}\left(\frac{\mathbf{r}_{\alpha Z}}{\rho_{\alpha Z}} - \frac{\mathbf{r}_{\beta\zeta}}{\rho_{\beta\zeta}}\right)\right], \tag{8.1.3}$$

$$-\omega_{\alpha Z}\mathbf{h}_{\alpha Z}\times\mathbf{k}+\frac{5}{2}\frac{kp_{\alpha Z}}{m_\alpha}\nabla T_{\alpha Z}=\frac{kT}{m_\alpha}\sum_{\beta,\zeta}\left[\frac{5}{2}\frac{\mu_{\alpha\beta}}{m_\alpha}G^{(2)}_{\alpha Z\beta\zeta}\right.$$
$$\times(\mathbf{w}_{\alpha Z}-\mathbf{w}_{\beta\zeta})+G^{(5)}_{\alpha Z\beta\zeta}\frac{\mathbf{h}_{\alpha Z}}{p_{\alpha Z}}+G^{(6)}_{\alpha Z\beta\zeta}\frac{\mathbf{h}_{\beta\zeta}}{p_{\beta\zeta}}$$
$$\left.+\frac{\mu_{\alpha\beta}}{kT}\left(G^{(9)}_{\alpha Z\beta\zeta}\frac{\mathbf{r}_{\alpha Z}}{p_{\alpha Z}}+G^{(10)}_{\alpha Z\beta\zeta}\frac{\mathbf{r}_{\beta\zeta}}{p_{\beta\zeta}}\right)\right],\tag{8.1.3'}$$

$$-\omega_{\alpha Z}\mathbf{r}_{\alpha Z}\times\mathbf{k}=\left(\frac{kT}{m_\alpha}\right)^2\sum_{\beta,\zeta}\left[\frac{35}{2}\left(\frac{\mu_{\alpha\beta}}{m_\alpha}\right)^2G^{(8)}_{\alpha Z\beta\zeta}(\mathbf{w}_{\alpha Z}-\mathbf{w}_{\beta\zeta})\right.$$
$$+7\frac{\mu_{\alpha\beta}}{m_\alpha}\left(G^{(9)}_{\alpha Z\beta\zeta}\frac{\mathbf{h}_{\alpha Z}}{p_{\alpha Z}}+G^{(10)}_{\alpha Z\beta\zeta}\frac{\mathbf{h}_{\beta\zeta}}{p_{\beta\zeta}}\right)$$
$$\left.+\frac{m_\alpha}{kT}G^{(11)}_{\alpha Z\beta\zeta}\frac{\mathbf{r}_{\alpha Z}}{p_{\alpha Z}}+\frac{m_\beta}{kT}G^{(12)}_{\alpha Z\beta\zeta}\frac{\mathbf{r}_{\beta\zeta}}{p_{\beta\zeta}}\right].\tag{8.1.3''}$$

The coefficients $G^{(n)}_{\alpha Z\beta\zeta}$ are defined by the expressions

$$G^{(1)}_{\alpha Z\beta\zeta}=-\lambda_{\alpha Z\beta\zeta},\qquad G^{(2)}_{\alpha Z\beta\zeta}=\frac{3}{5}\lambda_{\alpha Z\beta\zeta},$$
$$G^{(5)}_{\alpha Z\beta\zeta}=-\left(\frac{13}{10}\frac{m_\beta}{m_\alpha}+\frac{8}{5}+3\frac{m_\alpha}{m_\beta}\right)\kappa_{\alpha\beta}\lambda_{\alpha Z\beta\zeta},$$
$$G^{(6)}_{\alpha Z\beta\zeta}=\frac{27}{10}\kappa_{\alpha\beta}\lambda_{\alpha Z\beta\zeta},\quad G^{(8)}_{\alpha Z\beta\zeta}=-\frac{3}{14}\lambda_{\alpha Z\beta\zeta},$$
$$G^{(9)}_{\alpha Z\beta\zeta}=\frac{3}{5}\left(\frac{23}{28}\frac{m_\beta}{m_\alpha}+\frac{8}{7}+3\frac{m_\alpha}{m_\beta}\right)\kappa_{\alpha\beta}\lambda_{\alpha Z\beta\zeta},\tag{8.1.4}$$
$$G^{(10)}_{\alpha Z\beta\zeta}=-\frac{45}{28}\kappa_{\alpha\beta}\lambda_{\alpha Z\beta\zeta},$$
$$G^{(11)}_{\alpha Z\beta\zeta}=-\left[\frac{433}{280}\left(\frac{m_\beta}{m_\alpha}\right)^2+\frac{139}{35}\frac{m_\beta}{m_\alpha}+\frac{459}{35}+\frac{32}{5}\frac{m_\alpha}{m_\beta}\right.$$
$$\left.+5\left(\frac{m_\alpha}{m_\beta}\right)^2\right]\kappa^2_{\alpha\beta}\lambda_{\alpha Z\beta\zeta},$$
$$G^{(12)}_{\alpha Z\beta\zeta}=\frac{75}{8}\kappa^2_{\alpha\beta}\lambda_{\alpha Z\beta\zeta},$$

where

$$\kappa_{\alpha\beta} = \frac{m_\alpha m_\beta}{(m_\alpha + m_\beta)^2},$$

$$\lambda_{\alpha Z\beta\zeta} = \lambda_{\beta\zeta\alpha Z} = \frac{1}{3}(2\pi)^{-3/2} n_{\alpha Z} n_{\beta\zeta} e^4 Z^2 \zeta^2 \mu_{\alpha\beta}^{1/2} \frac{\ln \Lambda_{\alpha\beta}}{(kT)^{3/2}\varepsilon_0^2}.$$

The coefficients (8.1.4) are calculated by means of the expressions for the bracket integrals based on Sonine polynomials (Appendix 2) and on the fact that for Coulomb interactions at formal cutting off the impact parameter at a distance of the order of the Debye radius the expressions for $\Omega^{lr}_{\alpha\beta}$ can be represented as

$$\Omega^{lr}_{\alpha\beta} = \sqrt{\pi} l (r-1)! \left(\frac{Z_\alpha Z_\beta e^2}{4\pi\varepsilon_0}\right)^2 \frac{\ln \Lambda_{\alpha\beta}}{\mu_{\alpha\beta}^{1/2}(2kT)^{3/2}}, \tag{8.1.5}$$

where

$$\Lambda_{\alpha\beta} = \frac{12\pi\varepsilon_0^{3/2} kT}{Z_{eff} e^2}\left[\frac{kT}{n_e e^2 (1 + Z_{eff})}\right]^{1/2}.$$

The relevant system of equations for defining the viscous-stress tensor of the species is written as

$$\begin{aligned}
&-\omega_{\alpha Z}\{\pi_{\alpha Z lr} e_{slm} k_m\} + p_{\alpha Z} W_{rs} \\
&= \sum_{\beta,\zeta} \frac{kT}{m_\alpha + m_\beta}\left[\frac{G^{(3)}_{\alpha Z\beta\zeta}\pi_{\alpha Zrs}}{p_{\alpha Z}} + \frac{G^{(4)}_{\alpha Z\beta\zeta}\pi_{\beta\zeta rs}}{p_{\beta Z}}\right. \\
&\left.+ \frac{\mu_{\alpha\beta}}{kT}\left(\frac{G^{(13)}_{\alpha Z\beta\zeta}\sigma_{\alpha Zrs}}{p_{\alpha Z}} + \frac{G^{(14)}_{\alpha Z\beta\zeta}\sigma_{\beta\zeta rs}}{p_{\beta Z}}\right)\right],
\end{aligned} \tag{8.1.6}$$

$$\begin{aligned}
&-\omega_{\alpha Z}\{\sigma_{\alpha Z lr} e_{slm} k_m\} \\
&= \sum_{\beta,\zeta} \frac{kT}{m_\alpha + m_\beta}\left[\frac{7}{2}\frac{\mu_{\alpha\beta}}{m_\alpha}\left(\frac{G^{(13)}_{\alpha Z\beta\zeta}\pi_{\alpha Zrs}}{p_{\alpha Z}}\right.\right. \\
&\left.\left.+ \frac{G^{(14)}_{\alpha Z\beta\zeta}\pi_{\beta\zeta rs}}{p_{\beta\zeta}}\right) + \frac{G^{(15)}_{\alpha Z\beta\zeta}\sigma_{\alpha Zrs}}{p_{\alpha Z}} + \frac{G^{(16)}_{\alpha Z\beta\zeta}\sigma_{\beta\zeta rs}}{p_{\beta\zeta}}\right].
\end{aligned} \tag{8.1.6'}$$

And for the coefficients $G^{(n)}_{\alpha Z\beta\zeta}$, we have

$$G^{(3)}_{\alpha Z\beta\zeta} = -2\left(1+\frac{3}{5}\frac{m_\beta}{m_\alpha}\right)\lambda_{\alpha Z\beta\zeta}, \quad G^{(4)}_{\alpha Z\beta\zeta} = \frac{4}{5}\lambda_{\alpha Z\beta\zeta},$$

$$G^{(13)}_{\alpha Z\beta\zeta} = \left(\frac{18}{35}\frac{m_\beta}{m_\alpha}+\frac{6}{5}\right)\lambda_{\alpha Z\beta\zeta}, \quad G^{(4)}_{\alpha Z\beta\zeta} = -\frac{24}{35}\frac{m_\alpha}{m_\beta}\lambda_{\alpha Z\beta\zeta},$$

$$G^{(15)}_{\alpha Z\beta\zeta} = -\left(\frac{51}{35}\frac{m_\beta^2}{m_\alpha^2}+\frac{37}{7}\frac{m_\beta}{m_\alpha}+\frac{22}{5}+4\frac{m_\alpha}{m_\beta}\right)\kappa_{\alpha\beta}\lambda_{\alpha Z\beta\zeta},$$

$$G^{(16)}_{\alpha Z\beta\zeta} = \frac{24}{7}\kappa_{\alpha\beta}\lambda_{\alpha Z\beta\zeta}. \tag{8.1.7}$$

The solution of the equation systems (8.1.3) and (8.1.6) enables us to determine all the transport coefficients of interest for any species of the plasma at an arbitrary value of the magnetic field. However, in the general case the results expressed in terms of the ratios of the determinants are awkward which excludes their possible analysis. Therefore, in what follows the case of arbitrary $\omega_\alpha\tau$ is considered only for a simple two-component plasma which permits us to compare the results derived with those in [2,3]. The transport phenomena in a multispecies plasma will be treated in detail for the limiting cases $\omega_\alpha\tau \gg 1$ and $\omega_\alpha = 0$ allowing us to obtain relatively simple expressions in practically important applications such as (i) transverse fluxes in a strongly magnetized plasma, (ii) longitudinal fluxes for electrons and light ions, (iii) longitudinal fluxes for some species of heavy ions in arbitrary charge states.

8.2 Transport phenomena in a simple plasma

Consider transport phenomena in the simplest case of a two-component fully ionized plasma of electrons and a single species of ions with the charge number Z. In keeping with the analysis carried out in the previous chapter we can separate the ion equations from the electron equations.

First we shall calculate the heat flux of ions. That would require Eqs. (8.1.3′) and (8.1.3″) written for ions which, after omitting small terms dependent on the parameters of the electron component and substituting the explicit expressions for $G^{(n)}_{ii}$, will look as

$$
\begin{aligned}
-\beta_i \mathbf{h}_i \times \mathbf{k} + \frac{5}{2}\frac{k}{m_i} p_i \tau_{ii} \nabla T_i &= -\frac{2}{5}\mathbf{h}_i + \frac{3}{35}\frac{m_i}{kT}\mathbf{r}_i, \\
-\beta_i \frac{m_i}{kT}\mathbf{r}_i \times \mathbf{k} &= \frac{3}{5}\mathbf{h}_i - \frac{9}{14}\frac{m_i}{kT}\mathbf{r}_i,
\end{aligned} \tag{8.2.1}
$$

where

$$
\beta_i = \omega_i \tau_{ii}, \qquad \omega_i = \frac{eZ|\mathbf{B}|}{m_i}.
$$

Solving (8.2.1) with respect to $\mathbf{h}_i$ gives

$$
\mathbf{h}_i = -\lambda_{i\|} T_i - \lambda_{i\perp}\nabla_\perp T_i - \lambda_{i\wedge}\nabla T_i \times \mathbf{k}, \tag{8.2.2}
$$

where

$$
\begin{aligned}
\lambda_{i\|} &= 7.81\frac{k}{m_i} p_i \tau_{ii}, \quad \lambda_{i\perp} = \frac{k}{m_i} p_i \tau_{ii} \frac{\beta_i^2 + 0.33}{\Delta_i}, \\
\lambda_{i\wedge} &= \frac{5}{2}\frac{k}{m_i} p_i \tau_{ii} \beta_i \frac{\beta_i^2 + 0.46}{\Delta_i}, \\
\Delta_i &= \beta_i^4 + 0.68\beta_i^2 + 0.043,
\end{aligned} \tag{8.2.3}
$$

$$
\tau_{ii}^{-1} = \frac{1}{6\epsilon_0^2} n_i \left[\frac{Z^4 e^4}{m_i^{1/2}(\pi k T_i)^{3/2}}\right] \ln\Lambda.
$$

The results obtained agree with the familiar results of Braginskii [2] taking into account the differences in determining the relaxation times of ions ($\tau_i = 2\tau_{ii}$).

Now let us look upon (8.1.3′) and (8.1.3″) written for electrons. Preserving the relative velocity of electrons and ions $\mathbf{w} = \mathbf{w}_e - \mathbf{w}_i$ in them, neglecting the crossed electron–ion terms and using the explicit expressions for $G_{ee}^{(n)}$ and $G_{ei}^{(n)}$, we come to the following system of equations to define $\mathbf{h}_e$ and $\mathbf{r}_e$

$$
\begin{aligned}
\beta_e \mathbf{h}_e \times \mathbf{k} + \frac{5}{2}\frac{k}{m_e} p_e \tau_{ei} \nabla T_e - \frac{3}{2} p_e \mathbf{w} &= -\alpha_{11}\mathbf{h}_e + \alpha_{12}\frac{m_e}{kT_e}\mathbf{r}_e, \\
\frac{m_e}{kT_e}\beta_e \mathbf{r}_e \times \mathbf{k} + \frac{15}{4} p_e \mathbf{w} &= \alpha_{21}\mathbf{h}_e - \alpha_{22}\frac{m_e}{kT_e}\mathbf{r}_e,
\end{aligned} \tag{8.2.4}
$$

where $\beta_e = \omega_e \tau_{ei}$, $\omega_e = e|\mathbf{B}|/m_e$ and the coefficients α_{mn} are the functions of ion charge number Z

$$\alpha_{11} = \frac{2\sqrt{2}}{5Z} + \frac{13}{10}, \quad 7\alpha_{12} = \alpha_{21} = \frac{3\sqrt{2}}{5Z} + \frac{69}{20},$$

$$\alpha_{22} = \frac{9\sqrt{2}}{14Z} + \frac{433}{280}.$$

To find the relative velocity of the components $\mathbf{w}$ or the current density $\mathbf{j} = -n_e e\mathbf{w}$, we shall use the first equation (8.1.3) which at $\alpha = e$ and when neglecting the small terms is written as

$$n_e e\mathbf{E}' + \nabla p_e - \rho_e \omega_e \mathbf{w} \times \mathbf{k} = \mathbf{R}_{ei}, \tag{8.2.5}$$

where

$$\mathbf{R}_{ei} = -m_e n_e \tau_{ei}^{-1} \left(\mathbf{w} - \frac{3}{5}\frac{\mathbf{h}_e}{p_e} + \frac{3}{14}\frac{m_e}{kT_e}\frac{\mathbf{r}_e}{p_e} \right). \tag{8.2.6}$$

After Eqs. (8.2.4) are solved with respect to $\mathbf{h}_e$ and $\mathbf{r}_e$, the value $\mathbf{R}_{ei}$ appears to be expressed in terms of $\mathbf{w}$ and ∇T_e. In fact, it is quite adequate so that $\mathbf{w}$ in the function $\mathbf{E}^*$ and ∇T_e is defined from (8.2.5). Below, we give the expressions for $\mathbf{R}_{ei}$ and $\mathbf{h}_e$ following from the solution of Eqs. (8.2.4):

$$\begin{aligned}
&\mathbf{R}_{ei} = \mathbf{R}_{ei}^w + \mathbf{R}_{ei}^T, \qquad \mathbf{h}_e = \mathbf{h}_e^w + \mathbf{h}_e^T, \\
&\mathbf{R}_{ei}^w = -m_e n_e \tau_{ei}^{-1} \left(\alpha_{\parallel} \mathbf{w}_{\parallel} + \alpha_{\perp} \mathbf{w}_{\perp} - \alpha_{\wedge} \mathbf{k} \times \mathbf{w} \right), \\
&\mathbf{R}_{ei}^T = -n_e \left(\beta_{\parallel} \nabla_{\parallel} T_e + \beta_{\perp} \nabla_{\perp} T_e + \beta_{\wedge} \mathbf{k} \times \nabla T_e \right), \\
&\mathbf{h}_e^w = p_e \left(\beta_{\parallel} \mathbf{w}_{\parallel} + \beta_{\perp} \mathbf{w}_{\perp} + \beta_{\wedge} \mathbf{k} \times \mathbf{w} \right), \\
&\mathbf{h}_e^T = -\frac{p_e}{m_e} \tau_{ei} \left(\gamma_{\parallel} \nabla_{\parallel} T_e + \gamma_{\perp} \nabla_{\perp} T_e + \gamma_{\wedge} \mathbf{k} \times \nabla T_e \right).
\end{aligned} \tag{8.2.7}$$

From here on, the value $\mathbf{R}_{ei}$ will be termed the "electron–ion friction force"[1].

The coefficients α, β, and γ are the functions of Z and parameter β_e

[1]Note that in the literature [2] the "friction force" implies only that part of $\mathbf{R}_{ei}$ which is related to the relative velocities of the component (i.e. $\mathbf{R}_{ei}^w$) in which case $\mathbf{R}_{ei}^T$ is called the "thermal force".

$$\alpha_{\|} = 1 - \frac{p_1}{p_0}, \quad \alpha_{\perp} = 1 - \frac{p_1 p_0 + p_2 \beta_e^2}{\Delta_e}, \quad \alpha_{\wedge} = \beta_e \frac{p_3 + 1.7\beta_e^2}{\Delta_e},$$
$$\beta_{\|} = \frac{p_4}{p_0}, \quad \beta_{\perp} = \frac{p_4 p_0 + p_5 \beta_e^2}{\Delta_e}, \quad \beta_{\wedge} = \beta_e \frac{p_6 + 1.5\beta_e^2}{\Delta_e},$$
$$\gamma_{\|} = \frac{p_7}{p_0}, \quad \gamma_{\perp} = \frac{p_7 p_0 + p_8 \beta_e^2}{\Delta_e}, \quad \gamma_{\wedge} = \beta_e \frac{p_9 + 2.5\beta_e^2}{\Delta_e},$$
$$\Delta_e = \beta_e^4 + p_{10}\beta_e^2 + p_0^2, \tag{8.2.8}$$

where

$$p_0 = 0.31 + 1.20Z^{-1} + 0.41Z^{-2}, \quad p_1 = 0.22 + 0.73Z^{-1},$$
$$p_2 = 4.6 + 1.8Z^{-1}, \quad p_3 = 0.095 + 0.31Z^{-1} + 0.37Z^{-2},$$
$$p_4 = 0.47 + 0.94Z^{-1}, \quad p_5 = 3.8 + 1.3Z^{-1},$$
$$p_6 = 0.88 + 1.45Z^{-1} + 0.73Z^{-2}, \quad p_7 = 3.9 + 2.3Z^{-1}, \tag{8.2.9}$$
$$p_8 = 3.2 + 1.4Z^{-1}, \quad p_9 = 10.2 + 9.1Z^{-1} + 2.3Z^{-2},$$
$$p_{10} = 7.5 + 6.0Z^{-1} + 1.2Z^{-2}.$$

At $Z = 1, 2, 3, 4$ and $Z \to \infty$, the coefficients α, β, and γ calculated by means of (8.2.8) agree with the values given in [2,3].

It is to be noted that the results valid for the electron transport coefficients can be easily generalized for an arbitrary number of ion species. Indeed, since the electron equations are separated from the equations for heavy particles, the parameters of the ion species enter the separated equations for electrons in the form of certain sums ($\alpha \neq e$)

$$\sum_{\alpha,Z} n_{\alpha Z} Z_\alpha^2 = n_e Z_{eff}, \quad \sum_{\alpha,Z} n_{\alpha Z} Z_\alpha^2 \mathbf{w}_{\alpha Z} = n_e Z_{eff} \overline{\mathbf{w}}_i. \tag{8.2.10}$$

This results in the fact that in a multispecies plasma the full electron–ion friction force against the ions of all species and the electron heat flux are defined by the same expressions which were used in the case of a simple plasma accompanied by the replacement of Z with Z_{eff} in the relevant formulae and another definition of $\mathbf{w}$ which is written as $\mathbf{w} = \mathbf{w}_e - \overline{\mathbf{w}}_i$.

Using the expression for $\mathbf{R}_{ei}$ (8.2.6) in Eq. (8.2.5) and solving the latter with respect to $\mathbf{w}$, we can obtain the generalized Ohm's law for a two-component fully ionized plasma in the well-known form

$$\mathbf{j} = \sigma_{\|}\mathbf{E}_{\|eff} + \sigma_{\perp}\mathbf{E}_{\perp eff} + \sigma_{\wedge}\mathbf{k} \times \mathbf{E}_{eff}, \tag{8.2.11}$$

where the effective electric field

$$\mathbf{E}_{eff} = \mathbf{E}^* - \frac{\mathbf{R}_{ei}^T}{n_e e} \tag{8.2.12}$$

is introduced, while the electrical conductivities are defined by

$$\sigma_{\|} = \frac{n_e e^2}{m_e \alpha_{\|}} \tau_{ei}, \quad \sigma_{\perp} = \frac{n_e e^2 \tau_{ei} \alpha_{\perp}}{m_e \left[\alpha_{\perp}^2 + (\alpha_{\wedge} + \beta_e)^2\right]},$$
$$\sigma_{\wedge} = \frac{n_e e^2 \tau_{ei} (\alpha_{\wedge} + \beta_e)}{m_e \left[\alpha_{\perp}^2 + (\alpha_{\wedge} + \beta_e)^2\right]}. \tag{8.2.13}$$

This is not a simple generalization of the Ohm's law for the case of a multispecies plasma; nevertheless, in the approximation of the quasi two-species medium ($\beta_e \beta_i \ll 1$) when $\mathbf{w} \cong \mathbf{w}_e$ and $\mathbf{j} = -n_e e \mathbf{w}_e$, the expressions (8.2.11) and (8.2.13) prove to be valid for this case with the accompanying replacement of Z with Z_{eff} in the formulae.

Now let us turn back to the calculation of the viscosity in a two-component plasma. The general representation of the viscous-stress tensor of both electrons and ions in the magnetic field by means of five viscosity coefficients is given by Eq. (6.1.8). To obtain expressions of these coefficients explicitly, it would be helpful to use the fact that the tensor $e_{rlm} k_m$ on the left-hand side of the equations transforms the tensors $W_{lm}^{(1)}$, $W_{lm}^{(2)}$ and $W_{lm}^{(3)}$, $W_{lm}^{(4)}$ into each other so that the relationships are valid [2]

$$\left\{e_{rlm} W_{lk}^{(0)} k_m\right\} = 0, \qquad \left\{e_{rlm} W_{lk}^{(1)} k_m\right\} = -2W_{kr}^{(3)},$$
$$\left\{e_{rlm} W_{lk}^{(2)} k_m\right\} = -W_{kr}^{(4)}, \qquad \left\{e_{rlm} W_{lk}^{(3)} k_m\right\} = 2W_{kr}^{(1)},$$
$$\left\{e_{rlm} W_{lk}^{(4)} k_m\right\} = W_{kr}^{(2)}.$$

With the help of these relations, the system of Eqs. (8.1.6) is subdivided into three independent scalar systems of equations to define the viscosity coefficients.

The ion viscosity is found from the equations which are fully separated from the equations for electrons. In doing so, the three systems of equations can be conveniently represented in the unified complex form

$$p_i \tau_{ii} = (\gamma_{11} - im\beta_i) X_i + \gamma_{12} Y_i,$$
$$0 = \gamma_{21} X_i + (\gamma_{22} - im\beta_i) Y_i, \tag{8.2.14}$$

where the coefficients γ_{mn} for the plasma with a single species of ions are merely numbers

$$\gamma_{11} = \frac{3}{5}, \quad \gamma_{12} = -\frac{9}{70}, \quad \gamma_{21} = -\frac{9}{20}, \quad \gamma_{22} = \frac{41}{56}.$$

The values of m, X_i, and Y_i change their magnitudes depending on the calculated coefficient $\eta^{(p)}$. Thus, to calculate $\eta^{(0)}$ it is necessary to substitute $m = 0$, where $X_i = \eta_i^{(0)}$. The coefficients $\eta_i^{(2)}$ and $\eta_i^{(4)}$ can be obtained taking $m = 1$ and $X_i = \eta_i^{(1)} + i\eta_i^{(3)}$, whereas the coefficients $\eta_i^{(1)}$ and $\eta_i^{(3)}$ are found by substituting $m = 2$ and $X_i = \eta_i^{(2)} + i\eta_i^{(4)}$. The general solution of the system (8.2.14) is written as

$$X_i = p_i \tau_{ii} \frac{0.28 + 0.6 m^2 \beta_i^2 + i m \beta_i \left(0.6 + m^2 \beta_i^2\right)}{0.146 + m^2 \beta_i^2 + m^4 \beta_i^4}.$$

Using the appropriate values of m and X_i, the expressions for all coefficients of ion viscosity can be obtained

$$\eta_i^{(0)} = 1.92 p_i \tau_{ii}, \quad \eta_i^{(2)} = p_i \tau_{ii} \frac{0.28 + 0.6\beta_i^2}{\delta_i},$$
$$\eta_i^{(4)} = p_i \tau_{ii} \beta_i \frac{0.6 + \beta_i^2}{\delta_i}, \quad \eta_i^{(1)} = \eta_i^{(2)} \left[2\beta_i\right], \quad \eta_i^{(3)} = \eta_i^{(4)} \left[2\beta_i\right],$$
$$\delta_i = 0.146 + \beta_i^2 + \beta_i^4. \tag{8.2.15}$$

The electron viscosity coefficients are calculated similarly. It should be pointed out that at $\beta_i \geq 1$ the contribution of small cross terms related to the ion viscosity becomes essential and, hence, the system of equations analogous to (8.2.14) is of the kind

$$\begin{aligned} p_e \tau_{ei} + \frac{4}{5} \frac{\rho_e}{\rho_i} X_i &= \left(\gamma_{11}^e + i m \beta_e\right) X_e + \gamma_{12}^e Y_e, \\ -\frac{12}{5} \frac{\rho_e}{\rho_i} X_i &= \gamma_{21}^e X_e + \left(\gamma_{22}^e + i m \beta_e\right) Y_e, \end{aligned} \tag{8.2.16}$$

where the coefficients γ_{mn}^e are the functions of the ion charge number

$$\gamma_{11}^e = \frac{3\sqrt{2}}{5Z} + \frac{6}{5}, \quad \gamma_{21}^e = 7\gamma_{12}^e = -\left(\frac{9\sqrt{2}}{20Z} + \frac{9}{5}\right),$$

$$\gamma_{22}^e = \frac{41\sqrt{2}}{56Z} + \frac{51}{35}.$$

As to X_e and m, all the remarks are valid which are pertinent to the definition of the coefficients of ion viscosity. The effect of cross ion terms apart, the solution of Eqs. (8.2.16) can be given as

$$X_e = p_e\tau_{ei}\frac{\left[s_1 s_0 + s_2 m^2\beta_e^2 - im\beta_e\left(s_3 + m^2\beta_e^2\right)\right]}{\delta_e^{(m)}},$$

$$\delta_e^{(m)} = s_0^2 + s_4 m^2\beta_e^2 + m^4\beta_e^4,$$

where

$$s_0 = 0.82 + 1.82Z^{-1} + 0.72Z^{-2}, \quad s_1 = 1.46 + 1.04Z^{-1},$$

$$s_2 = 1.20 + 0.85Z^{-1}, \quad s_3 = 3.05 + 3.70Z^{-1} + 1.17Z^{-2},$$

$$s_4 = 5.32 + 6.36Z^{-1} + 2.02Z^{-2}.$$

Ultimately, for the coefficients of electron viscosity we have

$$\eta_e^{(0)} = p_e\tau_{ei}\frac{s_1}{s_0}, \qquad \eta_e^{(2)} = p_e\tau_{ei}\frac{s_1 s_0 + s_2\beta_e^2}{\delta_e},$$

$$\eta_e^{(4)} = -\,p_e\tau_{ei}\beta_e\frac{s_3 + \beta_e^2}{\delta_e}, \tag{8.2.17}$$

$$\eta_e^{(1)} = \eta_e^{(2)}\left[2\beta_e\right], \qquad \eta_e^{(3)} = \eta_e^{(4)}\left[2\beta_e\right],$$

$$\delta_e = s_0^2 + s_4\beta_e^2 + \beta_e^4$$

Note that at $Z = 1$ the expressions (8.2.17) as well as (8.2.15) for the coefficients of ion viscosity agree with the results [2]. Corrections to the electron viscosity due to X_i in Eqs. (8.2.16) are essential only for the coefficients $\eta_e^{(1)}$and $\eta_e^{(2)}$ provided that $\beta_i \geq 1$ or $\beta e \geq (m_i/m_e)^{1/2}$.

The general expressions for the corrections are of the kind

$$\Delta_{\eta_e}^{(2)} = \frac{4}{5}\frac{\rho_e}{\rho_i}\frac{\beta_e^3}{\delta_e}\eta_i^{(1)}, \quad \Delta_{\eta_e}^{(1)} = \Delta_{\eta_e}^{(2)}\left[2\beta_e, 2\beta_i\right].$$

At $\beta_e \gg 1$ and $\beta_i \sim 1$ the corrections are of the same order of value $(-p_e/\omega^2\tau_{ei})$ as the coefficients $\eta_e^{(1)}$ and $\eta_e^{(2)}$ proper. If $\beta_i \gg 1$, their contribution to the coefficients of electron viscosity is described by (7.1.7) obtained earlier when the $13N$-moment approximation used.

Finally, we should make the following conclusions: the requirement to use additional equations for the moments $\mathbf{r}_\alpha$ and $\sigma_{\alpha rs}$ arises mainly due to the inadequate accuracy in the calculation of longitudinal transport coefficients along with the transverse coefficients unless the plasma is strongly magnetized. If $\omega\tau \gg 1$, it is not necessary to account for higher approximations when calculating the transport coefficients across the magnetic field which means that the approximation of $13N$-moment used in two previous chapters works satisfactory when defining the transverse transport properties in a magnetized plasma. In particular, for a two-component fully ionized plasma the expressions derived in this section for the transverse transport coefficients at $\beta_e \gg 1$, $\beta_i \gg 1$ are transformed into the expressions obtained earlier for a partially ionized gas (see Chapter 7) if they are written in the limit of a fully ionized gas ($\alpha = 1$).

8.3 Transverse transport in a magnetized plasma

In a magnetized plasma, for which the condition $\omega\tau \gg 1$ is assumed to be satisfied for particles of all species, the transverse constituents of the diffusion velocities, heat fluxes and viscous-stress tensors can be defined from the general systems of equations (8.1.3) and (8.1.6) by expanding over $(\omega\tau)^{-1}$. Thus, for $\mathbf{w}_{\alpha Z\perp}$ and $\mathbf{h}_{\alpha Z\perp}$ in the zero approximation and neglecting collisions, we have

$$\mathbf{w}^{(0)}_{\alpha Z\perp} = \frac{1}{B}\mathbf{E}' \times \mathbf{k} + \frac{1}{\rho_{\alpha Z}\omega_{\alpha Z}}\mathbf{k} \times \nabla p_{\alpha Z} + \frac{1}{\omega_{\alpha Z}}\mathbf{k}\times\frac{d\mathbf{u}}{dt},$$
$$\mathbf{h}^{(0)}_{\alpha Z\perp} = \frac{5}{2}k\frac{p_{\alpha Z}}{m_\alpha\omega_{\alpha Z}}\mathbf{k} \times \nabla T_{\alpha Z}, \quad \mathbf{r}^{(0)}_{\alpha Z\perp} = 0. \tag{8.3.1}$$

The diffusion and heat fluxes appear in the first approximation with respect to $(\omega\tau)^{-1}$ and substituting the expressions (8.3.1) into the right-hand side of Eqs. (8.1.3). In this case we should remark that the account of the moment $\mathbf{r}_{\alpha Z}$ in the expansion, as in the magnetized plasma $\mathbf{r}^{(0)}_{\alpha Z\perp} \to 0$, does not lead to any improvements as compared with the $13N$-moment approximation. This reasoning explains why the values of the transverse fluxes can be deduced from the general expressions considered in Section 3.6 though formally the separating various charge states were not performed. To avoid cumbersome expressions, we shall confine ourselves to considering plasmas with nearly equal temperatures of heavy particles as was already mentioned earlier. Neglecting the forces of non-electromagnetic nature for simplicity, for $\mathbf{w}_{\alpha Z\perp}$ and $\mathbf{h}_{\alpha Z\perp}$ in the first approximation we have [1]

$$\mathbf{w}^{(1)}_{\alpha Z\perp} = -\frac{Z}{m^2_\alpha\omega^2_\alpha}\sum_\beta \mu_{\alpha\beta}\tau^{-1}_{\alpha\beta}\left\{\frac{\nabla_\perp p_{\alpha Z}}{Zn_{\alpha Z}} - \frac{kT}{n_\beta\overline{Z}^2_\beta}\right.$$
$$\left.\times\left[\nabla_\perp\left(n_\beta\overline{Z}_\beta\right) + \frac{n_\beta\overline{Z}_\beta}{T}\nabla_\perp T^{(1)}_\beta\right]\right.$$

$$+\left(\frac{m_\alpha}{Z}-\frac{m_\beta\overline{Z}_\beta}{\overline{Z_\beta^2}}\right)\frac{d\mathbf{u}_\perp}{dt}-\frac{3}{2}\mu_{\alpha\beta}k\left(\frac{\nabla_\perp T_{\alpha Z}}{Zm_\alpha}\right.$$
$$\left.\left.-\frac{\overline{Z}_\beta}{\overline{Z_\beta^2}m_\beta}\nabla_\perp T_\beta^{(1)}\right)\right\}, \tag{8.3.2}$$

$$\begin{aligned}\mathbf{h}_{\alpha Z\perp}^{(1)} &= \frac{3}{2}\frac{Zp_{\alpha Z}}{m_\alpha^2\omega_\alpha^2}\sum_\beta \mu_{\alpha\beta}^2\tau_{\alpha\beta}^{-1}\left\{\frac{\nabla_\perp p_{\alpha Z}}{Zn_{\alpha Z}}\right.\\ &-\frac{kT}{n_\beta\overline{Z_\beta^2}}\left[\nabla_\perp\left(n_\beta\overline{Z}_\beta\right)+\frac{n_\beta\overline{Z}_\beta}{T}\nabla_\perp T_\beta^{(1)}\right]\\ &\left.+\left(\frac{m_\alpha}{Z}-\frac{m_\beta\overline{Z}_\beta}{\overline{Z_\beta^2}}\right)\frac{d\mathbf{u}_\perp}{dt}\right\}-\frac{2p_{\alpha Z}k}{m_\alpha^2\omega_\alpha^2}\sum_\beta\frac{\mu_{\alpha\beta}^2}{m_\alpha+m_\beta}\\ &\times\tau_{\alpha\beta}^{-1}\left[\left(\frac{13}{8}\frac{m_\beta}{m_\alpha}+2+\frac{15}{2}\frac{m_\alpha}{m_\beta}\right)\nabla_\perp T_{\alpha Z}^{(1)}\right.\\ &\left.-\frac{27}{8}\frac{Z\overline{Z}_\beta}{\overline{Z_\beta^2}}\nabla_\perp T_\beta^{(1)}\right].\end{aligned} \tag{8.3.3}$$

Here $\omega_{\alpha Z}=eZ\left|\mathbf{B}\right|/m_\alpha$, T is the plasma temperature (whenever $T_{\alpha Z}$ is not under the gradient sign, it is taken to equal T). For convenience in Eqs. (8.3.2) and (8.3.3) the summation over the charge states of ions has been performed, therefore the sums in the given expressions imply the species of particles different in masses. When summing up, the following notations are introduced

$$\overline{Z_\alpha^k}=\sum_Z\frac{n_{\alpha Z}Z^k}{n_\alpha},\quad n_\alpha=\sum_Z n_{\alpha Z},\quad p_\alpha=\sum_Z p_{\alpha Z},$$

$$\tau_{\alpha\beta}^{-1}=\sum_{Z,\zeta}\frac{n_{\alpha Z}\tau_{\alpha Z\beta\zeta}^{-1}}{n_\alpha},\quad kT=\sum_\alpha\frac{p_\alpha}{n},$$

$$\nabla T_\alpha^{(k)}=\sum_Z\frac{n_{\alpha Z}Z^k\nabla T_{\alpha Z}}{n_\alpha\overline{Z_\alpha^k}},\quad \omega_\alpha^2=\frac{e^2\overline{Z_\alpha^2}B^2}{m_\alpha^2}.$$

It is appropriate to use the expressions of the transverse friction force $\mathbf{R}_{\alpha Z\perp}$ written in terms of diffusion velocities and temperature gradient

$$\begin{aligned}\mathbf{R}_{\alpha Z\perp} &= -\frac{n_{\alpha Z}Z^2}{\overline{Z_\alpha^2}}\sum_\beta \mu_{\alpha\beta}\tau_{\alpha\beta}^{-1}\Bigg[\mathbf{w}_{\alpha Z\perp} - \overline{\mathbf{w}}_{\beta\perp} - \frac{3}{2}\frac{k\mu_{\alpha\beta}}{eB} \\ &\times\left(\frac{1}{Zm_\alpha}\mathbf{k}\times\nabla_\perp T_{\alpha Z} - \frac{\overline{Z}_\beta}{m_\beta\overline{Z_\beta^2}}\mathbf{k}\times\nabla_\perp T_\beta^{(1)}\right)\Bigg],\end{aligned} \tag{8.3.4}$$

where

$$\overline{\mathbf{w}}_{\beta\perp} = \sum_Z \frac{n_{\beta Z}\mathbf{w}_{\beta Z\perp}Z^2}{n_\beta \overline{Z_\beta^2}}.$$

Now let us derive the expressions of the current density $\mathbf{j}_\perp = \sum\limits_{\alpha,Z} n_{\alpha Z}Ze\mathbf{w}_{\alpha Z\perp}$ in a magnetized plasma. Summing up the diffusion velocities of the species with the weight $n_{\alpha Z}Ze$ (for electrons, $Z = -1$), we get

$$\mathbf{j}_\perp = \frac{1}{B}\left(\mathbf{k}\times\nabla p + \rho\mathbf{k}\times\frac{d\mathbf{u}}{dt}\right), \tag{8.3.5}$$

where

$$p = \sum_\alpha p_\alpha, \qquad \rho = \sum_\alpha \rho_\alpha.$$

Using the same procedure which was adopted in Section 3.6, $d\mathbf{u}/dt$ can be expressed in terms of $\mathbf{E}'$ if the evident condition $\sum\limits_{\alpha,Z}\rho_{\alpha Z}\mathbf{w}_{\alpha Z\perp} = 0$ is applied and $\mathbf{w}_{\alpha Z\perp} = \mathbf{w}_{\alpha Z\perp}^{(0)} + \mathbf{w}_{\alpha Z\perp}^{(1)}$ is substituted into it. The ultimate expression for $\mathbf{j}_\perp$ is

$$\begin{aligned}\mathbf{j}_\perp &= \frac{1}{B}\left[\mathbf{k}\times\nabla(p - \langle p\rangle) + \rho e\left\langle\frac{m}{Z}\right\rangle^{-1}\mathbf{k}\times\mathbf{E}'\right] + \frac{1}{eB}\left\langle\frac{m}{Z}\right\rangle^{-1} \\ &\times\sum_{\alpha,Z}\frac{\rho_\alpha\mu_{\alpha\beta}\overline{Z}_\alpha}{\overline{Z_\alpha^2}}\tau_{\alpha\beta}^{-1}\Bigg[\frac{\nabla_\perp p_\alpha}{n_\alpha\overline{Z}_\alpha} - \frac{kT}{n_\beta\overline{Z_\beta^2}}\Bigg(\nabla_\perp n_\beta\overline{Z}_\beta \\ &+ \frac{n_\beta\overline{Z}_\beta}{T}\nabla_\perp T_\beta^{(1)}\Bigg) + \left(\frac{m_\alpha}{\overline{Z}_\alpha} - \frac{m_\beta\overline{Z}_\beta}{\overline{Z_\beta^2}}\right)\Bigg(\left\langle\frac{m}{Z}\right\rangle^{-1}e\mathbf{E}_\perp' \\ &- \frac{1}{\rho}\nabla_\perp\langle p\rangle\Bigg) - \frac{3}{2}k\mu_{\alpha\beta}\left(\frac{\nabla_\perp T_\alpha}{m_\alpha\overline{Z}_\alpha} - \frac{\overline{Z}_\beta}{m_\beta\overline{Z_\beta^2}}\nabla_\perp T_\beta^{(1)}\right)\Bigg],\end{aligned} \tag{8.3.6}$$

where

$$\left\langle \frac{m}{Z} \right\rangle = \sum_{\alpha,Z} \frac{m_\alpha \rho_{\alpha Z}}{\rho Z}, \qquad \langle p \rangle = \sum_{\alpha,Z} \frac{m_\alpha p_{\alpha Z}}{Z} \left\langle \frac{m}{Z} \right\rangle^{-1}.$$

The expression (8.3.6) generalizes the formula (3.6.6) for the case of different charge states of ions of every species. In addition, it takes into account the effect of thermal diffusion on the transverse current in plasma.

Not let us define the transverse ion thermal conductivity. Consider the case when the plasma comprises electrons, main ions of $Z_i = 1$ and impurities of charge number Z_I. Supposing $m_I \gg m_i$ and $\nabla_\perp T_i = \nabla_\perp T_I$, we obtain [1]

$$\lambda_\perp = \lambda_{i\perp} + \lambda_{I\perp} = \frac{k p_i}{m_i \omega_i^2 \tau_{ii}} \left\{ 1 + \frac{n_I Z_I^2}{n_i} \left[2.3 + \frac{n_I}{n_i} \left(\frac{m_I}{m_i} \right)^{1/2} \right] \right\}. \tag{8.3.7}$$

The coefficient in front of the brace brackets is the transverse ion thermal conductivity without impurity $\lambda_{i\perp}^{(0)}$, the term proportional to $(m_I/m_i)^{1/2}$ is an addition due to the thermal conductivity of heavy impurity ions and the rest is the correction to the thermal conductivity of the main plasma ions. It is convenient to express the transverse conductivity in terms of Z_{eff}

$$\lambda_\perp = \lambda_{i\perp}^{(0)} \left\{ 1 + (Z_{eff} - 1) \left[2.3 + \left(\frac{m_I}{m_i} \right)^{1/2} \frac{Z_{eff} - 1}{Z_I^2} \right] \right\}, \tag{8.3.7'}$$

with an implicit assumption of $Z_I \gg 1$.

In conclusion of the section, we shall treat the behavior of the viscosity coefficients in a magnetized plasma. When we are limited to the first approximation in expanding over $(\omega\tau)^{-1}$, the tensor $\sigma_{\alpha Z rs}$ will not appear in calculations and it will be sufficient to use the equations obtained in the $13N$-moment approximation. In the zero approximation over $(\omega\tau)^{-1}$ fully neglecting the collisions, we can find expressions for the coefficients $\eta_\alpha^{(3)}$ and $\eta_\alpha^{(4)}$. Substituting them into the right-hand side of Eqs. (8.1.6), we can obtain the expressions for $\eta_\alpha^{(1)}$ and $\eta_\alpha^{(2)}$ in the first approximation, the coefficients $\eta_\alpha^{(3)}$ and $\eta_\alpha^{(4)}$ being unchanged. Let us write out the values thus obtained for the transverse coefficients of viscosity

$$\eta_{\alpha Z}^{(4)} = 2\eta_{Z\alpha}^{(3)} = \frac{p_{\alpha Z}}{\omega_{\alpha Z}},$$

$$\eta_{\alpha Z}^{(2)} = 4\eta_{Z\alpha}^{(1)} = \frac{p_{\alpha Z}}{\omega_{\alpha Z}^2} \sum_\beta \kappa_{\alpha\beta} \tau_{\alpha\beta}^{-1} \left(\frac{6}{5} \frac{m_\beta}{m_\alpha} + 2 - \frac{4}{5} \frac{m_\beta}{m_\alpha} \frac{\bar{Z}_\beta Z}{\overline{Z_\beta^2}} \right). \quad (8.3.8)$$

Note that for the electron viscosity coefficient ($Z = -1$) the contribution of the last term in brackets of the expressions for $\eta_e^{(1)}$ and $\eta_e^{(2)}$ at $\beta \neq e$ is related to the influence of ion viscosity which we have discussed.

8.4 Longitudinal constituents of the friction forces and heat fluxes. Longitudinal viscosity

Expressions of the longitudinal particles and heat fluxes may be found by solving the system of equations (8.1.3) and the longitudinal constituent of the coefficients in the viscous-stress tensor by resolving (8.1.6) if $\omega_{\alpha Z} = 0$ is assumed in them. First let us consider the diffusion velocities and heat fluxes of individual species. Formally, to define $\mathbf{w}_{\alpha Z}$ and $\mathbf{h}_{\alpha Z}$ it is necessary to solve the total system of equations which comprises three equations (8.1.3) for each species. Then the diffusion velocities and heat fluxes appear to be expressed in terms of the gradients of the partial pressures and temperatures as well as the external force (in particular, the electric field E'). However, another solution is more convenient, namely, when the heat flux $\mathbf{h}_{\alpha Z}$ is represented in the form dependent on the temperature gradient and the differences of the diffusion velocities of species. With such an approach, defining $\mathbf{h}_{\alpha Z}$ and $\mathbf{r}_{\alpha Z}$ will require only the two last equations of the system (8.1.3) for each of the species. After substituting the values thus derived into the right-hand side of the first equations of the system, we can actually determine $\mathbf{R}_{\alpha Z\|}$ which corresponds to the longitudinal friction force for the particles of species α in the Z charge state. Then $\mathbf{R}_{\alpha Z\|}$ is also expressed in terms of the differences of the species velocities and temperature gradient. The diffusion velocities of the species proper are then found on the basis of the system of diffusion equations with the right-hand side determined in this way [1,4].

Even the system of Eqs. (8.1.3), which does not include the diffusion equations, in the general case of the plasma of electrons and an arbitrary number of ion species in various charge states turns out to be too cumbersome. One of the possible simplifications of this system requires separating equations for light particles (in particular, electrons) from it. We shall consider it later. Another way to simplify the system and reduce the number of equations is related to the equal mass of ions of the same species but differing ionization multiplicity. In the latter case the original equations can be summarized with respect to various charge states of the particles employing the fact that for the coefficients $G^{(n)}_{\alpha Z\beta\zeta}$ the following relations are valid

$$G^{(n)}_{\alpha Z\beta\zeta} = I_{\alpha Z} I_{\beta\zeta} \overline{G}^{(n)}_{\alpha\beta},$$

where

$$I_{\alpha Z} = \frac{n_{\alpha Z} Z^2}{n_\alpha \overline{Z^2_\alpha}}, \qquad \overline{G}^{(n)}_{\alpha\beta} = \sum_{Z,\zeta} G^{(n)}_{\alpha Z\beta\zeta}.$$

Taking such relations into account, Eqs. (8.1.3′) and (8.1.3″) in the case of interest ($\omega_{\alpha Z} = 0$) after summation over ζ can be represented as

$$\frac{5}{2} n_{\alpha Z} k \nabla T_{\alpha Z} = I_{\alpha Z} \sum_\beta \left[\frac{5}{2} \frac{\mu_{\alpha\beta}}{m_\alpha} \overline{G}^{(2)}_{\alpha\beta} (\mathbf{w}_{\alpha Z} - \overline{\mathbf{w}}_\beta) + \overline{G}^{(5)}_{\alpha\beta} \frac{\mathbf{h}_{\alpha Z}}{p_{\alpha Z}} \right.$$

$$\left. + \overline{G}^{(6)}_{\alpha\beta} \frac{\overline{\mathbf{h}}_\beta}{p_\beta} + \frac{\mu_{\alpha\beta}}{kT} \left(\overline{G}^{(9)}_{\alpha\beta} \frac{\mathbf{r}_{\alpha Z}}{p_{\alpha Z}} + \overline{G}^{(10)}_{\alpha\beta} \frac{\overline{\mathbf{r}}_\beta}{p_\beta} \right) \right.,$$

$$0 = I_{\alpha Z} \sum_\beta \left[\frac{35}{2} \left(\frac{\mu_{\alpha\beta}}{m_\alpha} \right)^2 \overline{G}^{(8)}_{\alpha\beta} (\mathbf{w}_{\alpha Z} - \overline{\mathbf{w}}_\beta) + 7 \frac{\mu_{\alpha\beta}}{m_\alpha} \left(\overline{G}^{(9)}_{\alpha\beta} \frac{\mathbf{h}_{\alpha Z}}{p_{\alpha Z}} \right. \right.$$

$$\left. \left. + \overline{G}^{(10)}_{\alpha\beta} \frac{\overline{\mathbf{h}}_\beta}{p_\beta} \right) + \frac{m_\alpha}{kT} \overline{G}^{(11)}_{\alpha\beta} \frac{\mathbf{r}_{\alpha Z}}{p_{\alpha Z}} + \frac{m_\beta}{kT} \overline{G}^{(12)}_{\alpha\beta} \frac{\overline{\mathbf{r}}_\beta}{p_\beta} \right], \tag{8.4.1}$$

where

$$\overline{\mathbf{w}}_\beta = \sum_Z I_{\beta Z} \mathbf{w}_{\beta Z}, \qquad \overline{\mathbf{h}}_\beta = \sum_Z \frac{p_\beta I_{\beta Z} \mathbf{h}_{\beta Z}}{p_{\beta Z}},$$

$$\overline{\mathbf{r}}_\beta = \sum_Z \frac{p_\beta I_{\beta Z} \mathbf{r}_{\beta Z}}{p_{\beta Z}}.$$

After summation of (8.4.1) over Z, we arrive to the following system of equations

$$
\begin{aligned}
\frac{5}{2}n_\alpha k\nabla T_\alpha =&\sum_\beta \left[\frac{5}{2}\frac{\mu_{\alpha\beta}}{m_\alpha}\overline{G}^{(2)}_{\alpha\beta}\left(\bar{\mathbf{w}}_\alpha - \bar{\mathbf{w}}_\beta\right) + \overline{G}^{(5)}_{\alpha\beta}\frac{\bar{\mathbf{h}}_\alpha}{p_\alpha}\right.\\
&\left. + \overline{G}^{(8)}_{\alpha\beta}\frac{\bar{\mathbf{h}}_\beta}{p_\beta} + \frac{\mu_{\alpha\beta}}{kT}\left(\overline{G}^{(9)}_{\alpha\beta}\frac{\bar{\mathbf{r}}_\alpha}{p_\alpha} + \overline{G}^{(10)}_{\alpha\beta}\frac{\bar{\mathbf{r}}_\beta}{p_\beta}\right)\right],\\
0 =&\sum_\beta \left[\frac{35}{2}\left(\frac{\mu_{\alpha\beta}}{m_\alpha}\right)^2 \overline{G}^{(8)}_{\alpha\beta}\left(\bar{\mathbf{w}}_\alpha - \bar{\mathbf{w}}_\beta\right) + 7\frac{\mu_{\alpha\beta}}{m_\alpha}\left(\overline{G}^{(9)}_{\alpha\beta}\frac{\bar{\mathbf{h}}_\alpha}{p_{\alpha Z}}\right.\right.\\
&\left.\left. + \overline{G}^{(10)}_{\alpha\beta}\frac{\bar{\mathbf{h}}_\beta}{p_\beta}\right) + \frac{m_\alpha}{kT}\overline{G}^{(11)}_{\alpha\beta}\frac{\bar{\mathbf{r}}_\alpha}{p_\alpha} + \frac{m_\beta}{kT}\overline{G}^{(12)}_{\alpha\beta}\frac{\bar{\mathbf{r}}_\beta}{p_\beta}\right].
\end{aligned}
\tag{8.4.2}
$$

Equations (8.4.2) are analogous to (8.1.3′) and (8.1.3″) but the number of equations is considerably smaller corresponding only to the species with different masses of particles. Formally, solution of this system provides expressing $\bar{\mathbf{r}}_\alpha$ and $\bar{\mathbf{h}}_\alpha$ in terms of $\bar{\mathbf{w}}_\beta$ and $\nabla_{||}T_\alpha$ and, after substitutions into the first equation (8.1.3) which must also be summed up over Z and ζ, results in an equation which will define $\overline{\mathbf{w}}_\alpha$. Yet, calculation of the mean values of $\overline{\mathbf{w}}_\alpha$, $\bar{\mathbf{r}}_\alpha$, and $\bar{\mathbf{h}}_\alpha$ would not give the whole solution, since to describe the plasma behavior would demand the knowledge of diffusion velocities and heat fluxes for each of the ion charge states. To calculate these values, divide Eq. (8.4.1) by $I_{\alpha Z}$ and subtract them from the appropriate equations of the system (8.4.2). Finally,

$$
\begin{aligned}
\frac{5}{2}n_\alpha k\left(\nabla T_\alpha - \frac{\overline{Z^2_\alpha}}{Z^2_\alpha}\nabla T_{\alpha Z}\right) =& S^{(2)}_\alpha\left(\overline{\mathbf{w}}_\alpha - \mathbf{w}_{\alpha Z}\right)\\
+ S^{(5)}_\alpha\left(\frac{\bar{\mathbf{h}}_\alpha}{p_\alpha} - \frac{\mathbf{h}_{\alpha Z}}{p_{\alpha Z}}\right) &+ \frac{m_\alpha}{kT}S^{(9)}_\alpha\left(\frac{\bar{\mathbf{r}}_\alpha}{p_\alpha} - \frac{\mathbf{r}_{\alpha Z}}{p_{\alpha Z}}\right),\\
0 = S^{(8)}_\alpha\left(\bar{\mathbf{w}}_\alpha - \mathbf{w}_{\alpha Z}\right) &+ 7S^{(9)}_\alpha\left(\frac{\bar{\mathbf{h}}_\alpha}{p_\alpha} - \frac{\mathbf{h}_{\alpha Z}}{p_{\alpha Z}}\right)\\
+ \frac{m_\alpha}{kT}S^{(11)}_\alpha\left(\frac{\bar{\mathbf{r}}_\alpha}{p_\alpha} - \frac{\mathbf{r}_{\alpha Z}}{p_{\alpha Z}}\right)&,
\end{aligned}
\tag{8.4.3}
$$

where

$$
S^{(2)}_\alpha = \frac{5}{2}\sum_\beta \frac{\mu_{\alpha\beta}}{m_\alpha}\overline{G}^{(2)}_{\alpha\beta}, \qquad S^{(5)}_\alpha = \sum_\beta \overline{G}^{(5)}_{\alpha\beta},
$$

$$S_\alpha^{(8)} = \frac{35}{2} \sum_\beta \left(\frac{\mu_{\alpha\beta}}{m_\alpha} \right)^2 \overline{G}_{\alpha\beta}^{(8)},$$

$$S_\alpha^{(9)} = \sum_\beta \frac{\mu_{\alpha\beta}}{m_\alpha} \overline{G}_{\alpha\beta}^{(9)}, \quad S_\alpha^{(11)} = \sum_\beta \overline{G}_{\alpha\beta}^{(11)}.$$

The solution of (8.4.3) for $(\mathbf{h}_{\alpha Z}/p_{\alpha Z}) - (\bar{\mathbf{h}}_\alpha/p_\alpha)$ can be written as

$$\frac{\mathbf{h}_{\alpha Z}}{p_{\alpha Z}} - \frac{\bar{\mathbf{h}}_\alpha}{p_\alpha} = n_\alpha \tau_\alpha \tau_{\alpha\alpha}^{-1} c_\alpha^{(5)} \left(\frac{\overline{Z_\alpha^2}}{Z_\alpha^2} \nabla T_{\alpha Z} - \nabla T_\alpha \right) + c_\alpha^{(6)} \left(\mathbf{w}_{\alpha Z} - \overline{\mathbf{w}}_\alpha \right), \tag{8.4.4}$$

where $\tau_\alpha = \sum_\beta (\mu_{\alpha\beta}/m_\alpha) \tau_{\alpha\beta}^{-1}$ and

$$c_\alpha^{(5)} = \frac{5}{2} \frac{\tau_\alpha \tau_{\alpha\alpha}^{-1} S_\alpha^{(11)}}{D_\alpha}, \quad c_\alpha^{(6)} = \frac{S_\alpha^{(2)} S_\alpha^{(11)} - S_\alpha^{(8)} S_\alpha^{(9)}}{D_\alpha},$$

$$D_\alpha = S_\alpha^{(5)} S_\alpha^{(11)} - 7 \left(S_\alpha^{(9)} \right)^2.$$

The relations of (8.4.4) together with the solution of (8.4.2) for $\bar{\mathbf{h}}_\alpha$ allow us to present $\mathbf{h}_{\alpha Z}$ as a function of the velocity differences $\mathbf{w}_{\alpha Z} - \overline{\mathbf{w}}_\beta$ and temperature gradients. Substitution of the solutions obtained into the right-hand side of (8.1.3) gives the expression of the longitudinal constituent of the friction force $\mathbf{R}_{\alpha Z}$. The ultimate expressions for these values can be written as [1]

$$\begin{aligned} \mathbf{R}_{\alpha Z} = & -\frac{n_{\alpha Z} Z^2}{\overline{Z_\alpha^2}} \left\{ \sum_\beta \left[c_{\alpha\beta}^{(1)} \mu_{\alpha\beta} \tau_{\alpha\beta}^{-1} \left(\mathbf{w}_{\alpha Z} - \overline{\mathbf{w}}_\beta \right) \right. \right. \\ & \left. \left. - c_{\alpha\beta}^{(2)} \tau_\beta \tau_{\alpha\beta}^{-1} \frac{\mu_{\alpha\beta}}{m_\beta} k \nabla T_\beta \right] + \frac{\overline{Z_\alpha^2}}{Z^2} c_\alpha^{(5)} k \nabla T_{\alpha Z} \right\}, \end{aligned} \tag{8.4.5}$$

$$\begin{aligned} \mathbf{h}_{\alpha Z} = & \frac{p_{\alpha Z} \tau_\alpha}{m_\alpha} \left\{ \sum_\beta \left[c_{\beta\alpha}^{(2)} \mu_{\alpha\beta} \tau_{\alpha\beta}^{-1} \left(\mathbf{w}_{\alpha Z} - \overline{\mathbf{w}}_\beta \right) \right. \right. \\ & \left. \left. - c_{\beta\alpha}^{(3)} \tau_\beta \tau_{\alpha\beta}^{-1} \frac{\mu_{\alpha\beta}}{m_\beta} k \nabla T_\beta \right] - \frac{\overline{Z_\alpha^2}}{Z^2} c_\alpha^{(6)} k \nabla T_{\alpha Z} \right\}, \end{aligned} \tag{8.4.6}$$

The numeric coefficients $c_{\alpha\beta}^{(n)}$ which are part of these expressions are defined as a result of solving the system of equations for the average values of fluxes (8.4.2), with the coefficients $c_\alpha^{(n)}$ given previously. The coefficients $c_{\alpha\beta}^{(n)}$ and $c_\alpha^{(n)}$ are related by the following expressions:

$$c_{\alpha\beta}^{(1)} = c_{\beta\alpha}^{(1)}, \quad c_{\alpha\beta}^{(3)} = c_{\beta\alpha}^{(3)}, \quad \sum_\beta c_{\alpha\beta}^{(1)} \tau_{\alpha\beta}^{-1} = c_\alpha^{(4)} \tau_\alpha^{-1},$$

$$\sum_\beta c_{\alpha\beta}^{(2)} \tau_{\alpha\beta}^{-1} = c_\alpha^{(5)} \tau_\alpha^{-1},$$

where

$$c_\alpha^{(4)} = 1 - \frac{2}{5} \frac{\tau_\alpha \tau_{\alpha\alpha}^{-1}}{D_\alpha} \left[\left(S_\alpha^{(2)}\right)^2 S_\alpha^{(11)} - 2 S_\alpha^{(2)} S_\alpha^{(8)} S_\alpha^{(9)} + \frac{1}{7} \left(S_\alpha^{(8)}\right)^2 S_\alpha^{(9)} \right].$$

Thus, the problem can be reduced to defining the coefficients $c_{\alpha\beta}^{(n)}$ for the species of different masses. The general expressions for these values are rather cumbersome though can be expressed in terms of the ratios of respective determinants. If a plasma comprises the particles whose mass is considerably smaller than the mass of the other particles (one of such species, viz. electrons, is always available), the explicit form of coefficients $c_{\alpha\beta}^{(n)}$ and $c_\alpha^{(n)}$ is found by expansion over the small parameter m_k/m_α (index k refers to light particles). In this case the system of equations for light particles is separated from the rest of the system and its solution brings the following expressions for the coefficients of interest

$$\begin{aligned}
c_{k\alpha}^{(1)} &= (1 + 0.24 Z_k^*) \frac{1 + 0.93 Z_k^*}{\Delta_k}, \quad c_{k\alpha}^{(2)} = 0, \\
c_{\alpha k}^{(2)} &= 1.56 (1 + 1.4 Z_k^*) \frac{1 + 0.52 Z_k^*}{\Delta_k}, \quad c_{k\alpha}^{(3)} = 0, \\
c_k^{(5)} - 0.5 c_{kk}^{(2)} \tau_k \tau_{kk}^{-1} &= 2.2 Z_k^* \frac{1 + 0.52 Z_k^*}{\Delta_k}, \\
c_k^{(6)} + 0.5 c_{kk}^{(3)} \tau_k \tau_{kk}^{-1} &= 3.9 (1 + 1.4 Z_k^*) \frac{1 + 1.7 Z_k^*}{\Delta_k}, \\
\Delta_k = (1 + 2.65 Z_k^*)(1 + 0.29 Z_k^*), &\quad Z_k^* = \sum_\alpha \frac{n_\alpha \overline{Z_\alpha^2}}{n_k \overline{Z_k^2}}.
\end{aligned} \tag{8.4.7}$$

Summation is carried over all sorts of particles for which $m_\alpha \gg m_k$. When writing down the coefficients, it is assumed that the light particles are those with a singularly possible charge state, hence $c_{kk}^{(n)}$ and $c_k^{(n)}$ are included into (8.4.5) and (8.4.6) in certain combinations described by the formulae of (8.4.7).

In the case $k = e$ (the light particles are electrons) the expressions (8.4.5)–(8.4.7) define the electron–ion friction force and electron heat flux in a multispecies plasma which, if $Z_k^* = Z_{eff}$ is taken, are coincident with the expressions discussed in Section 8.2. Since separation of equations for light particles is also valid for $T_k \neq T$ (on condition that $T_k/m_k \gg T_\alpha/m_\alpha$); instead of T their own temperature can be used in the equations (in particular, T_e for electrons).

The coefficients $c_{\alpha\beta}^{(n)}$ and $c_\alpha^{(n)}$ for heavy particles are found from the systems (8.4.2) and (8.4.3) written without the light species. Its influence can be exhibited only in the friction force. In fact, besides adding the force $\mathbf{R}_{\alpha k}$ to the friction force of heavy particles, $\mathbf{R}_\alpha$, the consideration of the light species slightly changes the coefficient $c_{\alpha\beta}^{(1)}$

$$c_{\alpha\beta}^{(1)} = c_{\alpha\beta}^{(1)*} + 1.8\left(\frac{m_k}{\mu_{\alpha\beta}}\right)^{1/2}\frac{1+0.3Z_k^*}{\Delta_k},$$

where $c_{\alpha\beta}^{(1)*}$ is the coefficient without consideration of the light species.

It is evident that if after separating equations for the lightest species of the plasma a species arises for which $m_i \ll m_\alpha$ at $\alpha \neq i, k$ (e.g. hydrogen ions in the plasma with heavy impurities), then the process of separation should be continued. In doing so, the coefficients in (8.4.5) and (8.4.6) for the friction force and heat flux of the species i will be defined by the formulae (8.4.7). Calculating the coefficients of the other species whose masses will not differ greatly turns out to be rather cumbersome even for two species of particles. Therefore, the numerical calculation becomes more advantageous.

Now we shall pass to the problem of longitudinal viscosity of a multicomponent plasma. The procedure of calculating the coefficient of the longitudinal viscosity in terms of Eqs. (5.1.6) in which $\omega_{\alpha Z} = 0$ closely resembles the technique used to derive the expression of the longitudinal heat flux. Accordingly, the general expression for $\eta_{\alpha Z}^{(0)}$ can be written as

$$\eta_{\alpha Z}^{(0)} = p_\alpha \tau_\alpha \sum_\beta \left(\frac{\mu_{\alpha\beta}}{m_\alpha}\tau_{\alpha\beta}^{-1}\tau_\beta c_{\alpha\beta} + \frac{\overline{Z^2}}{Z^2}c_\alpha\right),$$

which means that the derivation of the longitudinal viscous coefficient for every charge ion state is reduced to calculating the coefficients $c_{\alpha\beta}$ and c_α for different species. The possibility to separate equations for lighter species allows us to calculate the coefficients independent of the others as previously. In this case, we have a relation

$$\frac{\tau_k}{2\tau_{kk}}c_{kk} + c_k = 0.96\frac{(1+1.4Z_k^*)^2}{(1+1.8Z_k^*)(1+0.6Z_k^*)}, \quad c_{k\alpha} = 0.$$

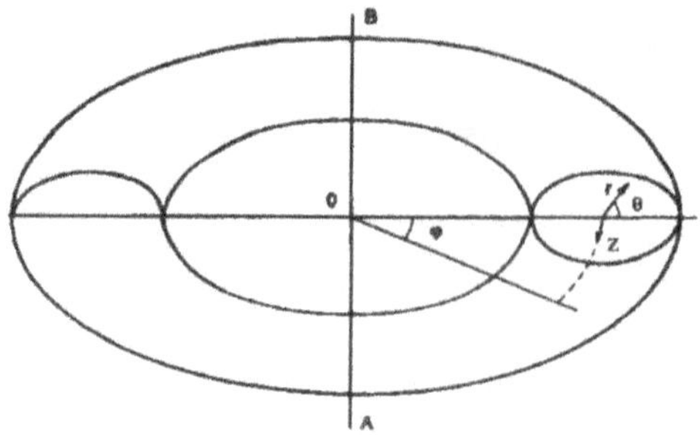

Figure 8.1. System of toroidal coordinates (r, θ, φ)

Provided only a single heavy species is kept after separating the equations for lighter particles,

$$c_{\alpha\alpha} = 0.167, \qquad c_{\alpha} = 0.793.$$

To determine the coefficients in the case of heavy species with slightly differing masses, it is appropriate to use numerical methods as we did earlier.

8.5 Diffusion and heat transfer in toroidal systems in Pfirsch–Schlüter regime

The particle and heat transport across a magnetic field in toroidal systems of magnetic plasma confinement (toroidal traps) essentially differs from the classical transport discussed in Section 8.3. The difference is due to the inhomogeneous and curved confining field which for the case of axial symmetrical geometry of a toroidal trap is represented by a set of magnetic field lines wound around the torus (see Fig. 8.1). The magnetic field in an axisymmetric tokamak-type confinement system can be described as

$$\mathbf{B} = \mathbf{B}_T + \mathbf{B}_\vartheta = \frac{B_0}{h}\left(\mathbf{e}_T + \Theta \mathbf{e}_\vartheta\right),$$

$$h = 1 + \varepsilon \cos\vartheta, \quad \varepsilon = \frac{r}{R} \ll 1.$$

Here $\mathbf{B}_T$ is the toroidal constituent of the magnetic field directed around the big axis of the torus and generated by the external currents, $\mathbf{B}_\vartheta$ is the poloidal field around the small axis and generated by the eigencurrents of plasma, $\Theta = B_\vartheta / B_T$ is the tangent angle between the direction **B** and the vector $\mathbf{e}_T$ parallel to the small axis of the torus, $\varepsilon = r/R$ is the ratio of the small and big radius of the magnetic surface. The value Θ does not usually exceed 0.1 and $\varepsilon \leq 0.2$ (for modern devices with the circular cross section [5]). An important parameter of the magnetic trap is the safety factor $q = \varepsilon/\Theta$, i.e. the number of revolutions of the magnetic force lines around the big toroidal axis per single revolution around the small axis. To stabilize most dangerous instabilities in plasma, it is necessary that the requirement $q > 1$ is met [6,7]. It is common knowledge that q is close to unity at the axis of the plasma column where the current density is maximal and grows when approaching the walls of the chamber to reach the value of about four or five [5].

In modern tokamaks, the plasma parameters are such that in central parts of the chamber the plasma particles move without collisions, i.e. the mean free path of the particles greatly exceeds the magnetic field connection length $L = qR$. It is evident that the transport relations obtained in the approximation $\lambda/L_0 \ll 1$ do not work when describing the phenomena in this area. However, in the vicinity of the chamber walls, a relatively cool layer of plasma arises in which the mean free path becomes less than qR at decreasing temperatures because of the abrupt increase of the average cross section of the Coulomb collisions. In the case of impurities as the mean free path of heavy multicharge impurity ions is considerably less than the mean free path of electrons and main ions of plasma, the condition $\lambda \ll qR$, corresponding to the so called Pfirsch–Schlüter regime [8], can be realized up to the central areas of the discharge.

The Pfirsch–Schlüter fluxes for a simple electron–ion plasma were historically the first fluxes to be calculated in a toroidal configuration. Galeev and Sagdeev showed [9] that in a rarefied plasma (at very small collision frequencies) the trapped particles with "banana" orbits are responsible for a significant enhancement of the diffusion coefficient and thermal conductivity. They also demonstrated the existence of an intermediate collision frequency regime in which the diffusion coefficients are independent of collision frequency (the "plateau regime").

Below, we will consider neoclassical diffusion and heat transfer in a multispecies plasma of toroidal traps in the Pfirsch–Schlüter regime when the use of hydrodynamic transport equations obtained in the previous chapters is possible. An important peculiarity of the transport phenomena in such conditions is the fact that the transport of particles and the heat transfer across the magnetic field is defined in terms of longitudinal components of the friction force and heat flux. Without going into detail (see [7–12]), we shall mention the quantitative considerations alone. The point is that the longitudinal transport of particles along the magnetic force lines must occur to compensate for the charge separation in the inhomogeneous toroidal magnetic field because of the gradient and centrifugal drifts of particles [11]. Such fluxes of particles are ensured by the constituent of the electrical field **E** along **B** which produces the force balancing the longitudinal friction force between the particles. The difference of drift velocities $\mathbf{E} \times \mathbf{k}/B$ at the internal and external lines of the torus due to the decreasing field **B** at the external profile creates a flux of plasma particles across the magnetic field which after averaging over the magnetic surface turns out to be q^2 times more than the flux due to the classical diffusion of particles across the field [8–10]. The growth of the heat fluxes across the magnetic field q^2 times is explained similarly but for the temperature gradient which acts as a space charge and which in the stationary state is compensated for by longitudinal heat fluxes.

Below, we give a procedure [13–15] in which the original equations for defining the neoclassical transport across a magnetic field are the equations of hydrodynamic equilibrium of every plasma species following from Eq. (8.1.3) when $d\mathbf{u}/dt = 0$ is assumed. In this case,

$$\nabla p_{\alpha Z} - n_{\alpha Z} eZ \left(\mathbf{E} + \mathbf{u}_\alpha \times \mathbf{B}\right) = \mathbf{R}_{\alpha Z}. \tag{8.5.1}$$

Equations (8.5.1) should be complemented by the requirement of the lack of particles and heat accumulation

$$\nabla \cdot \left(n_{\alpha Z} \mathbf{u}_{\alpha Z}\right) = 0, \qquad \nabla \cdot \mathbf{h}_{\alpha Z} = Q_{\alpha Z}, \tag{8.5.2}$$

where $Q_{\alpha Z}$ is the averaged heat transfer of particles of species α in the Z charge state resulting from collisions with other particles.

The radial flux of particles can be found from the poloidal component of the equilibrium equations (8.5.1)

$$n_{\alpha Z} \left(u_{\alpha Z}\right)_r = \frac{h}{eZB_0} \left(- R_{\alpha Z \perp} + \frac{\partial p_{\alpha Z}}{r \partial \vartheta} - n_{\alpha Z} eZE_\vartheta \right). \tag{8.5.3}$$

Here and below only the terms of the zero and first orders with respect to ε parameter are retained.

The requirement of the plasma species equilibrium across the magnetic field

$$\Theta \left(\frac{\partial p_{\alpha Z}}{r \partial \vartheta} - n_{\alpha Z} eZE_\vartheta \right) = R_{\alpha Z \parallel}$$

allows us to represent the averaged flux of particles across the magnetic surface $\Gamma_{\alpha Z}$ as [15]

$$\Gamma_{\alpha Z} = \langle h n_{\alpha Z} (u_{\alpha Z})_r \rangle = \frac{1}{eZB_0} \left\langle h^2 \left(\frac{R_{\alpha Z\|}}{\Theta} - R_{\alpha Z\perp} \right) \right\rangle, \qquad (8.5.4)$$

where $\langle A \rangle = \int A d\vartheta / 2\pi$, and the multiplier h in averaging appears due to toroidal geometry and dependence of B on r and ϑ. It is to be noted that the formula (8.5.4) ensures ambipolarity of the fluxes ($\sum_{\alpha,Z} \Gamma_{\alpha Z} = 0$ as $\sum_{\alpha,Z} R_{\alpha Z} = 0$).

The part of the flux correspondent to the transverse component of the friction force $\mathbf{R}_{\alpha Z\perp}$ defines the common classical transport of particles across the magnetic field. It can be calculated if the expression for $\mathbf{R}_{\alpha Z\perp}$ (8.3.4) is taken into account. Yet, as the neoclassical toroidal transport is q^2 times more than the classical one and as $q^2 \geq 10$ in the area of where the Pfirsch–Schlüter regime is usually realized, only the part of the flux which is associated with $\mathbf{R}_{\alpha Z\|}$ will be considered.

To calculate the radial drift following Eq. (8.5.4), it is necessary to define $\mathbf{R}_{\alpha Z\|}$. To this end, we will use expression (8.4.5) in which the longitudinal velocities $\mathbf{u}_{\alpha Z}$ and temperature gradients should be expressed in terms of the transverse gradients of temperature and density with the help of Eqs. (8.5.2). Indeed, the first of the equations written in toroidal coordinates looks as

$$\frac{1}{rh} \frac{\partial}{\partial \vartheta} \left\{ h \left[n_{\alpha Z} \left(u_{\alpha Z\perp} + \Theta u_{\alpha Z\|} \right) \right] \right\} = 0, \qquad (8.5.5)$$

where

$$u_{\alpha Z\perp} = -\frac{h}{eB_0 n_{\alpha Z} Z} \frac{\partial p_{\alpha Z}}{\partial r}.$$

The terms arising because of the derivatives along the toroidal coordinate are not considered because of the azimuthal symmetry and the derivatives $\partial/\partial r$ describing the radial flux of the particles which is essential only if $\omega\tau\Theta^2 \leq 1$ are also omitted. If the dependence of h on ϑ is taken into account, Eq. (8.5.5) permits us to find the longitudinal constituent of the species velocity

$$u_{\alpha Z\|} = \frac{2q\cos\vartheta}{eB_0 n_{\alpha Z}} \frac{\partial p_{\alpha Z}}{\partial r}. \qquad (8.5.6)$$

The longitudinal gradient of temperatures cannot be found that simply because the right-hand side of the second equation in the relations (8.5.2) is not equal to zero. In toroidal coordinates at the same assumptions as Eq. (8.5.5), the equation is written as

$$\frac{1}{rh}\frac{\partial}{\partial \vartheta}\left[h\left(-\frac{5}{2}\frac{k^2T}{eB_0}h\frac{\partial T}{\partial r}+\Theta h_{\alpha Z\|}\right)\right]=Q_{\alpha Z}, \tag{8.5.7}$$

where $h_{\alpha Z\|}$ is defined by Eq. (8.4.6).

Equation (8.5.7) is conveniently solved by substituting the temperature of each species in the form

$$T_{\alpha Z}=T(r)+T^{(1)}_{\alpha Z}(r)\sin\vartheta,$$

where $T^{(1)}_{\alpha Z}(r)$ is a small correction to the general temperature of the plasma $T(r)$ [16]. Then after substitution of $u_{\alpha Z\|}$ (8.5.6) into $h_{\alpha Z\|}$ (8.4.6) we have instead of Eq. (8.5.7) a system of equations for $T^{(1)}_{\alpha Z}$. To avoid cumbersome formulae, we shall put the temperatures of various charged states of ions of a certain species equal, i.e. $T^{(1)}_{\alpha Z}\equiv T^{(1)}_{\alpha}$. In fact, it can be justified by solving the whole system of equations for $T^{(1)}_{\alpha Z}$ and using the equality of the mass of the particles of a certain species and the requirement $\lambda \ll qr$. Besides, we shall assume approximately that $\overline{Z^k_\alpha}=\left(\overline{Z}_\alpha\right)^k$, and the averaging line over Z_α will be omitted. As a result, we come to the following system of equations for the corrections $T^{(1)}_\alpha$:

$$\begin{aligned}
&\frac{2}{eB_0R}\left[\frac{5}{2}\frac{k}{Z_\alpha}\frac{\partial T}{\partial r}+\sum_\beta\frac{\mu_{\alpha\beta}}{m_\alpha}\tau_\alpha\tau^{-1}_{\alpha\beta}c^{(2)}_{\beta\alpha}\right.\\
&\left.\times\left(\frac{1}{n_\beta Z^2_\beta}\frac{\partial(p_\beta Z_\beta)}{\partial r}-\frac{1}{n_\alpha}\frac{\partial p_\alpha Z^{-1}_\alpha}{\partial r}\right)\right]\\
&=\sum_\beta\frac{\tau^{-1}_{\alpha\beta}}{m_\alpha+m_\beta}\left(3\mu_{\alpha\beta}\frac{T^{(1)}_\beta-T^{(1)}_\alpha}{T}-\frac{\Theta^2}{r^2}\tau_\alpha\tau_\beta c^{(3)}_{\alpha\beta}T^{(1)}_\beta\right)\\
&-\frac{\Theta^2}{r^2}\frac{c^{(6)}_\alpha\tau_\alpha}{m_\alpha}T^{(1)}_\alpha.
\end{aligned} \tag{8.5.8}$$

The general solution of the system expressed in terms of determinants is rather clumsy and is not subject to a simple analysis; therefore, we intend to keep the dependence on $T^{(1)}_\alpha$ in the final result. The flux of particles of species α in the charged state Z across the magnetic surface with allowance made for Eqs. (8.4.5), (8.5.4), and (8.5.6) looks in this case as

$$\Gamma_{\alpha Z}=-\frac{\rho^2_{B\alpha}q^2}{kT}n_{\alpha Z}Z\left[\sum_\beta\frac{\mu_{\alpha\beta}}{m_\alpha}\tau^{-1}_{\alpha\beta}c^{(1)}_{\alpha\beta}\left(\frac{1}{Zn_{\alpha Z}}\frac{\partial p_{\alpha Z}}{\partial r}\right.\right.$$

$$
-\frac{1}{n_\beta Z_\beta^2}\frac{\partial p_\beta Z_\beta}{\partial r}\Big) + \frac{eB_0 k}{2q^2 m_\alpha R}\Big(\sum_\beta \frac{\mu_{\alpha\beta} n_\beta}{m_\alpha n_\alpha}
$$

$$
\times \tau_\beta \tau_{\beta\alpha}^{-1} c_{\alpha\beta}^{(2)} T_\beta^{(1)} - c_\alpha^{(5)} T_\alpha^{(1)}\Big)\Big], \tag{8.5.9}
$$

where

$$
\rho_{B\alpha}^2 = \frac{2kT}{m_\alpha \omega_\alpha^2}.
$$

The radial heat flux is directly expressed by means of the poloidal temperature gradient

$$
H_\alpha = \frac{5}{2}\frac{p_\alpha k}{eB_0 Z_\alpha}\left\langle h^2 \frac{\partial T_\alpha}{r\partial\vartheta}\right\rangle = \frac{5}{2}\frac{p_\alpha k}{eB_0 Z_\alpha R} T_\alpha^{(1)}. \tag{8.5.10}
$$

Thus, the final determination of the radial particle fluxes and heat fluxes in the Pfirsch–Schlüter regime is reduced to solving the system of linear algebraic equations (8.5.8). As the analysis is too complicated for the general case, we shall confine ourselves to some special cases which anyway allow us to reveal the main physical results.

First let us analyse the diffusion and heat transport in a simple plasma of electrons and single species of Z charge ions. The system (8.5.8) for corrections of temperature amounts to the following system

$$
\frac{2}{eB_0 R}\left(-\frac{5}{2}k\frac{\partial T}{\partial r} + \frac{a_1}{n_e}\frac{\partial p}{\partial r}\right) = -\frac{\Theta^2}{r^2}\frac{kT_e^{(1)}}{m_e} a_2 \tau_e + 3\frac{m_e}{m_i}\tau_{ei}^{-1}\frac{T_i^{(1)} - T_e^{(1)}}{T},
$$

$$
\frac{2}{eB_0 R}\left(\frac{5}{2}\frac{k}{Z}\frac{\partial T}{\partial r}\right) = -\frac{\Theta^2}{r^2}\frac{kT_i^{(1)}}{m_i} a_3 \tau_i + 3\frac{m_e}{m_i}\tau_{ie}^{-1}\frac{T_e^{(1)} - T_i^{(1)}}{T}, \tag{8.5.11}
$$

where $p = p_e + p_i$ and the coefficients a_n are equal to

$$
a_1 = \tau_e \tau_{ei}^{-1} c_{ie}^{(2)} = c_e^{(5)} - \frac{1}{2}\tau_e \tau_{ee}^{-1} c_{ee}^{(2)} = 2.2Z\,(1 + 0.52Z)\,\Delta^{-1},
$$

$$
a_2 = c_e^{(6)} + \tau_e \tau_{ee}^{-1} c_{ee}^{(3)} = 3.9\,(1 + 1.4Z)\,(1 + 1.7Z)\,\Delta^{-1}
$$

$$
a_3 = 3.9, \quad \Delta = 1 + 3Z + 0.75Z^2,
$$

$$
\tau_e = \tau_{ei}, \qquad \tau_i = 2\tau_{ii}.
$$

In the general case of arbitrary heat exchange between electrons and ions, the solution is represented as [15,16]

$$T_e^{(1)} = \frac{1}{1+s^2}\frac{B_e}{\lambda_{e\|}} + \frac{s^2}{1+s^2}\frac{B_e+B_i}{\lambda_{e\|}+\lambda_{i\|}},$$
$$T_i^{(1)} = \frac{1}{1+s^2}\frac{B_i}{\lambda_{i\|}} + \frac{s^2}{1+s^2}\frac{B_e+B_i}{\lambda_{e\|}+\lambda_{i\|}}, \tag{8.5.12}$$

where $\lambda_{e\|}$ and $\lambda_{i\|}$ are the coefficients of longitudinal thermal conductivity of electrons and ions

$$\lambda_{e\|} = a_2\frac{k}{m_e}p_e\tau_e, \qquad \lambda_{i\|} = 3.9\frac{k}{m_i}p_i\tau_i,$$

and B_e and B_i correspond to the left-hand sides of Eqs. (8.5.11)

$$B_e = -\frac{2q^2k^2TR}{m_e\omega_e}\left(-\frac{5}{2}n_e\frac{\partial T_e}{\partial r} + \frac{a_1}{k}\frac{\partial p}{\partial r}\right),$$

$$B_i = -\frac{2q^2k^2TR}{m_i\omega_i}\frac{5}{2}n_i\frac{\partial T_i}{\partial r}.$$

The parameter

$$s^2 = \frac{\lambda_{e\|}+\lambda_{i\|}}{\lambda_{e\|}\lambda_{i\|}}\frac{3m_en_e}{m_i\tau_{ei}}\left(\frac{r}{\Theta}\right)^2$$

characterizes the relative role of the heat exchange or the temperature equilibration.

Countering that $\lambda_{e\|} \gg \lambda_{i\|}$, we can express s^2 in terms of the ratio of the magnetic field connection length qR to the ion mean free path l_i

$$s^2 = 2\left(\frac{m_e}{m_i}\right)^{1/2}\frac{q^2R^2}{l_i^2Z}.$$

In the regime of diffusion under study ($l_i \ll qR$) the parameter s^2 can be either small or big. The case of the weak heat exchange ($s^2 \ll 1$) is implemented in the vicinity of the boundary of the Pfirsch–Schlüter regime at $l_i \gg qR(m_e/m_i)^{1/4}Z^{1/2}$. If $s^2 \gg 1$ (the intense heat exchange), the temperatures of electrons and ions are assumed to be practically equal.

Let us consider the case of a weak temperature equilibration ($s^2 \ll 1$). Substituting $T_e^{(1)} = B_e/\lambda_{e\|}$ and $T_i^{(1)} = B_i/\lambda_{i\|}$ into Eq. (8.5.10), we have

$$H_e = -\frac{\rho_{Be}^2 q^2}{\tau_{ei}} \left(b_1 n_e k \frac{\partial T_e}{\partial r} - b_2 \frac{\partial p}{\partial r} \right),$$

$$H_i = -0.8 \frac{\rho_{Bi}^2 q^2}{\tau_{ii}} n_i k \frac{\partial T_i}{\partial r}, \tag{8.5.13}$$

where

$$b_1 = 1.13 \frac{1 + 3Z + 0.75Z^2}{Z(1 + 1.7Z)}, \qquad b_2 = \frac{1 + 0.52Z}{1 + 1.7Z}. \tag{8.5.14}$$

The diffusion flux across the magnetic surface can be obtained from the expression (8.5.9)

$$\Gamma_e = -\Gamma_i Z = -\frac{\rho_{Be}^2 q^2}{kT} \tau_{ei}^{-1} \left(b_3 \frac{\partial p}{\partial r} - b_2 n_e k \frac{\partial T}{\partial r} \right), \tag{8.5.15}$$

where

$$b_3 = \frac{1 + 3.75Z + 3.1Z^2 + 0.62Z^3}{(1 + 1.7Z)(1 + 3Z + 0.75Z^2)}. \tag{8.5.16}$$

Consider how the expressions for the fluxes change when the parameter s^2 grows. As $\lambda_{e\|} \gg \lambda_{i\|}$, and $B_e \sim B_i$, the correction to the electron temperature $T_e^{(1)}$ will be slightly changed from $T_e^{(1)} = B_e/\lambda_{e\|}$ in the limit $s^2 \ll 1$ to $T_e^{(1)} = (B_e + B_i)/\lambda_{e\|}$ in the limit $s^2 \gg 1$. Therefore, both the electron heat flux and the fluxes of particles will be expressed by the formulae of the form (8.5.13) and (8.5.15) in which the temperature gradient $\partial T_e/\partial r$ should by replaced by

$$\frac{\partial T_e}{\partial r} - \left(\frac{n_i}{Zn_e} \frac{s^2}{1 + s^2} \right) \frac{\partial T_i}{\partial r}.$$

This is not the case for the correction of $T_i^{(1)}$ at increasing s^2. At $s^2 \geq 1$, $T_i^{(1)}$ decreases proportionally to s^{-2} reaching $(B_e + B_i)/\lambda_{e\|}$ at $s^2 \gg (m_i/m_e)^{1/2}$ [with $l_i \ll (m_e/m_i)^{1/2} qR$]. The ion heat flux compares by the order of value with the electron flux and becomes less (to be precise, at $q < (m_i/m_e)^{1/4}$) than the classical flux across the magnetic field and, hence, negligible.

Now turn back to the case of two-ion species, one of which corresponds to the main ions of plasma ($Z_i = 1$) and the other to the ions of the impurity which are in the singularly possible charge state of Z_I. In this case the condition $m_I \gg m_i$ is assumed to be fulfilled. Since the friction force of impurity ions against the main ions of the plasma exceeds the friction force of impurity ions against electrons [$(m_i/m_e)^{1/2}$ times], the latter can influence the ion diffusion of impurity indirectly via the heat exchange. If $(l/qR) \ll (m_e/m_i)^{1/2}$, the effect of the electrons can be neglected at all. In this case, the calculation of the radial impurity ion flux wholly repeats the calculation of diffusion in a simple plasma already discussed. Even the coefficients $c_{\alpha\beta}^{(n)}$ and $c_\alpha^{(n)}$, because of the big ratio of ion masses, are defined by the same expressions with the simultaneous substitution of Z for $Z^* = Z_I^2 n_I/n_i$. Then for the impurity ion flux, we have [15]

$$\Gamma_I = \frac{\rho_{Bi}^2 q^2}{ZkT}\tau_{iI}^{-1}\left[b_3\left(\frac{\partial p}{\partial r} - \frac{n_i}{n_I Z_I}\frac{\partial p_I}{\partial r}\right)\right.$$
$$\left. - b_2 n_i k\left(\frac{\partial T_i}{\partial r} + \frac{n_I}{n_i Z_I}\frac{s_I^2}{1+s_I^2}\frac{\partial T_I}{\partial r}\right)\right], \qquad (8.5.17)$$

where b_2 and b_3 are given by the relations (8.5.14) and (8.5.16) where $Z = Z^*$. The parameter s_I^2 is defined by the formula

$$s_I^2 = 2\left(\frac{m_i}{m_I}\right)^{1/2}\frac{q^2R^2}{l_I^2 Z_I}.$$

Since the ratio of m_i/m_I is not as small as the ratio of the electron mass to the ion mass, the parameter s_I^2 in the Pfirsch–Schlüter regime cannot be considerably less than unity, i.e. there is always an intense heat exchange between the ions. This results in a difference between expression (8.5.17) and the well-known Rutherford result [14] obtained in the assumption of the weak temperature equilibration ($s_I^2 \ll 1$).

Some practical applications of the results presented in this section will be considered for the case of a multispecies plasma composed of electrons, primary single charged ions and multicharged ions, in the next chapter.

Chapter 9

Some practical applications

9.1 Electrical conductivity of the MHD generator plasma

One of the power devices where the multicomponent plasma is used is the magnetohydrodynamic (MHD) power generator. Its simplified scheme corresponding to the so called continuous electrode Faraday generator is presented in Fig. 9.1. In the MHD generator the vector of the magnetic field **B** is directed across the motion of an electrically conducting gas which flows along an insulated duct at a speed **u**. The gas charged particles are under the influence of an induced electrical field $\mathbf{u} \times \mathbf{B}$ which generates an electric current directed perpendicular to vectors **u** and **B**. With a pair of electrodes on opposite sides of the duct in contact with the conducting gas and closed at the external load, there is a chance to transform the energy of the plasma flowing across the magnetic field into the energy delivered to the external load [1,2].

A partially ionized plasma which is a mixture of gas combusting products and an easily ionized admixture of alkali metal vapours which ensures a required degree of ionization at temperatures of 2000–2500 K is commonly used as the working fluid of open cycle MHD generators. The working fluid of the closed cycle generator is, a rule, a mixture of an inert gas (helium, argon) with cesium or potassium.

To estimate the current density obtained and the electrical power delivered to the load by the gas volume unit of the MHD generator with allowance for the Hall effect and ion-slip, it is necessary to have an expression for the generalized Ohm's law which takes into consid-

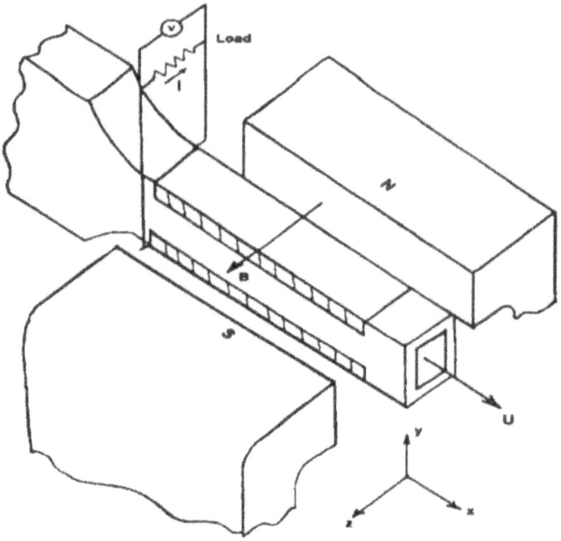

Figure 9.1. Simplified scheme of MHD generator

eration the complex chemical composition of the working fluid. The structure of the generalized Ohm's law is seen to be the same as it was for the three-component partially ionized gas (see Section 7.3). The expressions for coefficients A and B which define the values of the transverse electrical conductivity of the plasma with an arbitrary number of neutral components will be given later. For the time being, we shall consider the relationships for the transverse (with respect to the magnetic field) constituents of the current density.

Let a partially ionized gas flow along the x axis whereas the magnetic field is directed along the z axis. Then, for the constituents of the current density j_y and j_x across the magnetic field, we have

$$
\begin{aligned}
j_y &= \sigma_0 \frac{AE_y' + B\beta_0 E_x}{A^2 + B^2\beta_0^2}, \\
j_x &= \sigma_0 \frac{AE_x - B\beta_0 E_y'}{A^2 + B^2\beta_0^2},
\end{aligned}
\tag{9.1.1}
$$

where $E_y' = E_y - uB_z = -\,uB_z\,(1-k)$, with $k = E_y/uB_z$ as the load parameter and $\beta_0 = \omega_e\tau_0$ as the Hall parameter. For the continuous electrode Faraday generator $\mathbf{E}_x = 0$ and in this case

$$j_y = -\frac{\sigma_0 A}{A^2 + B^2\beta_0^2} uB_z (1-k), \quad j_x = -\frac{B}{A}\beta_0 j_y. \tag{9.1.2}$$

The Hall effect resulting in a current arising along the axes x can considerably reduce the current j_y collected from the electrodes and ultimately the power density extracted into the load. To minimize the unfavorable effect, the electrodes are segmented so that each pair of oppositely charged electrodes closes on a separate load. In an ideal segmented Faraday generator $j_x = 0$, due to which an electrical field $E_x = (B/A)\beta_0 E_y'$ arises. Substituting this value into j_y (9.1.1), we obtain

$$j_y = -\frac{\sigma_0}{A} uB_z (1-k), \quad A = 1 - \frac{\Delta_0}{1+\beta_e^{*2}} + \delta_0\beta_0^2, \tag{9.1.3}$$

which means that under the given conditions the Hall effect is inessential, yet at great values of $\beta_0 \sim \delta_0^{-1/2}$ the effect of ion slip becomes appreciable.

If the Hall parameter is small ($\beta_0 \ll 1$), then in all cases

$$j_y = -\sigma uB_z (1-k),$$

where $\sigma = \sigma_0/(1-\Delta_0)$ is the electrical conductivity of the plasma defined by the expressions correspondent to the second approximation of the theory.

The electric energy transferred into the load by the plasma unit volume per unit time, or the specific electrical power of the MHD generator is equal to

$$P = -\mathbf{j}\cdot\mathbf{E}. \tag{9.1.4}$$

For the continuous electrode Faraday generator

$$P = \frac{\sigma_0 A}{A^2 + B^2\beta_0^2} u^2 B_z^2 k (1-k) \tag{9.1.5}$$

while for the ideal segmented Faraday generator

$$P = \frac{\sigma_0}{A} u^2 B_z^2 k (1-k). \tag{9.1.6}$$

Now consider the problem of generalization of the relevant expressions for the coefficients A and B in the case of partially ionized plasma with several species of neutrals [3]. Let K be the number of neutral species of various types, so that the general number of the plasma species is $N = K + 2$ (for simplicity, ions are assumed to arise from an easily ionized admixture). Since the number of independent diffusion equations is equal to $N-1$, an equation for electrons ($\alpha = e$) and K equations for the neutral species are taken as initial expressions. If the variables $\mathbf{S}_\beta = n_e e\,(\mathbf{w}_i - \mathbf{w}_e)$ for $\beta = 1, ..., K$ and $\mathbf{X}_e = n_e e \mathbf{h}_e / p_e$ are introduced, then the electron equation (6.2.1), taking into account (3.2.4), can be written as

$$\mathbf{j} + \omega_e \tau_0 \mathbf{j} \times \mathbf{k} = \sigma_0 \mathbf{E}^* + \tau_0 \sum_{\beta=1}^{K} \tau_{e\beta}^{-1} \mathbf{S}_\beta + \omega_e \tau_0 \sum_{\beta=1}^{K} \frac{\rho_\beta}{\rho} \mathbf{S}_\beta \times \mathbf{k} + \nu_0 \tau_0 \mathbf{X}_e. \tag{9.1.7}$$

In turn, $\mathbf{X}_e$ is found from the independent equation for the electron heat flux (6.2.3) which can be written as

$$\mathbf{X}_e + \beta_e^* \mathbf{X}_e \times \mathbf{k} = \frac{5}{2} \nu_0 \tau_e^* \mathbf{j} - \frac{5}{2} \tau_e^* \sum_{\beta=1}^{K} \xi_{e\beta} \tau_{e\beta}^{-1} \mathbf{S}_\beta, \tag{9.1.8}$$

(for simplicity, the terms related to the temperature gradient are omitted).

To define $\mathbf{S}_\beta$ at $\alpha = 1, ...K$, K independent diffusion equations (6.1.1) are given. As earlier, the terms comprising heat fluxes of ions and neutrals are neglected and the terms will temperature gradients are omitted. In this case, the equations for the neutral species are

$$\sum_{\beta=1}^{K} c_{\alpha\beta} \mathbf{S}_\beta = \lambda_{e\alpha} \mathbf{j} + n_e e \frac{\rho_\alpha}{\rho} \mathbf{j} \times \mathbf{B}, \tag{9.1.9}$$

where $c_{\alpha\beta}$ and $\lambda_{\alpha\beta}$ are defined by Eqs. (3.3.3) and (3.3.4). Solving Eq. (9.1.9) brings

$$\mathbf{S}_\alpha = c_\alpha \mathbf{j} + f_\alpha \mathbf{j} \times \mathbf{k},$$

$$c_\alpha = \sum_{\beta=1}^{K} \frac{|c|_{\beta\alpha}}{|c|} \lambda_{e\beta}, \quad f_\alpha = \rho_e \omega_e \sum_{\beta=1}^{K} \frac{\rho_\beta}{\rho} \frac{|c|_{\beta\alpha}}{|c|}. \tag{9.1.10}$$

It is to be noted that the non-diagonal elements of the determinants depend on the values characterizing the interaction of neutral particles alone. The diagonal terms $c_{\alpha\alpha} = \sum_\gamma \lambda_{\alpha\gamma}$ along with other elements contain $\lambda_{\alpha e}$ and $\lambda_{\alpha i}$, with $\lambda_{\alpha e}/\lambda_{\alpha i} \sim (m_e/\mu_{i\alpha})^{1/2} \ll 1$. Under these conditions, the estimation of the coefficients c_α and f_α gives

$$c_\alpha \ll \left(\frac{m_e}{m_k}\right)^{1/2}, \qquad f_\alpha \le \left(\frac{m_e}{m_k}\right)^{1/2} \omega_e \tau_0,$$

where the k index refers to the lightest neutral species. Substituting (9.1.7) into (9.1.4) and (9.1.5) and neglecting the terms $\sim (m_e/m_k)^{1/2}$, we come to the expression for the current density $\mathbf{j}$ which is coincident in form with the expression for the generalized Ohm's law (7.3.4), in which case, due to the assumed simplification, $\mathbf{E}_{eff} = \mathbf{E}^*$. The coefficients A and B as part of the expressions for the constituent of the current density $\mathbf{j}_y$ and $\mathbf{j}_x$ (9.1.1) can be, as previously, represented as

$$A = 1 - \frac{\Delta_0}{1+\beta_e^{*2}} + \delta_0\beta_0^2,$$

$$B = 1 + \frac{\gamma\Delta_0}{1+\beta_e^{*2}},$$

where

$$\beta_e^* = \omega_e\tau_e^*,$$

$$\Delta_0 = \frac{5}{2}\nu_0^2\tau_0\tau_e^*, \qquad \gamma = \tau_0^{-1}\tau_e^*,$$

and coefficients τ_0, ν_0 and τ_e^* are defined as

$$\tau_0^{-1} = \sum_{\beta=1,\beta\neq e}^{N} \tau_{e\beta}^{-1}, \qquad \nu_0 = \sum_{\beta=1,\beta\neq e}^{N} \xi_{e\beta}\tau_{e\beta}^{-1},$$

$$(\tau_e^*)^{-1} = 0.4\tau_{ee}^{-1} + \sum_{\beta=1,\beta\neq e}^{N} \left(2.5 - 1.2B_{e\beta}^*\right)\tau_{e\beta}^{-1}. \tag{9.1.11}$$

As to the coefficient δ_0, which is related to the effect of ion-slip, it can be written in the case under consideration as

$$\delta_0 = \rho_e\tau_0^{-1}\sum_{\alpha}\sum_{\beta}\frac{\rho_\alpha\rho_\beta}{\rho^2}\frac{|c|_{\beta\alpha}}{|c|}. \tag{9.1.12}$$

It is helpful to give δ_0 for the case when the mixture includes one or two species of neutral particles. Thus, for the case of three-component plasma ($K = 1$) we have

$$\delta_0 = \rho_e\tau_0^{-1}\left(\frac{\rho_1}{\rho}\right)^2\lambda_{i1}^{-1} = \left(\frac{m_e}{\mu_{i1}}\right)^{1/2}\left(\frac{\rho_1}{\rho}\right)^2\frac{n_1Q_{e1}+n_iQ_{ei}}{n_1Q_{i1}},$$

and for the case of two species of neutrals ($K = 2$)

$$\delta_0 = \frac{n_em_e}{\rho^2\tau_0}\frac{\rho_1^2\lambda_{i2}+\rho_2^2\lambda_{i1}+(\rho_1+\rho_2)^2\lambda_{12}}{\lambda_{i1}\lambda_{i2}+\lambda_{12}(\lambda_{i1}+\lambda_{i2})}.$$

In the case of a weakly ionized plasma ($\alpha \ll 1$), the latter expression may be simplified

$$\delta_0 = \frac{m_e^{1/2}\left(n_1\overline{Q}_{e1} + n_2\overline{Q}_{e2} + n_i\overline{Q}_{ei}\right)}{\mu_{i1}^{1/2} n_1\overline{Q}_{i1} + \mu_{i2}^{1/2} n_2\overline{Q}_{i2}}.$$

The second approximation to the electrical conductivity of an MHD generator plasma is satisfactory in many cases, e.g., for the potassium-seeded combustion product plasma. For the cesium-or potassium-seeded argon plasma, the convergence of successive approximations to the electrical conductivity is worse and the use of the Lorentz approximation to σ at a low degree of ionization is preferable. Note that in the calculations of σ at an arbitrary degree of ionization, the interpolation expressions are frequently used [4,5]. Let us consider, e.g., the well-known Lin–Kantrowitz formula [4]

$$\frac{1}{\sigma} = \frac{1}{\sigma_{en}} + \frac{1}{\sigma_{ei}} \tag{9.1.13}$$

based on the assumption of the addition of specific resistances related to electron–neutral and electron–ion collisions. Here σ_{en} is the electrical conductivity of a weakly ionized gas in the Lorentz approximation and σ_{ei} is the well-known Spitzer result for a fully ionized plasma. But the fact is that this rule is valid for the effective collision frequency only

$$\tau_\alpha^{-1} = \tau_{en}^{-1} + \tau_{ei}^{-1},$$

or for the specific resistances calculated in the first approximation. Then the electrical conductivity (9.1.13) expressed through the values of $[\sigma_{en}]_2$ and $[\sigma_{ei}]_2$ calculated in the second approximation may be written as

$$\frac{1}{[\sigma]_2} = \frac{1}{[\sigma_{en}]_2}\frac{1-\Delta_0}{1-\Delta_{en}} + \frac{1}{[\sigma_{ei}]_2}\frac{1-\Delta_0}{1-\Delta_{ei}}. \tag{9.1.14}$$

The difference between (9.1.13) and (9.1.14) is small at the limits of weak and full ionization but it rises to 50–100% at intermediate degrees of ionization.

The best accuracy is given by using the Frost mixture rules [5] in which the modified collision frequency is introduced into the expression for $\sigma_{\parallel}$ in the Lorentz approximation. However, the corresponding formulae for $\sigma_{\perp}$ and $\sigma_{\wedge}$ are less satisfactory [6], therefore the application of the expressions for the electrical conductivity calculated in the second approximation of the exact kinetic theory is preferable in many cases.

9.2 Separation of the mixture species in a rotating plasma

Consider a plasma in the space between two coaxial cylindrical electrodes. If an electric field E_r is applied to it, then, as is known, azimuthal rotation of the plasma arises when there is a longitudinal magnetic field B_z directed along the cylindrical axis. Assuming that the main plasma parameters keep changing only in the radial direction and the requirement $\Omega/\omega_i \ll 1$ is met (where $\Omega = u_\varphi/r$ is the angular velocity of plasma rotation), for the acceleration components in the cylindrical frame of reference (r, φ, z), we have [7]

$$\frac{du}{dt} \equiv \left\{-\frac{u_\varphi^2}{r}, u_r\left(\frac{\partial u_\varphi}{\partial r} + \frac{u_\varphi}{r}\right), 0\right\}.$$

For the case of a fully ionized magnetized plasma according to Section 3.6 (Eqs. (3.6.3) and (3.6.4)), the expression for the azimuthal velocity of the plasma species can be written as

$$u_{\alpha\varphi}^{(0)} = -\frac{E_r}{B} + \frac{1}{\rho_\alpha \omega_\alpha}\left(\frac{\partial p_\alpha}{\partial r} - \frac{\rho_\alpha u_\varphi^2}{r}\right), \tag{9.2.1}$$

and for the radial diffusion velocities as

$$\begin{aligned} w_{\alpha r}^{(1)} &= -\frac{1}{m_\alpha \omega_\alpha^2}\sum_\beta \mu_{\alpha\beta}\tau_{\alpha\beta}^{-1}\left[\frac{1}{n_\alpha}\left(\frac{\partial p_\alpha}{\partial r} - \frac{Z_\alpha}{Z_\beta n_\beta}\frac{\partial p_\beta}{\partial r}\right)\right. \\ &\left. -\left(m_\alpha - \frac{Z_\alpha}{Z_\beta}m_\beta\right)\frac{u_\varphi^2}{r}\right] + \frac{1}{\omega_\alpha}u_r\left(\frac{\partial u_\varphi}{\partial r} + \frac{u_\varphi}{r}\right). \end{aligned} \tag{9.2.2}$$

The last term in the expression accounts for the Coriolis force. When the condition $\Omega/\omega_i \ll 1$ is satisfied, its effect on the radial diffusion of the ion species can be neglected [7].

For the multispecies plasma from Eq. (9.2.2) it follows that the ions of different species in the field of the centrifugal force which occurs in the rotating plasma can be radially separated [7]. In particular, for plasma of two ion species i and I, setting the radial flux of particles equal to zero provides

$$\frac{dn_i}{Z_i n_i} - \frac{dn_I}{Z_I n_I} = \frac{m_i u_\varphi^2}{kT}\left(\frac{1}{Z_i} - \frac{m_I}{m_i Z_i}\right)\frac{dr}{r}. \tag{9.2.3}$$

Assuming for simplicity that u_φ and T do not depend on r and integrating Eq. (9.2.3) on the assumption $Z_i = Z_I = Z$, we arrive at the expression which defines the coefficient of separation of the ion species $\alpha = (n_i/n_I)\,/\,(n_{i0}/n_{I0})$

$$\ln \alpha = \frac{m_I - m_i}{kT} u_\varphi^2 \ln \frac{r_0}{r}, \tag{9.2.4}$$

where n_α and $n_{\alpha 0}$ is the density of the particles of species α at distances r and r_0, respectively.

Now we shall refer to a case of a weakly ionized plasma in the crossing electrical and magnetic fields. The rotation of the weakly ionized gas mixture in the space between two cylindrical electrodes brings in the centrifugal effect of separating the mixture analogues to the common separation in the gas centrifuge. One can show that along with this phenomenon there occurs an additional effect of separation due to the "ion wind" [8] which means that the average momentum transfer during the collisions of neutrals and ions moving under the applied electric field can be treated as an extra external force which, together with the gradients of the partial pressure of neutral species, determines their relative diffusion. Since the average momentum transfer depends on the reduced mass of the colliding particles, the effect of the ion wind will result in the separation of the neutral species of the mixture. To analyse this phenomenon will require the diffusion equations (3.1.19) written for the neutral species (recall that in this derivation thermal diffusion was neglected). These equations take the form

$$n_k \sum_{i=1}^{S} \mu_{ki} \tau_{ki}^{-1} (\mathbf{w}_k - \mathbf{w}_i) + n_k \sum_{k' \neq k}^{S} \mu_{kk'} \tau_{kk'}^{-1} (\mathbf{w}_k - \mathbf{w}_{k'}) = -\,\mathbf{G}_k, \tag{9.2.5}$$

where the interactions between neutrals of given species k and ions of various species i and neutrals of different species k' are separated.

Using the expression for the diffusion ion velocity $\mathbf{w}_i$ (3.5.4) in the limit of the weak ionization and substituting it into (9.2.5), we can define the real effective friction force affecting the neutral species by ions. For the steady plasma rotation $d\mathbf{u}/dt = 0$ and $\mathbf{G}_\alpha = \nabla p_\alpha$. The plasma pressure gradient in the radial direction is related to the Hall current j_y via the equilibrium condition[1]

$$j_y B = \frac{dp}{dx}. \tag{9.2.6}$$

[1]If $d \ll R$ (where d is the distance between the coaxial cylinders, R is the radius of one of the cylinders) the problem can be dealt with in the Cartesian system of coordinates (the x axis is directed along the radius, while the y axis corresponds to the azimuthal direction).

To develop an adequate speed of rotation, it is required that $\beta_e^2 \gg 1$. Then, substituting the expression for j_y into Eq. (9.2.6) and neglecting the gradients of the partial pressure of the charged species along x as compared with dp/dx, we come to the relation connecting the gradient of total pressure and the value of field $E_x' = E_x + u_y B_z$

$$\frac{dp}{dx} = E_x' \sum_{i=1}^{S} \frac{n_i e_i}{1+\beta_i^2}. \tag{9.2.7}$$

If $\beta_i^2 \ll 1$, then the radial velocity of ions of k species following (3.5.4) is defined by the expression

$$(\mathbf{w}_{i_k})_x = \frac{b_{i_k}}{n_e e} \frac{dp}{dx} \tag{9.2.8}$$

under the condition of $dp_i/dx \ll dp/dx$. Note that with a single species of ions available in the plasma, the expression (9.2.8) is also valid at an arbitrary β_i.

Let us assume that in the steady state there is no full radial flux of heavy particles of every species

$$(\mathbf{u}_k)_x + \gamma_k (\mathbf{u}_{i_k})_x = 0, \qquad \gamma_k = \frac{n_{i_k}}{n_k} \ll 1. \tag{9.2.9}$$

Since at $m_{ik} = m_k$ the mass-average velocity along the x axis is $u_x \approx (\rho_e/\rho)\, u_e$ and it is less than the velocity of ions, we can take $(\mathbf{u}_{i_k})_x = (\mathbf{w}_{i_k})_x$. The indices i_k belong to ion i of k species.

Substituting into the diffusion equations for the neutral species (9.2.5) the velocities of ions and neutrals written as projected onto the x axis and expressed in terms of Eqs. (9.2.8) and (9.2.9), we have

$$\frac{dp_k}{dx} = \frac{1}{n_e}\frac{dp}{dx}\left[\sum_{m\neq k}\left(\frac{n_{i_k} n_m a_{km}}{\sum_l n_l a_{i_k l}} - \frac{n_{i_m} n_k a_{km}}{\sum_l n_l a_{i_m l}}\right) + \sum_{i,m} \frac{n_{i_m} n_k a_{ki_m}}{\sum_l n_l a_{i_k l}}\right], \tag{9.2.10}$$

where the coefficients $a_{\alpha\beta}$ are introduced with the help of the relation $n_\beta a_{\alpha\beta} = \mu_{\alpha\beta} \tau_{\alpha\beta}^{-1}$. Integrating Eqs. (9.2.10) results in the distribution of the density of neutral species with respect to the radius (x axis) depending on the gradient of total pressure dp/dx or the field E_x'.

As an example, look at the binary isotopic gas mixture with an addition of a different easily ionized component. Taking the relative concentration of the addition small ($n_3 \ll n_1 + n_2$) as well as the relative difference of isotopes masses $[(m_2 - m_1)/m_1]$, the resulting solution of Eqs. (9.2.10) will give the following expression for the enrichment coefficient [8]

$$\varepsilon = \alpha - 1 = \frac{\left(\frac{c}{1-c}\right)_{R_I}}{\left(\frac{c}{1-c}\right)_{R_{II}}} - 1 = \left[\frac{a_{i_3 2} - a_{i_3 1} + a_{31} - a_{32}}{a_{i_3 2} c + a_{i_3 1}(1-c)}\right] \ln \frac{P_{R_I}}{P_{R_{II}}}, \tag{9.2.11}$$

where indices 1 and 2 refer to the species of the mixture being separated and indices i_3 and 3 to the ions and atoms of the addition; R_I and R_{II} are the radii of the external and internal cylinders; $c = n_2/(n_1 + n_2)$ is the concentration of the heavy isotope. Using the expression for $\tau_{\alpha\beta}^{-1}$ in the form of (3.1.12), we have

$$a_{\alpha\beta} = \frac{4}{3}\mu_{\alpha\beta}\langle v_{\alpha\beta}\rangle Q_{\alpha\beta}, \quad \langle v_{\alpha\beta}\rangle = \left(\frac{8kT}{\pi\mu_{\alpha\beta}}\right)^{1/2}.$$

Substituting the expressions above into (9.2.11) and considering the small value of the relative difference of isotope masses, we obtain [8]

$$\varepsilon = \frac{(m_2 - m_1) m_3}{2m_1(m_1 + m_3)}\left(1 - \frac{Q_{13}}{Q_{1i_3}}\right) \ln \frac{P_{R_I}}{P_{R_{II}}}, \tag{9.2.12}$$

where Q_{13} and Q_{1i_3} are the mean effective cross sections of collisions of isotopic ions and ions of addition, respectively. It follows from (9.2.12) that at $Q_{13}/Q_{1i_3} < 1$ the heavy isotopes center in the area of the higher pressure. Therefore, in the general case, with the positive polarity of the voltage applied (when the internal cylinder is an anode) the centrifugal and ion effects of the enrichment supplement each other and with the negative polarity they are subtracted.

It should be mentioned that the results obtained here cannot be directly applied in the case when the current in the plasma is due to the ion transfer of the separated isotopic mixture proper. As the dominant process in the interaction between an ion and an atom of the same chemical species is the resonance charge transfer when the transferred momentum is proportional to the ion mass m_i rather than the reduced mass μ_{in}, the collisions of this type do not result in the species separation as the analysis proves. Hence, the coefficient ε for such a mixture depends not only on the relative difference of the isotope mass but appears to be proportional to the ratio of the averaged effective cross sections of the ion–neutral elastic collision and the cross section of the resonance charge transfer. A detailed analysis of this case is presented elsewhere [8].

9.3 Temperature screening of impurities in a fully ionized multispecies plasma

In Chapter 3, one of the well-known results referring to the diffusion of ions in a fully ionized multispecies plasma has already been dealt with [9]. The analysis of diffusion fluxes of particles across a strong magnetic field brings us to the conclusion that after a certain time there occurs an equilibrium distribution of ion densities over radius (3.6.11) from which it follows that ions of greater charge concentrate in the area of greater plasma density. This was obtained as resulting from the expressions for the diffusion fluxes of particles defined in the quasihydrodynamic approach, i.e. without allowance for the thermal force or the thermal diffusion in the diffusion equations. It is worth to consider a more general case when, together with the species density gradients, there is a temperature gradient in the plasma.

Let plasma be composed of electrons, primary single charged ions i and multicharged ions I. The expression for the equilibrium ion density n_I in the cylindrically symmetric case (when the magnetic field $\mathbf{B}$ is directed along the cylinder axis) can be derived from the general equation (8.3.2) setting the radial constituents of the particle flux $\boldsymbol{\Gamma}_I = \sum_Z n_{IZ}\mathbf{w}_{IZ}$ and the ion current density equal to zero. Ignoring the collisions between the ions and electrons and inertial terms (which is possible when considering slow processes in the plasma) and assuming approximately $\overline{Z_I^2} = Z_I^2$ (the average line is further omitted), we have

$$\frac{1}{n_i}\frac{\partial p_i}{\partial r} - \frac{1}{n_I Z_I^2}\frac{\partial p_i Z_I}{\partial r} - \frac{3}{2}\mu_{iI}\left(\frac{1}{m_i} - \frac{1}{m_I Z_I}\right)\frac{\partial T}{\partial r} = 0,$$
$$\frac{\partial Z_I}{\partial r} = 0. \qquad (9.3.1)$$

An important sequence of these equations is the fact that equilibrium is reached only at $Z_I = \text{const}$, i.e. in the areas where the impurity ions are ionized to a maximum charged state plausible at a given electron temperature. Otherwise, because of the collisions of impurity ions of different charges, a flux

$$\Gamma_I = \frac{1}{4}\rho_{BI}^2 \frac{n_i}{Z_I \tau_{II}}\frac{\partial Z_I}{\partial r}, \quad \rho_{BI}^2 = \frac{2kT}{m_I \omega_I^2} \qquad (9.3.2)$$

directed along the growing Z_I, can occur[2].

In the areas where $Z_I = \text{const}$ unless thermal diffusion is taken into account ($\partial T/\partial r = 0$), the condition (9.3.1) is the familiar result (3.6.11). Allowance for the terms of temperature gradient shows that

[2]It is to be noted that a comparatively simple expression for $\boldsymbol{\Gamma}_I$ is explained by an assumption made as to the approximate equality between $\overline{Z_I^2}$ and Z_I^2. Generally, this expression is more complicated and, in particular, at a great difference between $\overline{Z_I^2}$ and Z_I^2 it can turn to zero.

under these conditions the temperature screening of impurities is possible. Indeed, from Eqs. (9.3.1) it follows that

$$\frac{n_I}{n_i} = An_i^{\frac{Z_I}{Z_i}-1} T^{\left[\frac{3}{2}\mu_{iI}\left(\frac{1}{m_I}-\frac{Z_I}{m_i Z_i}\right)+\left(\frac{Z_I}{Z_i}-1\right)\right]}. \tag{9.3.3}$$

In the characteristic case when $m_I \gg m_i$ and $Z_i = 1$ this relationship implies that the impurity ions do not concentrate in the area of maximum density if the density of primary ions of plasma grows slower than $T^{(Z_I+2)/2(Z_I-1)}$. For the plasma of two ion species with equal charges ($Z_I = Z_i = 1$) the formula (9.3.3) shows that in the area of maximum temperature ions of lighter species should be concentrated. In particular, the densities of the ions of deuterium and tritium in equilibrium in the D–T plasma are related by the expression

$$\frac{n_T}{n_D} = AT^{-0.3}.$$

Now let us see what conclusions will be made when analyzing the impurity diffusion in toroidal systems when the requirements corresponding to the so called Pfirsch–Schlüter regime are satisfied. For this purpose, we shall use the expression for the impurity ion flux (8.5.17) obtained in Section 8.5 at $m_I \gg m_i$.

In the absence of the temperature gradient, the formula (8.5.17) gives the same equilibrium distribution of impurity concentrations (3.6.11) as in the classical case, i.e. the maxima in the centre of the plasma column. For simplicity, we assume that the requirement $n_I Z_I^2 \gg n_i$ ($Z^* \gg 1$) is met, in which case $b_2 \approx 1/2$, $b_3 \approx 1/3$. Then the full flux can be presented as

$$\begin{aligned}\Gamma_I &= \frac{\rho_{Bi}^2 q^2}{2Z} n_i \tau_{iI}^{-1} \left[\frac{\partial \ln n_i}{\partial r} - \frac{1}{Z_I}\frac{\partial \ln n_I}{\partial r}\right. \\ &\quad \left. - \left(\frac{1}{Z_I} - \frac{1}{3} + \frac{2}{3}\frac{n_I}{n_i Z_I}\right)\frac{\partial \ln T}{\partial r}\right].\end{aligned} \tag{9.3.4}$$

In principle, the temperature screening of impurities or their removal to the periphery of the plasma column due to thermal diffusion is possible if the factor in front of $\partial \ln T/\partial r$ is negative. It is realizable when

$$\frac{n_I}{n_i} > \frac{Z_I - 3}{2},$$

from which it follows that the temperature screening of impurities becomes possible only for the impurities with small Z_I, in fact only at

$Z_I \leq 3$ in the areas of abrupt decreases of temperatures. When the characteristic scales of changing densities and temperatures are similar, screening is possible only if $n_I > (2Z_I - 3)\, n_i$, i.e. only for the impurity with the charge $Z_I = 1$.

It is notable that an interesting effect in screening impurities occurs when the multicharge state of ions is considered. In the analysis of the temperature screening the impurity ions were presumed to be in a single fixed charged state. Now we shall take the particle transport across the magnetic field allowing for the multicharged impurity. For simplicity we shall confine ourselves to the case when the temperature gradient is negligible which is the case, e.g., in the intense heat exchange between ions and light particles when $l/qR \ll (m_k/m_i)^{1/4}$. In the expression for the flux of the particles of species α in the Z charge state it corresponds to the case of neglecting $T_\alpha^{(1)}$ and $T_\beta^{(1)}$ and preserving only the first term in braces. The resulting full flux of the particles of species α in the multispecies plasma is found by simple summation of the partial fluxes ($\Gamma_\alpha = \sum_\beta \Gamma_{\alpha\beta}$) where

$$\Gamma_{\alpha\beta} = \frac{\rho_{B\alpha}^2 q^2}{kT} \frac{\mu_{\alpha\beta}}{m_\alpha} c_{\alpha\beta}^{(1)} \tau_{\alpha\beta}^{-1} \left[\frac{n_\alpha Z_\alpha}{n_\beta Z_\beta} \frac{\partial p_\beta}{\partial r} - \frac{\partial p_\alpha}{\partial r} + \frac{p_\alpha Z_\alpha}{Z_\beta^2} \frac{\partial Z_\beta}{\partial r} \right]. \tag{9.3.5}$$

It is to be remembered that in all the expressions the averaging lines above Z_α, Z_β and Z_β^2 are omitted for simplicity.

It is seen from Eq. (9.3.5) that every flux of particles, in addition to the conventional part proportional to the difference of relative gradients of the species partial pressures, has a part proportional to $\partial Z_\beta/\partial r$ because of the multicharged ions. The transport of particles due to it is always directed to the growing temperature and, therefore, prevents from temperature screening of the impurities. The contribution of the additional flux to the total flux of the impurity of species α is mainly associated with collisions of the impurity–impurity type. It is especially great in the area of abruptly growing Z_β, i.e. in the areas of the plasma column where the temperature corresponds to a denser arrangement of ionization potentials of specific impurity ions. The impurity ions ionized up to Z and $Z+1$ charges at different temperature and, hence, in different spacial areas of the column start mixing as a result of mutual diffusion. In this case, their diffusion fluxes are related by the ambipolarity condition $Z\Gamma_{I,Z} = -(Z+1)\,\Gamma_{I,Z+1}$ so that the flux of ions with $Z+1$ charge is somewhat smaller than the flux of ions of Z charge and on average there is the diffusion flux of $\Gamma_I \sim Z^{-1}\Gamma_{I,Z}$ directed to the growing Z_I.

9.4 Radial particle and heat fluxes of the impurities at Pfirsch–Schlüter regime

Consider particle and heat fluxes in a multispecies plasma of electrons, protons (with index p) and impurity ions of species i and I in an arbitrary charge state. Let the particle masses satisfy the relations $m_e \ll m_p \ll m_i, m_I$ which is nearly always the case. As to the ratio of m_I/m_i, it can be arbitrary. Owing to the assumed relationships of masses, we can use the generalization of the expressions in Section 8.5 when there are several heavy components in the plasma.

To avoid writing too clumsy results, we shall limit ourselves to the investigation of the most important case when the requirement $l/qR \geq 1$ is met. Then $s_p^2 \ll 1$ and $s_{i,I} \ll 1$, i.e. the heat exchange between electrons and ions as well as between proton and ions of the impurities is inessential and therefore the system of Eqs. (8.5.8) has to be resolved only for two ion species: i and I. The resulting expressions for radial particle and heat fluxes of the species of the smallest mass (i.e. electrons and proton) are an obvious generalization of the formulae (8.5.13) and (8.5.15) for the case when there are several heavy components in the plasma

$$\Gamma_k = \sum_{\alpha} \Gamma_{k\alpha},$$

$$\Gamma_{k\alpha} = -\frac{\rho_{Bk}^2 q^2}{kT}\left[Z_k b_3 \tau_{k\alpha}^{-1}\left(\frac{\partial p_k}{\partial r} - \frac{n_k}{n_\alpha Z_\alpha^2}\frac{\partial p_\alpha Z_\alpha}{\partial r}\right) - b_5 n_k \tau_k^{-1}\frac{n_\alpha Z_\alpha^2}{\sum_\beta n_\beta Z_\beta^2}\frac{\partial T_k}{\partial r}\right], \tag{9.4.1}$$

$$H_k = -\rho_{Bk}^2 q^2\left[b_4 n_k \tau_k^{-1}\frac{\partial T_k}{\partial r} + b_2 \sum_\beta \tau_{k\alpha}^{-1}\left(\frac{\partial p_k}{\partial r} - \frac{n_k}{n_\alpha Z_\alpha^2}\frac{\partial p_\alpha Z_\alpha}{\partial r}\right)\right], \tag{9.4.2}$$

where b_1 and b_3 are defined by Eqs. (8.5.14) and (8.5.16) in which Z must be substituted for $Z_k^* = \sum_\alpha n_\alpha Z_\alpha^2/n_k$ and for b_4 and b_5 one has

$$b_4 = 1.6\frac{1 + 3Z_k^* + 0.75Z_k^{*2}}{(1 + 1.4Z_k^*)(1 + 1.7Z_k^*)},$$

$$b_5 = \frac{1.4Z_k^*(1 + 0.52Z_k^*)}{(1 + 1.4Z_k^*)(1 + 1.7Z_k^*)}.$$

Note that expressions (9.4.1) and (9.4.2) are of less importance than the corresponding expressions for the impurity ions becuse, from the viewpoint of the collision parameter, electrons and protons are often beyond the scope of the Pfirsch–Schlüter regime. As to the ions of i and I species, after omitting the cumbersome calculations, the final expressions can be represented as

$$\Gamma_I = \Gamma_{Ie} + \Gamma_{Ip} + \Gamma_{Ii}, \quad \Gamma_i = \Gamma_{ie} + \Gamma_{ip} + \Gamma_{iI},$$

where

$$\Gamma_{Ie} = \frac{\Gamma_{eI}}{Z_I}, \qquad \Gamma_{Ip} = \frac{\Gamma_{pI}}{Z_I},$$
$$\Gamma_{ie} = \frac{\Gamma_{ei}}{Z_i}, \qquad \Gamma_{ip} = \frac{\Gamma_{pi}}{Z_i}.$$

The values $\Gamma_{ei}, \Gamma_{eI}, \Gamma_{pi}, \Gamma_{pI}$ are defined by Eqs. (9.4.1) and for Γ_{iI} we have

$$\begin{aligned} \Gamma_{iI} \;=\; & -\frac{\rho_{Bi}^2 q^2}{kT}\frac{\mu_{iI}}{m_i}\tau_{iI}^{-1}\left[A_1\left(\frac{\partial p_i}{\partial r} - \frac{n_i Z_i}{n_I Z_I}\frac{\partial p_I}{\partial r}\right)\right. \\ & \left. + n_i Z_i k\left(A_2\frac{\partial T}{\partial r} + A_3\frac{T\partial Z_i^{-1}}{\partial r} + A_4\frac{T\partial Z_I^{-1}}{\partial r}\right)\right], \end{aligned} \tag{9.4.3}$$

where

$$\begin{aligned} A_1 &= c_{iI}^{(1)} + \frac{1}{A_0}\left(\frac{\tau_I}{m_I}c_{iI}^{(2)} - \frac{\tau_i}{m_i}c_{Ii}^{(2)}\right)^2\frac{m_i m_I}{\tau_i \tau_I}, \\ A_2 &= \frac{5}{2}\left(\frac{n_i}{Z_i} + \frac{n_I}{Z_I}\right)\frac{m_i m_I}{\mu_{iI} n_i A_0}\frac{\tau_{iI}}{\tau_i \tau_I}\left(\frac{\tau_I}{m_I}c_{iI}^{(2)} - \frac{\tau_i}{m_i}c_{Ii}^{(2)}\right), \\ A_3 &= \frac{m_i}{2\mu_{iI}}\frac{\tau_{iI}}{\tau_{ii}}c_{ii}^{(1)} + \frac{1}{A_0}\left(\frac{m_I}{m_i}\frac{\tau_i}{\tau_I}c_{Ii}^{(2)} - c_{iI}^{(2)}\right) \\ &\quad \times\left(\frac{m_i}{\mu_{iI}}\frac{\tau_{iI}}{\tau_{ii}}c_{ii}^{(2)} + c_{Ii}^{(2)} + \frac{m_i}{m_I}\frac{\tau_I}{\tau_i}c_{iI}^{(2)}\right), \\ A_4 &= c_{iI}^{(1)} + \frac{1}{A_0}\left(\frac{m_I}{m_i}\frac{\tau_i}{\tau_I}c_{Ii}^{(2)} - c_{iI}^{(2)}\right) \\ &\quad + \left(\frac{m_i}{\mu_{iI}}\frac{\tau_{iI}}{\tau_{ii}}c_{ii}^{(2)} + \frac{\tau_{Ii}}{\tau_{iI}}c_{Ii}^{(2)} + \frac{m_i \tau_I}{m_I \tau_i}\frac{\tau_{Ii}}{\tau_{iI}}c_{iI}^{(2)}\right), \end{aligned} \tag{9.4.4}$$

$$A_0 = 2c_{iI}^{(3)} + \frac{m_I}{2\mu_{iI}} \frac{\tau_{iI}}{\tau_{ii}} \frac{\tau_i}{\tau_I} c_{ii}^{(3)} + \frac{m_i}{2\mu_{iI}} \frac{\tau_{Ii}}{\tau_{II}} \frac{\tau_I}{\tau_i} c_{II}^{(3)} + \frac{m_I}{\mu_{iI}} \frac{\tau_{iI}}{\tau_I} c_i^{(6)} + \frac{m_i}{\mu_{iI}} \frac{\tau_{Ii}}{\tau_i} c_I^{(6)}.$$

The total heat flux of heavy ions is defined as

$$\begin{aligned} H = &-\frac{5}{2} \frac{\rho_{BI}^2 q^2}{A_0} \left(\frac{n_i}{Z_i} + \frac{n_I}{Z_I} \right) \frac{m_i Z_I^2}{n_i \mu_{iI}} \frac{\tau_{iI}}{\tau_i \tau_I} \\ &\times \left[\frac{5}{2} \left(\frac{n_i}{Z_i} + \frac{n_I}{Z_I} \right) \frac{\partial T}{\partial r} - kT \left(A_5 n_i \frac{\partial Z_i^{-1}}{\partial r} - A_6 n_I \frac{\partial Z_I^{-1}}{\partial r} \right) \right. \\ &\left. + n_I \left(\frac{\mu_{iI}}{m_i} \frac{\tau_i}{\tau_{Ii}} c_{Ii}^{(2)} - \frac{\mu_{iI}}{m_I} \frac{\tau_i}{\tau_{Ii}} c_{iI}^{(2)} \right) \left(\frac{1}{n_I Z_I} \frac{\partial p_I}{\partial r} - \frac{1}{n_i Z_i} \frac{\partial p_i}{\partial r} \right) \right], \end{aligned} \quad (9.4.5)$$

where

$$\begin{aligned} A_5 &= \frac{\tau_i}{\tau_{ii}} c_{ii}^{(2)} + \frac{\mu_{iI}}{m_i} \frac{\tau_i}{\tau_{iI}} c_{Ii}^{(2)} + \frac{\mu_{iI}}{m_I} \frac{\tau_I}{\tau_{iI}} c_{iI}^{(2)}, \\ A_6 &= \frac{\tau_I}{\tau_{II}} c_{II}^{(2)} + \frac{\mu_{iI}}{m_i} \frac{\tau_i}{\tau_{Ii}} c_{Ii}^{(2)} + \frac{\mu_{iI}}{m_I} \frac{\tau_I}{\tau_{Ii}} c_{iI}^{(2)}. \end{aligned} \quad (9.4.6)$$

The formulae (9.4.3)–(9.4.6) permit one to calculate the values of radial ion fluxes of the impurities of an arbitrary mass and the radial heat flux related to the heavy particles in the explicit form if the coefficients $c_{\alpha\beta}^{(n)}$ and $c_\alpha^{(n)}$ are known. The latter can be calculated with the help of the approximated formulae

$$c^{(n)} = P_1^{(n)} + \frac{P_2^{(n)}}{Z_{iI} + P_3^{(n)}},$$

where

$$Z_{iI} = \frac{n_I \overline{Z_I^2}}{n_i \overline{Z_i^2}}.$$

Index I corresponds to heavier species of impurities. The values P_1, P_2 and P_3 calculated by the numerical method for either pair of ion impurities from the set of carbon, oxygen, iron and tungsten are given in Table 9.1. The estimated error in the calculation is about 2%.

Table 9.1. The coefficients $c_{\alpha\beta}^{(n)}$ and $c_{\alpha}^{(n)}$

c	P	C–O	Fe–W	O–Fe	C–Fe	O–W	C–W
	P_1	0.74	0.63	0.63	0.63	0.62	0.62
$c_{ii}^{(1)}$	P_2	0.09	0.11	0.11	0.10	0.09	0.09
	P_3	0.86	0.53	0.53	0.49	0.45	0.45
	P_1	0.77	0.55	0.54	0.48	0.38	0.36
$c_{iI}^{(1)}$	P_2	0.10	0.27	0.27	0.29	0.30	0.31
	P_3	0.87	0.63	0.62	0.57	0.50	0.49
	P_1	0.84	0.84	0.84	0.84	0.84	0.84
$c_{II}^{(1)}$	P_2	0.12	0.36	0.36	0.38	0.33	0.31
	P_3	0.88	0.65	0.63	0.58	0.49	0.48
	P_1	0.51	0.38	0.38	0.40	0.47	0.49
$c_{ii}^{(2)}$	P_2	0.07	0.11	0.1	0.06	0.01	0.01
	P_3	0.92	0.55	0.51	0.32	0.10	0.10
	P_1	0.37	–0.07	–0.08	–0.12	-0.11	–0.09
$c_{iI}^{(2)}$	P_2	0.07	0.27	0.27	0.27	0.18	0.14
	P_3	1.00	1.16	1.15	1.15	1.12	1.10
	P_1	0.75	1.14	1.16	1.23	1.29	1.29
$c_{Ii}^{(2)}$	P_2	0.04	0.02	0.02	0.01	0.00	0.00
	P_3	0.80	0.10	0.10	0.08	0.00	0.00
	P_1	0.59	0.58	0.59	0.59	0.58	0.58
$c_{II}^{(2)}$	P_2	0.04	–0.12	–0.45	–0.30	–0.32	–0.31
	P_3	0.86	13.00	8.00	2.90	1.40	1.20
	P_1	0.75	1.17	1.19	1.26	1.37	1.39
$c_{i}^{(5)}$	P_2	–0.17	–0.78	–0.82	–0.95	–1.19	–1.26
	P_3	1.09	1.36	1.37	1.43	1.56	1.59
	P_1	0.59	0.58	0.58	0.58	0.57	0.57
$c_{I}^{(5)}$	P_2	–0.15	–0.52	–0.53	–0.56	–0.50	–0.46
	P_3	1.09	1.26	1.25	1.21	0.95	0.86
	P_1	1.54	2.57	2.69	3.41	6.79	8.00
$c_{ii}^{(3)}$	P_2	–0.31	–4.00	–4.65	–8.80	–38.00	-52.00
	P_3	1.28	3.10	3.29	4.10	6.80	7.64
	P_1	1.35	0.64	0.56	0.20	–0.50	–0.55
$c_{iI}^{(3)}$	P_2	–0.28	–0.17	–0.07	1.20	4.00	4.50
	P_3	1.26	1.43	0.90	6.00	6.00	6.60

Chapter 10

Transport phenomena in a polyatomic gas mixture

10.1 Introduction

We come again to the general theory of transport phenomena in a multicomponent gas mixture. The point of discussion is a partially ionized gas whose heavy particles are polyatomic molecules or molecular ions. A characteristic feature of such a molecular gas are inelastic collisions in it which result in a change of translational, rotational and vibrational states of colliding particles.

A variety of inelastic processes are inherent in a partially ionized gas in the general case. They include excitation of rotational and vibrational degrees of freedom of molecules and molecular ions in addition to dissociation, ionization and recombination, electronic excitation as well as emission and absorption of radiation related to them. The latter processes play the crucial role in the establishment of the level of electron concentration and temperature in the plasma that is characterized by the significance of the relevant terms dependent on the inelastic collision in the continuity and energy equations for electrons.

The derivation of conservation equations for electrons as well as the equations for the rates of the processes governing the population of different excited levels of atoms, given inelastic collisions, is dealt with elsewhere [1–3]. This chapter concentrates on the transport equations and expressions for the kinetic coefficients due to rotational and vibrational excitations of heavy particles, though formally electrons can be included into this scheme since their interaction with heavy particles can also bring in the excitation of the internal degrees of freedom of molecules. Hence, the treatment given below largely corresponds to the case of a weakly ionized molecular gas at the stage when the dissociation of molecules has not become essential yet and the prevailing part of the molecules is still in the ground electron state.

A consistent kinetic theory of gases with rotational and vibrational degrees of freedom should be developed in terms of quantum mechanical kinetic equation of Waldmann–Snider [4]. It is written for the single-particle Wigner distribution function corresponding to the conventional distribution function in the phase space of translational degrees of freedom of a molecule and density matrix in the space of quantum numbers describing its internal state. By the way, prior to works on a rigorous derivation of this equation, Wang Chang, Uhlenbeck and de Bour had suggested a formal kinetic theory of polyatomic gases based on the modified kinetic Boltzmann equation with the distribution function dependent on the velocity and discrete values of the internal energy of molecules which corresponds to the semiclassical description [5,6]. This equation was considered in Chapter 1 [see Eq. (1.2.23)].

Traditionally, the Wang Chang and Uhlenbeck scheme, in addition to the well-known dependence of the distribution function on $\mathbf{v}, \mathbf{r}$ and t, takes into account the dependence of f_i on the internal energy of the molecule in an ith quantum state E_i, i.e. $f_i = f(\mathbf{v}, \mathbf{r}, t, E_i)$. Thus set, the distribution function is quite adequate when describing vibrationaly excited molecules. Yet, to be precise, for a gas with rotational degrees of freedom such an assumption related to the independence of the density matrix for the internal states on the orientation of the vector of the angular momentum $\mathbf{M}$. The necessity to account for the dependence of the distribution function on the mutual orientation of $\mathbf{M}$ and $\mathbf{v}$ was first pointed out by Kagan and Afanasiev [7] and Kagan and Maximov [8]. The numerical estimates of the effect of the factor mentioned on the kinetic coefficients for some models of nonspherical molecules proved that the corrections derived are small, as a rule [9–11]. However, accounting for the peculiar problem becomes principally essential when analyzing the transport phenomena in a molecular gas with electric and magnetic fields applied. The fact that expansion of the distribution function must generally include all the tensors of the appropriate rank with the components of the independent vectors $\mathbf{M}$ and $\mathbf{v}$ and the allowance for the dependence of the nonspherical part of the potential of molecular interaction upon the angle between these vectors leads to an explanation of a number of new physical phenomena resulting from the precession of the rotational angular momentum of the molecules in an external field, the effect of Senftleben–Beenakker in particular [8,4]. Within the framework of this book, we shall not consider these problems described in detail in a monograph [4]; therefore, our analysis will be based on the application of the distribution function dependent on E_i alone.

The formal kinetic theory of transport and relaxation phenomena developed for the case of a pure polyatomic gas [5,6] and generalized for a polyatomic gas mixture [12,13] is usually based upon the generalized Chapman–Enskog method. It is well known that this method is not capable of describing processes which involve essentially relaxation and as a result of which one establishes time-independent linear relations between fluxes and gradients of thermodynamic quantities. This drawback of the method becomes tangible in the case of a polyatomic gas when, apart from the typical time τ describing the relaxation to equilibrium in translational degrees of freedom, yet another parameter appears, namely, the characteristic time τ_E of energy exchange between the translational and internal degrees of freedom. In practice, a consistent generalization of the Chapman–Enskog method can be realized only if $\tau_E \sim \tau$. It is in this case of the so called "easy" energy exchange that the term of the volume viscosity proportional to the divergence of the gas velocity occurs in the pressure tensor. In the case of "slow" exchange ($\tau_E \gg \tau$), when τ_E is of the same order as the characteristic time of the problem τ_L, the linear relation, by means of which the coefficient of the volume viscosity is introduced, becomes invalid and the relevant term in the pressure tensor can be found only from the relaxation equation for the translational or internal energy of the gas.

The relaxation type of the transport equations in terms of the moment method turns out to be more suitable to the peculiar problem under consideration as it allows us to describe the behaviour of a polyatomic gas in the whole range of the parameter values $Z = \tau_E/\tau$, i.e. in the cases of both easy and slow energy exchanges. Accordingly, the transport and relaxation phenomena in a gas mixture with the internal degrees of freedom will be treated in what follows in terms of the Grad's moment method the generalization on which was carried out elsewhere [14–16].

10.2 Macroscopic parameters and conservation equations for a polyatomic gas mixture

In accordance with the assumptions taken above we intend to describe the state of a partially ionized gas mixture, which comprises molecules and ions with internal degrees of freedom, with the help of the distribution function

$$f_{\alpha i} = f_\alpha (\mathbf{v}_\alpha, \mathbf{r}, t, E_{\alpha i}), \tag{10.2.1}$$

where $E_{\alpha i}$ is the internal energy of the particles of species α in an ith quantum state. The generalized kinetic equation, which the function $f_{\alpha i}$ satisfies, is written as

$$\frac{\partial f_{\alpha i}}{\partial t} + \mathbf{v}_\alpha \cdot \nabla f_{\alpha i} + \frac{1}{m_\alpha} \mathbf{F}_\alpha \cdot \nabla_v f_{\alpha i}$$
$$= \sum_\beta \sum_{jkl} \iint \left(f'_{\alpha k} f'_{1\beta l} - f_{\alpha i} f_{1\beta j} \right) g \sigma_{\alpha\beta} (ij \mid kl, g, \chi) \, d\Omega d\mathbf{v}_{1\beta}. \tag{10.2.2}$$

(see Chapter 1, Eq. (1.2.33)).

The macroscopic parameters of species and the mixture as a whole are some moments of $f_{\alpha i}$, defined in addition to integrating over velocities by summing over the internal states of the particles (see Eqs. (1.1.13)–(1.1.23)).

The local one-temperature Maxwell–Boltzmann distribution may be used as the distribution function of the zero approximation in the Chapman–Enskog method or as a weight function in expansion over polynomials in the generalized moment method

$$f_{\alpha i}^{(0)} = n_\alpha \left(\frac{m_\alpha}{2\pi kT} \right)^{3/2} Q_\alpha^{-1} \exp\left(-\frac{m_\alpha}{2kT} c_\alpha^2 - \varepsilon_{\alpha i} \right), \tag{10.2.3}$$

where

$$Q_\alpha = \sum_i \exp(-\varepsilon_{\alpha i}), \qquad \varepsilon_{\alpha i} = \frac{E_{\alpha i}}{kT}. \tag{10.2.4}$$

Note that the requirements of applicability of both methods demand an assumption that the basic hydrodynamic variables such as the number density of a particle n_α, the mass-average velocity $\mathbf{u}$ and the total mean energy of the mixture per unit volume nE should be defined similarly in the full nonequilibrium distribution function $f_{\alpha i}$ as well as in the function $f_{\alpha i}^{(0)}$

$$n_\alpha = \sum_i \int f_{\alpha i} d\mathbf{v} = \sum_i \int f_{\alpha i}^{(0)} d\mathbf{v}, \tag{10.2.5}$$

$$\rho \mathbf{u} = \sum_\alpha \sum_i \int m_\alpha \mathbf{v} f_{\alpha i} d\mathbf{v} = \sum_\alpha \sum_i \int m_\alpha \mathbf{v} f_{\alpha i}^{(0)} d\mathbf{v}, \tag{10.2.6}$$

$$nE = \sum_\alpha \sum_i \int \left(\frac{m_\alpha c^2}{2} + E_{\alpha i} \right) f_{\alpha i} d\mathbf{v}$$
$$= \sum_\alpha \sum_i \int \left(\frac{m_\alpha c^2}{2} + E_{\alpha i} \right) f_{\alpha i}^{(0)} d\mathbf{v}. \qquad (10.2.7)$$

The relationships (10.2.5)–(10.2.7) are, in fact, additional constraints which interpose on the small deviations of the distribution function from the local Maxwell–Boltzmann distribution. Within the framework of the moment method that would lead to vanishing of some first coefficients of expansion of the distribution function in series over the system of the generalized polynomials.

It is seen that the condition (10.2.7) serves in reality as a definition of the temperature of a polyatomic gas mixture T which leads to

$$nE = \sum_\alpha \left(n_\alpha E_\alpha^{tr} + n_\alpha E_\alpha^{int} \right) = \sum_\alpha \left(n_\alpha E_{\alpha 0}^{tr} + n_\alpha E_{\alpha 0}^{int} \right), \qquad (10.2.8)$$

where

$$E_{\alpha 0}^{tr} = \frac{3}{2} kT, \qquad E_{\alpha 0}^{int} = \langle \varepsilon_\alpha \rangle kT,$$
$$\langle \varepsilon_\alpha \rangle = Q_\alpha^{-1} \sum_i \varepsilon_{\alpha i} \exp\left(-\varepsilon_{\alpha i}\right). \qquad (10.2.9)$$

Let us introduce deviations of mean translational and internal energies of species α from their equilibrium values

$$\Delta E_\alpha^{tr} = E_\alpha^{tr} - \frac{3}{2} kT, \quad \Delta E_\alpha^{int} = E_\alpha^{int} - \langle \varepsilon_\alpha \rangle kT. \qquad (10.2.10)$$

Then the condition (10.2.8) can be rewritten as

$$\sum_\alpha n_\alpha \left(\Delta E_\alpha^{tr} + \Delta E_\alpha^{int} \right) = 0. \qquad (10.2.11)$$

It is to be recalled that the partial pressure tensor is defined by the expression (1.1.15)

$$P_{\alpha rs} = P_\alpha \delta_{rs} + \pi_{\alpha rs}, \qquad (10.2.12)$$

where $\pi_{\alpha rs}$ corresponds to the partial viscous-stress tensor, whereas the diagonal part of the tensor is related to the mean translational energy of particles of species α per unit volume by the expression

$$P_\alpha = \frac{2}{3} n_\alpha E_\alpha^{tr}. \qquad (10.2.13)$$

As the equilibrium value $E^{tr}_{\alpha 0}$ corresponds to the partial hydrostatic pressure of particles of species α

$$p_\alpha = \frac{2}{3} n_\alpha E^{tr}_{\alpha 0}$$

the expression for $P_{\alpha rs}$ can be written as

$$P_{\alpha rs} = \left(p_\alpha + \frac{2}{3} n_\alpha \Delta E^{tr}_\alpha \right) \delta_{rs} + \pi_{\alpha rs}. \tag{10.2.14}$$

For the mixture as a whole

$$P_{rs} = \left(p + \frac{3}{2} n \Delta E^{tr} \right) \delta_{rs} + \pi_{rs}, \tag{10.2.15}$$

where

$$n \Delta E^{tr} = \sum_\alpha n_\alpha \Delta E^{tr}_\alpha .$$

Of fundamental importance are the macroscopic parameters such as the energy flux densities (heat fluxes) of species α, $\mathbf{q}^{tr}_\alpha$ and $\mathbf{q}^{int}_\alpha$, which correspond to the translational and internal degrees of freedom of particles. Further, we shall need expressions for the reduced partial heat fluxes

$$\mathbf{h}^{tr}_\alpha = \mathbf{q}^{tr}_\alpha - \frac{5}{2} p_\alpha \mathbf{w}_\alpha, \quad \mathbf{h}^{int}_\alpha = \mathbf{q}^{int}_\alpha - \langle \varepsilon_\alpha \rangle kT \mathbf{w}_\alpha. \tag{10.2.16}$$

Thus, the basic macroscopic parameters describing the state of a polyatomic gas mixture include $n_\alpha, \mathbf{u}$ and T, i.e. the parameters of the local equilibrium distribution function, and $\mathbf{w}_\alpha, \pi_{\alpha rs}, \Delta E^{tr}_\alpha, \Delta E^{int}_\alpha$, $\mathbf{h}^{tr}_\alpha, \mathbf{h}^{int}_\alpha$ to be defined by the nonequilibrium part of the distribution function. $17N$ values (where N is the number of species) are independent because the additional requirements $\sum_\alpha \rho_\alpha \mathbf{w}_\alpha = 0$ and (10.2.11) are met.

The parameters $n_\alpha, \mathbf{u}$ and T should satisfy the conservation equations of the particle number, momentum and energy similarly to the case of a monoatomic gas mixture. Such equations can be derived from the kinetic equation (10.2.2) when it is multiplied by $1, m_\alpha \mathbf{v}_\alpha$ and $(m_\alpha c_\alpha^2/2) + E_{\alpha i}$, respectively, integrated over velocities and summed over the internal states. As the result,

$$\frac{\partial n_\alpha}{\partial t} + \nabla \cdot n_\alpha \mathbf{u}_\alpha = 0, \tag{10.2.17}$$

$$\rho\frac{du_r}{dt}+\frac{\partial P_{rs}}{\partial x_s}-\sum_\alpha n_\alpha X_{\alpha r}-(\mathbf{j}\times\mathbf{B})_r=0, \tag{10.2.18}$$

$$\frac{d(nE)}{dt}+nE\nabla\cdot\mathbf{u}+P_{rs}\frac{\partial u_r}{\partial x_s}+\nabla\cdot\mathbf{q}-\sum_\alpha n_\alpha\mathbf{w}_\alpha\cdot\mathbf{X}_\alpha-\mathbf{j}\cdot\mathbf{E}' = 0. \tag{10.2.19}$$

When obtaining the last two equations, summation over all the species was carried out, with

$$\mathbf{q}=\sum_\alpha\left(\mathbf{q}_\alpha^{tr}+\mathbf{q}_\alpha^{int}\right). \tag{10.2.20}$$

The energy equation (10.2.19) can be transformed into the equation of temperature T by using the continuity equation and relationship (10.2.8)

$$nc_v\frac{dT}{dt}+P_{rs}\frac{\partial u_r}{\partial x_s}+\nabla\cdot\mathbf{q}-kT\sum_\alpha\left(\frac{3}{2}+\langle\varepsilon_\alpha\rangle\right)\nabla\cdot n_\alpha\mathbf{w}_\alpha$$
$$-\sum_\alpha n_\alpha\mathbf{w}_\alpha\cdot\mathbf{X}_\alpha-\mathbf{j}\cdot\mathbf{E}'=0, \tag{10.2.21}$$

where

$$c_v=c_v^{tr}+c_v^{int}=\frac{3}{2}k+k\sum_\alpha\left(\langle\varepsilon_\alpha^2\rangle-\langle\varepsilon_\alpha\rangle^2\right) \tag{10.2.22}$$

is the heat capacity of a gas mixture, including both the translational and internal degrees of freedom.

10.3 Expansion of the distribution function and the transport equations

As in the case of a gas mixture of monoatomic structureless particles, we shall expand the distribution function in series over generalized tensorial orthogonal polynomials. In doing so, we use the one-temperature local Maxwell–Boltzmann distribution (10.2.3) as the distribution function of the zero approximation which corresponds to the choice of the weight function in the form

$$\omega_{\alpha i}=(2\pi)^{-3/2}Q_\alpha^{-1}\exp\left(-\frac{\xi_\alpha^2}{2}\right)\exp\left(-\varepsilon_{\alpha i}\right), \tag{10.3.1}$$

where

$$\boldsymbol{\xi}_\alpha = \left(\frac{m_\alpha}{kT}\right)^{1/2} \mathbf{c}.$$

The choice of one-temperature approximation implies that, in fact, we do not consider large deviations of the mean translational and internal energies of a species from the equilibrium values corresponding to them and defined at temperature T. When treating a monoatomic gas mixture such method of attack corresponds to the approximation of Section 4.4.

Following the considerations given in Section 4.1, the part of the weight function (10.3.1) dependent on $\boldsymbol{\xi}_\alpha$ generates the irreducible tensorial Hermite polynomials $H^{mn}_{\alpha r_1 \dots r_m}(\boldsymbol{\xi}_\alpha)$ in expansion of the distribution function. An extra Boltzmann multiplier $Q_\alpha^{-1}\exp(-\varepsilon_{\alpha i})$ in (10.3.1) results in the new polynomials in the expansion whose form is readily defined from the orthogonalization procedure with respect to the set $1, \varepsilon_{\alpha i}, \varepsilon_{\alpha i}^2$ with the same part of the weight function. Direct application of the procedure results in the polynomials

$$R_\alpha^{(0)} = 1, \qquad R_\alpha^{(1)} = \varepsilon_{\alpha i} - \langle \varepsilon_\alpha \rangle,$$

$$R_\alpha^{(q)} = \varepsilon_{\alpha i} R_\alpha^{(q-1)} - \sum_{s=0}^{q-1} \left\langle \varepsilon_{\alpha i} R_\alpha^{(q-1)} R_\alpha^{(s)} \right\rangle \frac{R_\alpha^{(s)}}{\left\langle R_\alpha^{(s)2} \right\rangle}, \tag{10.3.2}$$

where $\langle \dots \rangle$ means averaging with the Boltzmann multiplier $Q_\alpha^{-1}\exp(-\varepsilon_{\alpha i})$. The first two polynomials were used by Wang Chang and Uhlenbeck in [5] (see also [17]). The expansion of $f_{\alpha i}$ can be represented in the form [14,15]

$$f_{\alpha i} = f_{\alpha i}^{(0)} \sum_{mnq} \sigma_{mn} a^{mnq}_{\alpha r_1 \dots r_m} H^{mn}_{\alpha r_1 \dots r_m}(\boldsymbol{\xi}_\alpha) R_\alpha^{(q)}(\varepsilon_{\alpha i}). \tag{10.3.3}$$

The expansion coefficients according to the orthogonality requirement for the polynomials meet the relationship

$$n_\alpha a^{mnq}_{\alpha r_1 \dots r_m} \left\langle R_\alpha^{(q)2} \right\rangle = \sum_i \int H^{mn}_{\alpha r_1 \dots r_m}(\boldsymbol{\xi}_\alpha) R_\alpha^{(q)}(\varepsilon_{\alpha i}) f_{\alpha i} d\mathbf{c}. \tag{10.3.4}$$

Similar to the mixture of structureless particles, the infinite system of equations for the coefficients $a^{mnq}_{\alpha r_1 \dots r_m}$ and parameters $n_\alpha, \mathbf{u}$ and T can be produced by multiplying the kinetic equation (10.2.2) (previously transformed into the variable $\mathbf{c}_\alpha = \mathbf{v}_\alpha - \mathbf{u}$ instead of $\mathbf{v}_\alpha$) by the product of polynomials $H^{mn}_{\alpha r_1 \ldots r_m}(\boldsymbol{\xi}_\alpha)\, R^{(q)}_\alpha(\varepsilon_{\alpha i})$ and subsequent integration over velocities and summing over all the values of the quantum numbers i.

The further procedure due to the truncation of the hierarchy of equations and the closed system of equations obtained for a finite number of macroscopic variables is entirely equivalent to the one given in Chapter 4. In particular, the $17N$-moment approximation corresponds to the $13N$-moment approximation in the case of a gas mixture of monoatomic particles; it is defined by the set of parameters derived above which describes the state of the mixture. In expansion (10.3.3) the terms which contain the products H^{mn}_α by $R^{(0)}_\alpha = 1$ for $m, n = 0, 1$ and $m = 2$, $n = 0$ and the products of H^{mn}_α by $R^{(1)}_\alpha = \varepsilon_{\alpha i} - \langle \varepsilon_\alpha \rangle$ for $m = 0, 1$ and $n = 0$ should be conserved. Then the appropriate expansion coefficients are expressed with the help of (10.3.4) in terms of the basic macroscopic parameters of the mixture as follows

$$
\begin{gathered}
a^{000}_\alpha = 1, \quad a^{010}_\alpha = 2\frac{\Delta E^{tr}_\alpha}{kT}, \quad a^{001}_\alpha = \frac{\Delta E^{int}_\alpha}{c^{int}_\alpha T}, \\
a^{100}_{\alpha r} = \gamma_\alpha^{1/2} w_{\alpha r}, \qquad a^{110}_{\alpha r} = 2\gamma_\alpha^{1/2}\frac{h^{tr}_{\alpha r}}{p_\alpha}, \\
a^{200}_{\alpha rs} = \frac{\pi_{\alpha rs}}{p_\alpha}, \qquad a^{101}_{\alpha r} = \gamma_\alpha^{1/2}\frac{h^{int}_{\alpha r}}{n_\alpha c^{int}_\alpha T},
\end{gathered}
\tag{10.3.5}
$$

where

$$
\begin{gathered}
h^{tr}_{\alpha r} = q^{tr}_{\alpha r} - \frac{5}{2} p_\alpha w_{\alpha r}, \quad h^{int}_{\alpha r} = q^{int}_{\alpha r} - \langle \varepsilon_\alpha \rangle p_\alpha w_{\alpha r}, \\
p_\alpha = n_\alpha kT, \qquad \gamma_\alpha = \frac{m_\alpha}{kT}, \\
c^{int}_\alpha = \frac{\partial E^{int}_{\alpha 0}}{\partial T} = k\left(\langle \varepsilon^2_\alpha \rangle - \langle \varepsilon_\alpha \rangle^2\right).
\end{gathered}
\tag{10.3.6}
$$

The expansion of $f_{\alpha i}$ in the $17N$-moment approximation ultimately looks as

$$
\begin{aligned}
f_{\alpha i} \;=\; & f^{(0)}_{\alpha i}\Bigg[1 + \gamma_\alpha^{1/2} w_{\alpha r}\xi_{\alpha r} + \frac{1}{3}\frac{n_\alpha \Delta E^{tr}_\alpha}{p_\alpha}\left(\xi^2_\alpha - 3\right) \\
& + \frac{k}{c^{int}_\alpha}\frac{n_\alpha \Delta E^{int}_\alpha}{p_\alpha}\left(\varepsilon_{\alpha i} - \langle \varepsilon_\alpha \rangle\right) + \frac{1}{2}\frac{\pi_{\alpha rs}}{p_\alpha}\left(\xi_{\alpha r}\xi_{\alpha s} - \frac{1}{3}\delta_{rs}\xi^2_\alpha\right) \\
& + \frac{1}{5}\frac{\gamma_\alpha^{1/2}}{p_\alpha} h^{tr}_{\alpha r}\xi_{\alpha r}\left(\xi^2_\alpha - 5\right) + \frac{k}{c^{int}_\alpha}\frac{\gamma_\alpha^{1/2}}{p_\alpha} h^{int}_{\alpha r}\xi_{\alpha r}\left(\varepsilon_{\alpha i} - \langle \varepsilon_\alpha \rangle\right)\Bigg].
\end{aligned}
\tag{10.3.7}
$$

The closed set of equations for the values $\rho_\alpha, \rho_\alpha w_{\alpha r}, n_\alpha \Delta E_\alpha^{tr}$, $n_\alpha \Delta E_\alpha^{int}, \pi_{\alpha rs}, h_{\alpha r}^{tr}$ and h_α^{int} corresponding to such an approximation is obtained after multiplying the kinetic equations by the appropriate polynomials available in expansion (10.3.7), integrating over velocities and summation over all the states

$$\frac{d\rho_\alpha}{dt} + \rho_\alpha \nabla \cdot \mathbf{u} + \nabla \cdot (\rho_\alpha \mathbf{w}_\alpha) = 0, \tag{10.3.8}$$

$$\begin{aligned} \frac{d}{dt}\rho_\alpha w_{\alpha r} + \rho_\alpha w_{\alpha s}\frac{\partial u_r}{\partial x_s} + \rho_\alpha w_{\alpha r}\nabla \cdot \mathbf{u} + \rho_\alpha \frac{du_r}{dt} + \frac{\partial p_\alpha}{\partial x_r} \\ + \frac{2}{3}\frac{\partial}{\partial x_r} n_\alpha \Delta E_\alpha^{tr} + \frac{\partial \pi_{\alpha rs}}{\partial x_s} = R_{\alpha r}^{100}, \end{aligned} \tag{10.3.9}$$

$$\begin{aligned} \frac{dn_\alpha \Delta E_\alpha^{tr}}{dt} + \frac{5}{3} n_\alpha \Delta E_\alpha^{tr} \nabla \cdot \mathbf{u} + \frac{3}{2} n_\alpha k \frac{dT}{dt} + p_\alpha \nabla \cdot \mathbf{u} + \pi_{\alpha rs}\frac{\partial u_r}{\partial x_s} \\ + \nabla \cdot \mathbf{q}_\alpha^{tr} - E_{\alpha 0}^{tr} \nabla \cdot (n_\alpha \mathbf{w}_\alpha) + \rho_\alpha w_{\alpha l}\frac{\partial u_l}{\partial t} = R_\alpha^{010}, \end{aligned} \tag{10.3.10}$$

$$\begin{aligned} \frac{dn_\alpha \Delta E_\alpha^{int}}{dt} + n_\alpha \Delta E_\alpha^{int} \nabla \cdot \mathbf{u} + n_\alpha c_\alpha^{int}\frac{dT}{dt} \\ + \nabla \cdot \mathbf{q}_\alpha^{int} - E_{\alpha 0}^{int} \nabla \cdot (n_\alpha \mathbf{w}_\alpha) = R_\alpha^{001}, \end{aligned} \tag{10.3.11}$$

$$\begin{aligned} \frac{d\pi_{\alpha rs}}{dt} + \frac{4}{5}\left\{\frac{\partial q_{\alpha r}^{tr}}{\partial x_s}\right\} + 2\rho_\alpha \left\{w_{\alpha r}\frac{du_r}{dt}\right\} + 2\left\{\pi_{\alpha rl}\frac{\partial u_s}{\partial x_l}\right\} \\ + \pi_{\alpha rs}\nabla \cdot \mathbf{u} + \frac{4}{3} n_\alpha \Delta E_\alpha^{tr} \varepsilon_{rs} + 2p_\alpha \varepsilon_{rs} = R_{\alpha rs}^{200}, \end{aligned} \tag{10.3.12}$$

$$\begin{aligned} \frac{dh_{\alpha r}^{tr}}{dt} + \frac{7}{5}h_{\alpha s}^{tr}\frac{\partial u_r}{\partial x_s} + \frac{2}{5}h_{\alpha s}^{tr}\frac{\partial u_s}{\partial x_r} + \frac{7}{5}h_{\alpha r}^{tr}\nabla \cdot \mathbf{u} + \frac{5}{2}n_\alpha k w_{\alpha r}\frac{dT}{dt} \\ + 2p_\alpha w_{\alpha s}\varepsilon_{rs} + \frac{5}{3}p_\alpha w_{\alpha s}\frac{\partial u_r}{\partial x_s} + \frac{kT}{m_\alpha}\frac{\partial \pi_{\alpha rs}}{\partial x_s} + \frac{k}{m_\alpha}\pi_{\alpha rs}\frac{\partial T}{\partial x_s} \\ + \frac{5}{3}\frac{k}{m_\alpha} n_\alpha \Delta E_\alpha^{tr}\frac{\partial T}{\partial x_r} + \frac{5}{3}\frac{kT}{m_\alpha}\frac{\partial n_\alpha \Delta E_\alpha^{tr}}{\partial x_r} + \frac{5}{3}n_\alpha \Delta E_\alpha^{tr}\frac{du_r}{dt} \\ + \pi_{\alpha rs}\frac{du_s}{dt} + \frac{5}{2}\frac{k}{m_\alpha}P_{\alpha rs}\frac{\partial T}{\partial x_s} = R_{\alpha r}^{110}, \end{aligned} \tag{10.3.13}$$

$$\frac{dh_{\alpha r}^{int}}{dt} + h_{\alpha s}^{int}\frac{\partial u_r}{\partial x_s} + h_{\alpha r}^{int}\nabla\cdot\mathbf{u} + n_\alpha c_\alpha^{int} w_{\alpha r}\frac{dT}{dt} + \frac{kT}{m_\alpha}\frac{\partial n_\alpha \Delta E_\alpha^{int}}{\partial x_r}$$
$$+ \frac{k}{m_\alpha} n_\alpha \Delta E_\alpha^{int}\frac{\partial T}{\partial x_r} + n_\alpha \Delta E_\alpha^{int}\frac{du_r}{dt} + \frac{c_\alpha^{int}}{m_\alpha}P_{\alpha rs}\frac{\partial T}{\partial x_s} = R_{\alpha r}^{101}. \tag{10.3.14}$$

The moments with respect to the collision integral on the right-hand sides of Eqs. (10.3.9)–(10.3.14) are defined as

$$R_{\alpha r_1\ldots r_m}^{mnq} = \sum_\beta\sum_{ijkl}\int\left[G_{\alpha r_1\ldots r_m}^{mn}\left(\xi_\alpha'\right)T_\alpha^{(q)}\left(\varepsilon_{\alpha k}\right) - G_{\alpha r_1\ldots r_m}^{mn}\left(\xi_\alpha\right)\right.$$
$$\left.\times T_\alpha^{(q)}\left(\varepsilon_{\alpha i}\right)\right] f_{\alpha i} f_{1\beta j}\, g\sigma\left(ij\mid kl, g, \chi\right)\sin\chi\, d\chi\, d\varphi\, d\mathbf{v}_\alpha d\mathbf{v}_{1\beta}, \tag{10.3.15}$$

where G_α^{mn} are related to H_α^{mn} via (4.2.6) and $T_\alpha^{(q)} = (kT)^q R_\alpha^{(q)}$.

After substituting the expansions (10.3.7) into (10.3.15), the values R_α^{mnq} are expressed in terms of a set of the same moments of the distribution function which occur on the left-hand sides of Eqs. (10.3.9)–(10.3.14). As previously, the right-hand sides comprise only moment linear terms which implies that the conditions

$$\Delta E_\alpha^{tr} \ll \frac{3}{2}kT, \quad \Delta E_\alpha^{int} \ll c_\alpha^{int}T, \quad \pi_{\alpha rs} \ll p_\alpha,$$
$$|\mathbf{w}_\alpha| \ll \left(\frac{kT}{m_\alpha}\right)^{1/2}, \quad |\mathbf{q}_\alpha^{tr}| \ll \left(\frac{kT}{m_\alpha}\right)^{1/2} n_\alpha kT, \tag{10.3.16}$$
$$|\mathbf{q}_\alpha^{int}| \ll \left(\frac{kT}{m_\alpha}\right)^{1/2} c_\alpha^{int} n_\alpha T$$

are met. The latter corresponds to considering the states of a gas mixture which does not differ from the equilibrium states too much.

When calculating $R_{\alpha r_1\ldots r_m}^{mnq}$, we pass to the variables $\mathbf{G}_\alpha, \mathbf{g}_{\alpha\beta}$ and $\mathbf{g}_{\alpha\beta}'$ in line with (1.2.6); in contrast to (1.2.8) the relative velocities before and after collisions are related by

$$\mathbf{g}' = g'\mathbf{i}\cos\chi + g'\mathbf{j}\sin\chi\cos\varphi + g'\mathbf{k}\sin\chi\sin\varphi,$$

where

$$g'^2 = g^2 - \frac{2kT}{\mu_{\alpha\beta}}\Delta\varepsilon_{\alpha\beta}, \quad \Delta\varepsilon_{\alpha\beta} = \varepsilon_{\alpha k} + \varepsilon_{1\beta l} - \varepsilon_{\alpha i} - \varepsilon_{1\beta j}.$$

After integrating over the angle φ and the variable $\mathbf{G}$ the coefficients at the dimensionless moments on the right-hand sides of Eqs. (10.3.9)–(10.3.14) are expressed in terms of the integral expressions of the kind

$$\begin{aligned}\langle F\rangle_{\alpha\beta} = & \left(\frac{kT}{2\pi\mu_{\alpha\beta}}\right)^{1/2} Q_\alpha^{-1} Q_\beta^{-1} \sum_{ijkl} \int_0^\infty d\zeta \int_0^{2\pi} d\varphi \int_0^\pi \sin\chi d\chi \\ & \times \left[F\zeta^3 \exp\left(-\zeta^2 - \varepsilon_{\alpha i} - \varepsilon_{1\beta j}\right) \sigma_{\alpha\beta}(ij \mid kl, \zeta, \chi)\right].\end{aligned} \tag{10.3.17}$$

When F is the function of ζ and ζ', these values are a generalization of the appropriate integrals $\Omega^{lr}_{\alpha\beta}$ introduced in the theory of monoatomic gases. However, the internal degrees of freedom bring in new integral values with which the characteristic relaxation times for the mean translational and internal energies are related in addition to corrections of the transport coefficients due to inelastic collisions.

We do not write the explicit expressions for R^{mnq}_α obtained as a result of the calculation procedure described above in the full form. They will be used later, in the analysis of the transport phenomena and relaxation in a pure polyatomic gas as well as when dealing with expressions for the kinetic coefficients in the case of a polyatomic gas mixture.

10.4 Transport and relaxation phenomena in a pure polyatomic gas

The characteristic features of the phenomena due to internal degrees of freedom of molecules can be revealed using an example of a pure polyatomic gas of molecules of one species. As in a weakly ionized gas, apart from electron and ion diffusion, the transport phenomena are also determined by the behaviour of its molecules, this case bears on the description of a weakly ionized molecular plasma.

Calculation of the right-hand sides of Eqs. (10.3.10)–(10.3.14) for a pure polyatomic gas results in

$$\begin{aligned} R^{010} = -R^{001} = -\frac{4}{3}\frac{c_v}{c^{int}} n^2 \Omega_E \Delta E^{tr}, \quad R^{200}_{rs} = -\frac{8}{5} n\Omega_\eta \pi_{rs}, \\ R^{110}_r = -\frac{16}{15} n\left(\Omega_\eta + \frac{25}{4}\Omega_E\right) q^{tr}_r + \frac{5}{3}\frac{k}{c^{int}} n\Omega_E q^{int}_r, \\ R^{101}_r = \frac{2}{3} n\Omega_E q^{tr}_r - \frac{8}{3} n\left(\Omega_D + \frac{k}{c^{int}}\Omega_E\right) q^{int}_r, \end{aligned} \tag{10.4.1}$$

where

$$\Omega_E = \langle \Delta\varepsilon^2 \rangle_{11}, \quad \Omega_\eta = \left\langle \zeta^2 \left(\zeta^2 - \zeta'^2 \cos^2\chi\right) - \frac{1}{6}\Delta\varepsilon^2 \right\rangle_{11},$$

$$\frac{c^{int}}{k}\Omega_D = \left\langle (\varepsilon_i - \langle\varepsilon\rangle)\left[(\varepsilon_i - \varepsilon_{1j})\zeta^2 - (\varepsilon_k - \varepsilon_{1l})\zeta\zeta'\cos\chi\right]\right\rangle_{11}, \tag{10.4.2}$$

and

$$\zeta = \left(\frac{m}{4kT}\right)^{1/2} \mathbf{g}, \quad \zeta'^2 = \zeta^2 - \Delta\varepsilon, \quad \Delta\varepsilon = \varepsilon_k + \varepsilon_{1l} - \varepsilon_i - \varepsilon_{1j},$$

$$c^{int} = k\left(\langle\varepsilon^2\rangle - \langle\varepsilon\rangle^2\right), \quad c_v = \frac{3}{2}k + c^{int}. \tag{10.4.3}$$

Consider first the homogeneous state of a polyatomic gas in which the basic hydrodynamic variables $\rho, \mathbf{u}$ and T do not depend on $\mathbf{r}$ and t, and the others are functions of time alone. Equations (10.3.10)–(10.3.14) are then reduced to

$$\frac{\partial \Delta E^{tr}}{\partial t} = -\frac{2}{3}\frac{c_v}{k}\tau_E^{-1}\Delta E^{tr}, \quad \frac{\partial \pi_{rs}}{\partial t} = -\tau_\eta^{-1}\pi_{rs},$$

$$\frac{\partial \mathbf{q}^{tr}}{\partial t} = -\frac{2}{3}\left(\tau_\eta^{-1} + \frac{5}{6}\frac{c^{int}}{k}\tau_E^{-1}\right)\mathbf{q}^{tr} + \frac{5}{6}\tau_E^{-1}\mathbf{q}^{int}, \tag{10.4.4}$$

$$\frac{\partial \mathbf{q}^{int}}{\partial t} = -\left(\tau_D^{-1} + \frac{1}{2}\tau_E^{-1}\right)\mathbf{q}^{int} + \frac{1}{3}\frac{c^{int}}{k}\tau_E^{-1}\mathbf{q}^{tr},$$

where

$$\tau_E^{-1} = \frac{2k}{c^{int}}n\Omega_E, \quad \tau_\eta^{-1} = \frac{8}{5}n\Omega_\eta, \quad \tau_D^{-1} = \frac{8}{3}n\Omega_D. \tag{10.4.5}$$

To estimate the order of magnitude of the characteristic times τ_η and τ_D, consider a limiting case when inelastic collisions can be neglected. Assuming that $g' = g$ and $\Delta\varepsilon = 0$ and that the effective cross section does not depend on the internal states of the colliding pair, we have

$$\Omega_\eta = \langle \zeta^4 \left(1 - \cos^2\chi\right)\rangle_{11} = \left[\sum_i \exp(-\varepsilon_i)\right]^{-2} \times \sum_i\sum_j \exp(-\varepsilon_i - \varepsilon_{1j})\,\Omega^{22} = \Omega^{22},$$

$$\frac{c^{int}}{k}\Omega_D = \left\langle \zeta^2 \left(\varepsilon_i - \langle\varepsilon\rangle\right)\left(\varepsilon_i - \varepsilon_{1j}\right)(1-\cos\chi)\right\rangle$$

$$= \left[\sum_i \exp(-\varepsilon_i)\right]^{-2} \sum_i\sum_j \left(\varepsilon_i^2 - \varepsilon_i\varepsilon_{1j}\right)$$

$$\times \exp\left(-\varepsilon_i - \varepsilon_{1j}\right)\Omega^{11} = \frac{c^{int}}{k}\Omega^{11},$$

where the relation $\langle\varepsilon^2\rangle - \langle\varepsilon\rangle^2 = c^{int}/k$ is used and Ω integrals corresponding to the familiar Chapman–Cowling integrals

$$\Omega^{lr} = \left(\frac{kT}{\pi m}\right)^{1/2} \int_0^\infty\int_0^\pi\int_0^{2\pi} \zeta^{2r+3}\left(1-\cos^l\chi\right)\sigma(g,\chi) \times \exp\left(-\zeta^2\right) d\zeta \sin\chi d\chi d\varphi$$

are introduced.

It will be recalled that

$$\Omega^{11} = \left(\frac{kT}{\pi m}\right)^{1/2}\pi d^2, \quad \Omega^{22} = 2\left(\frac{kT}{\pi m}\right)^{1/2}\pi d^2$$

for the model of elastic hard spheres. Then

$$\tau_\eta^{-1} = \frac{4}{5}\sqrt{2}n\langle v\rangle\pi d^2, \quad \tau_D^{-1} = \frac{2}{3}\sqrt{2}n\langle v\rangle\pi d^2,$$

$$\langle v\rangle = \left(\frac{8kT}{\pi m}\right)^{1/2}.$$

Thus, τ_η^{-1} and τ_D^{-1} are of the same order as the mean frequency of elastic collisions $\nu = \sqrt{2}n\langle v\rangle\pi d^2$ that is used in the elementary kinetic theory of gases. As for the characteristic time of the energy exchange between the translational and internal degrees of freedom of molecules τ_E, this parameter describes the result of inelastic collisions and, therefore, the ratio τ_E/τ_η can be arbitrary.

Let us represent the total pressure tensor P_{rs} as

$$P_{rs} = \left(p + \frac{2}{3}n\Delta E^{tr}\right)\delta_{rs} + \pi_{rs}, \quad p = nkT. \tag{10.4.6}$$

Then, from the solutions of Eqs. (10.4.4) for ΔE^{tr} and π_{rs} it follows that the off-diagonal components of this tensor and nonequilibrium additions to the diagonal part decrease. Parameters τ_η^{-1} and $(2c_v/3k)\,\tau_E^{-1}$

in the equations are, in fact, respective decrements of attenuation. In the case of slow energy exchange $\tau_E \gg \tau_\eta$, i.e. π_{rs} decreases much faster than ΔE^{tr}.

In the general case, the relaxation of the heat flux $\mathbf{q} = \mathbf{q}^{tr} + \mathbf{q}^{int}$ is described by a linear combination of exponents with decrements of attenuation defined by the roots of the characteristic equation of the system of related equations for $\mathbf{q}^{tr}$ and $\mathbf{q}^{int}$. In the case of slow energy exchange the crossed terms on the right-hand sides of Eqs. (10.4.4) for $\mathbf{q}^{tr}$ and $\mathbf{q}^{int}$ can be neglected, and the relaxation of the heat flux $\mathbf{q}$ is described by the expression

$$\mathbf{q}(t) = \mathbf{q}^{tr}(0)\exp\left(-\frac{2}{3}\frac{t}{\tau_\eta}\right) + \mathbf{q}^{int}(0)\exp\left(-\frac{2}{3}\frac{t}{\tau_D}\right).$$

Now let us examine the slightly inhomogeneous and slowly changing gas state. Their description implies, as was previously mentioned, that the macroscopic parameters of a gas slightly change with distance and time correspondent to the mean free path and time between the collisions of the molecules.

As $\tau_\eta \sim \tau_D \sim \tau \ll \tau_L$, the derivatives $d\pi_{rs}/dt$, $d\mathbf{q}^{tr}/dt$ and $d\mathbf{q}^{int}/dt$ are negligible on the left-hand sides of the equations of the viscous-stress tensor and heat fluxes. In the case of easy energy exchange, there is $\tau_E \sim \tau_\eta$ which permits us to neglect the derivative $d\Delta E^{tr}/dt$ in the equation for ΔE^{tr}. The left-hand inhomogeneous sides contain some nonlinear terms including the products of nonequilibrium parameters by spatial derivatives of $\mathbf{u}$ and T which can also be omitted in the slightly inhomogeneous states ($\lambda \ll L$) under consideration and as a consequence of (10.3.16). Finally, we come to the relationships

$$\frac{c^{int}}{c_v}p\nabla\cdot\mathbf{u} + \frac{c^{int}}{c_v}\nabla\cdot\mathbf{q}^{tr} - \frac{3}{2}\frac{k}{c_v}\nabla\cdot\mathbf{q}^{int} = -\frac{2}{3}\frac{c_v}{k}\tau_E^{-1}n\Delta E^{tr}, \quad (10.4.7)$$

$$2p\epsilon_{rs} + \frac{4}{5}\left\{\frac{\partial q_r^{tr}}{\partial x_s}\right\} = -\tau_\eta^{-1}\pi_{rs}, \quad (10.4.8)$$

$$\frac{5}{2}\frac{k}{m}p\frac{\partial T}{\partial x_r} + \frac{5}{3}\frac{p}{m}\frac{\partial\Delta E^{tr}}{\partial x_r} + \frac{kT}{m}\frac{\partial\pi_{rs}}{\partial x_s}$$
$$= -\frac{2}{3}\left(\tau_\eta^{-1} + \frac{5}{6}\frac{c^{int}}{k}\tau_E^{-1}\right)q_r^{tr} + \frac{5}{6}\tau_E^{-1}q_r^{int}, \quad (10.4.9)$$

$$\frac{c^{int}}{m}p\frac{\partial T}{\partial x_r} + \frac{p}{m}\frac{\partial\Delta E^{int}}{\partial x_r} = -\left(\tau_D^{-1} + \frac{1}{2}\tau_E^{-1}\right)q_r^{int} + \frac{1}{3}\frac{c^{int}}{k}\tau_E^{-1}q_r^{tr}. \quad (10.4.10)$$

On the left-hand sides of Eqs. (10.4.7)–(10.4.10), the terms with the derivations of $\mathbf{q}^{tr}, \mathbf{q}^{int}, \pi_{rs}$ and ΔE^{tr} are left. Some of them may be important when analyzing essentially non-one-dimensional problems and the problem of sound propagation in a polyatomic gas. In most cases, however, these terms can be omitted and Eqs. (10.4.7)–(10.4.10) result in common linear transport relations

$$\frac{2}{3}n\Delta E^{tr} = -\zeta\nabla\cdot\mathbf{u}, \quad \pi_{rs} = -2\eta\varepsilon_{rs}, \quad \mathbf{q} = -\lambda\nabla T, \qquad (10.4.11)$$

where η and ζ are the shear and volume viscosities, $\lambda = \lambda^{tr} + \lambda^{int}$ is the thermal conductivity, defined by

$$\eta = p\tau_\eta, \qquad \zeta = \frac{c^{int}k}{c_v^2}p\tau_E, \qquad (10.4.12)$$

$$\frac{\lambda^{tr}m}{\eta} = \frac{15}{4}k - \frac{5}{\pi Z}c^{int}\left(\frac{5}{2} - \frac{\rho D^{int}}{\eta}\right)\delta^{-1},$$

$$\frac{\lambda^{int}m}{\eta} = \frac{\rho D^{int}}{\eta}c^{int}\left[1 + \frac{2}{\pi Z}\left(\frac{5}{2} - \frac{\rho D^{int}}{\eta}\right)\delta^{-1}\right], \qquad (10.4.13)$$

$$\delta = 1 + \frac{2}{\pi Z}\left(\frac{5}{3}\frac{c^{int}}{k} + \frac{\rho D^{int}}{\eta}\right)$$

and

$$Z = \frac{4}{\pi}\frac{\tau_E}{\tau_\eta}, \qquad (10.4.14)$$

where Z is the collisional number which may be considered as the number of collisions required for the equilibration of mean internal and translational energies of molecules. The diffusion coefficient for the internal energy D^{int} is defined by the relation

$$\rho D^{int} = p\tau_D = \frac{3}{8}\frac{kT}{\Omega_D}. \qquad (10.4.15)$$

The expressions obtained fully agree with the results of the first Chapman–Enskog approximation for easy energy exchange [5,6].

Equations (10.4.8)–(10.4.10) and the resulting expressions of the viscosity η and thermal conductivity λ also remain essential in the case of slow energy exchange between translational and internal degrees of freedom. In particular, at $Z \gg 1$

$$\lambda = \left(\frac{15}{4}k + \frac{\rho D^{int}}{\eta}c^{int}\right)\frac{\eta}{m}. \qquad (10.4.16)$$

Assuming $D^{int} = D_{11}$, where D_{11} is the self-diffusion coefficient we come to the well-known modified Eucken expression for thermal conductivity [18].

Yet, the coefficient of volume viscosity ζ cannot be introduced in a straightforward manner in the case of slow energy exchange because at $\tau_E \gg \tau_\eta$ a full relaxation equation for $n\Delta E^{tr}$ (10.3.10) must be used instead of (10.4.7). Neglecting the non-linear terms, it is written in the case of a pure gas as

$$\frac{d\Delta E^{tr}}{dt} + \frac{c^{int}}{c_v}kT\nabla\cdot\mathbf{u} + \frac{1}{nc_v}\left(c^{int}\nabla\cdot\mathbf{q}^{tr} - \frac{3}{2}k\nabla\cdot\mathbf{q}^{int}\right)$$
$$= -\frac{2}{3}\frac{c_v}{k}\tau_E^{-1}\Delta E^{tr}. \qquad (10.4.17)$$

10.5 Two-temperature approximation. Relaxation equation of internal energy

Let us show that the energy relaxation equation (10.4.17) can be obtained if the two-temperature Maxwell–Boltzmann distribution function of the kind

$$f_i^{(0)} = \left(\frac{m}{2\pi kT^{tr}}\right)^{3/2} Q^{-1}\left(T^{int}\right)\exp\left(-\frac{mc^2}{2kT^{tr}} - \frac{E_i}{kT^{int}}\right) \qquad (10.5.1)$$

is used as a zero approximation (the weight function).

When inelastic collisions in the gas are relatively rare, it would be natural to consider that not only the total mean energy nE, but each of constituents nE^{tr} and nE^{int} separately are defined in the same way both in the full distribution function f_i and in the distribution function of the zero approximation (10.5.1). Hence,

$$E^{tr} = E_0^{tr} = \frac{3}{2}kT^{tr}, \quad E^{int} = E_0^{int} = kT^{int}\langle\varepsilon\rangle. \qquad (10.5.2)$$

In this approach, the moments of the distribution function of interest are $n, \mathbf{u}, T^{tr}$ and T^{int} as part of the weight function and $\pi_{rs}, \mathbf{q}^{tr}$ and $\mathbf{q}^{int}$ occurring in the expansion. The moment equations correspondent to them at $\tau \ll \tau_L$ and $\lambda \ll L_i$ bring us to the linear relationships for π_{rs} and $\mathbf{q}$ which are coincident with Eqs. (10.4.11) if we stick to the assumption used earlier as to the small deviations of the translational and internal energies from their equilibrium values at temperature T. Because of Eqs. (10.5.2), the pressure tensor is determined as

$$P_{rs} = nkT^{tr}\delta_{rs} + \pi_{rs}$$

and it does not include the explicit term of the volume viscosity. Instead, the full system of equations includes, together with the equation of conservation of the total energy, the relaxation equation for T^{tr} and T^{int}.

As we are interested in energy relaxation only, for simplicity $\pi_{rs}, \mathbf{q}^{tr}$ and $\mathbf{q}^{int}$ are assumed to be zero. In that case, expansion for f_i coincides with $f_i^{(0)}$. Expanding Eq. (10.5.1) into series over small relative differences $(T^{tr} - T)/T$ and $(T^{int} - T)/T$ and neglecting the squared terms, we find

$$\begin{aligned} f_i &= n\left(\frac{m}{2\pi kT}\right)^{3/2} Q^{-1}(T)\exp\left(-\frac{mc^2}{2kT} - \varepsilon_i\right) \\ &\times \left[1 + \frac{1}{2}\frac{T^{tr} - T}{T}\left(\xi^2 - 3\right) + \frac{T^{int} - T}{T}\left(\varepsilon_i - \langle\varepsilon\rangle\right)\right]. \end{aligned} \tag{10.5.3}$$

It is easily seen that for small energy deviations considered the relations

$$\Delta E^{tr} = E^{tr} - E_0^{tr} \approx \left(\frac{\partial E_0^{tr}}{\partial T^{tr}}\right)_{T^{tr}=T}\left(T^{tr} - T\right) = \frac{3}{2}k\left(T^{tr} - T\right),$$

$$\Delta E^{int} = E^{int} - E_0^{int} \approx \left(\frac{\partial E_0^{int}}{\partial T^{int}}\right)_{T^{int}=T}\left(T^{int} - T\right)$$

$$= c^{int}\left(T^{int} - T\right) \tag{10.5.4}$$

are approximately valid and the expansion (10.5.3) becomes equivalent to an earlier expansion (10.3.7) written for the case of a pure gas.

In the theory of two-temperature relaxation, the relaxation equations of translational or internal energy are commonly used instead of the equation of the energy deviations. They follow from the kinetic equation (1.2.23), when it is multiplied by $mc^2/2$ and E_i, respectively, integrated over velocities and summed over the internal states

$$\frac{dnE^{tr}}{dt} + nE^{tr}\nabla\cdot\mathbf{u} + P_{rs}\frac{\partial u_r}{\partial x_s} + \nabla\cdot\mathbf{q}^{tr} = R^{tr}, \tag{10.5.5}$$

$$\frac{dnE^{int}}{dt} + nE^{int}\nabla\cdot\mathbf{u} + \nabla\cdot\mathbf{q}^{int} = R^{int}. \qquad (10.5.6)$$

It is evident that summation of these equations brings in the conservation equation of total energy (10.2.19) or the equation for temperature T (10.2.21). It is seen that substituting E^{tr} for $(3/2)\,kT + \Delta E^{tr}$ and dT/dt of (10.2.20) into Eq. (10.5.5) gives the relaxation equation written earlier (10.4.7), as the right-hand side of Eq. (10.5.5) coincides with R^{010} calculated in the previous section. As, from the condition $\left(\Delta E^{tr} + \Delta E^{int}\right) = 0$ and relations (10.5.4), it follows that

$$T^{int} - T = -\frac{3}{2}\frac{k}{c^{int}}\left(T^{tr} - T\right)$$

the expression of R^{010} (10.4.1) can be represented as

$$R^{010} = -nc^{int}\tau_E^{-1}\left(T^{tr} - T^{int}\right). \qquad (10.5.7)$$

Equation (10.5.6) with allowance for the continuity equation (10.2.17) can be written as

$$n\frac{dE^{int}}{dt} + \nabla\cdot\mathbf{q}^{int} = -nc^{int}\tau_E^{-1}\left(T^{tr} - T^{int}\right), \qquad (10.5.8)$$

which is in agreement with the well-known equation of relaxation of the internal energy obtained in the semi-phenomenological theory of two-temperature relaxation [19].

So far only one kind of degrees of freedom has been considered for simplicity. If both rotational and vibrational degrees of freedom are simultaneously excited, a change of the scheme should be introduced. In particular, in the expansion of the distribution function, along with $\Delta E^{tr}, \Delta E^{rot}$ and $\mathbf{q}^{tr}, \mathbf{q}^{rot}$ there occur new variables ΔE^{vib} and $\mathbf{q}^{vib}$ for which special moment equations must be written. Correspondingly, the $21N$-moment approximation arises instead of $17N$-moment approximation [20]. It is essential that in the equations of energy relaxation and the equations of heat fluxes new crossed terms occur which depend on the characteristic collision frequencies $\left(\tau_E^{-1}\right)_{rv}$ and $\left(\tau_D^{-1}\right)_{rv}$ defined in terms of probabilities of transition with simultaneous changes of both rotational and vibrational quantum numbers.

The analysis of relaxation processes, with the conventional requirement met

$$\left(\tau_E\right)_{rot} \ll \left(\tau_E\right)_{vib} \ll \left(\tau_E\right)_{rv}, \qquad (10.5.9)$$

is given in [20]. It turns out that, if the relaxation of rotational and vibrational temperatures is considered at fixed T^{tr}, the separation of the relaxation equations describing independent relaxation processes of internal energies for every degree of freedom with respect to their equilibrium values defined by their translational temperature can be justified. In terms of these relationships the dispersion equation and the expressions of the phase velocity of the sound propagation c and the coefficient of sound absorption α following from it can be derived for the case when both rotational and vibrational degrees of freedom are taken into account.

The analysis proves that these expressions in the whole range of sound frequencies ω satisfying the requirement $\omega\tau \ll 1$ are well approximated by the relations resulting from the theory of sound propagation based on the phenomenological equations of relaxation. In particular, the dispersion of the phase velocity of sound propagation is described by [19,20]

$$c^2 = c_0^2\left[1+\frac{kc^{vib}}{c_p\left(c_v-c^{vib}\right)}\frac{\left(\omega\tau_{vib}''\right)^2}{1+\left(\omega\tau_{vib}''\right)^2} + \frac{2}{3\gamma}\frac{c^{rot}}{\left(c_v-c^{vib}\right)}\frac{\left(\omega\tau_{rot}'\right)^2}{1+\left(\omega\tau_{rot}'\right)^2}\right], \tag{10.5.10}$$

where c_0 is the sound velocity without dispersion, $\gamma = c_p/c_v$. For the relaxation part of the absorption coefficient, we have

$$\frac{2\alpha_{rel}c_0}{\omega} = \left(\frac{c_0}{c}\right)^3\left[\frac{kc^{vib}}{c_p\left(c_v-c^{vib}\right)}\frac{\omega\tau_{vib}''}{1+\left(\omega\tau_{vib}''\right)^2} + \frac{2}{3\gamma}\frac{c^{rot}}{\left(c_v-c^{vib}\right)}\frac{\left(\omega\tau_{rot}'\right)^2}{1+\left(\omega\tau_{rot}'\right)^2}\right]. \tag{10.5.11}$$

The times τ_{rot}' and τ_{vib}'' are related to the conventional times of the rotational and vibrational relaxation τ_{rot} and τ_{vib}

$$\tau_{rot}' = \frac{3}{2}\frac{k}{\left(c_v-c^{vib}\right)}\tau_{rot}, \quad \tau_{vib}'' = \frac{3}{2}\frac{\left(c_v-c^{vib}\right)}{c_v}\tau_{vib}.$$

Experimental data of the absorption of sound in molecular gases are often used to determine rotational and vibrational numbers of collisions Z_{rot} and Z_{vib} the knowledge of which, in turn, is necessary for theoretical estimates of the Eucken factor or thermal conductivity of polyatomic gases.

10.6 Estimates of contributions of inelastic collisions to kinetic coefficients

The expressions (10.4.12)–(10.4.13) allow us to calculate the values of all kinetic coefficients of interest if the quantum mechanical cross sections of scattering $\sigma(kl \mid ij, g, \chi)$ under the integral sign of the expressions for Ω_η, Ω_E and Ω_D are known. In the general case, the problem of establishing such cross sections by methods of quantum mechanics is rather complicated. Some advances have been made only in the analysis of cross sections of the rotational excitation of simpler molecules, e.g. H_2 [21,22]. For diatomic and linear polyatomic molecules the expressions for the matrix element T of the scattering matrix directly related to the cross section of the inelastic scattering were treated in [23]. Recently, a number of quantum mechanical calculations of effective cross sections for atom–diatomic molecule interactions was carried out for He–H_2, He–N_2, He–CO_2, Ar–H_2 systems (see [24] for a review of the results).

Some information on the contribution of inelastic collisions to the kinetic coefficients can be derived from fairly simple classical models of molecular interaction. The passage to the classical variables is accompanied by the replacement of summation over internal states with integrating over classical variables.

Let us consider, for example, a model of rough spheres of diameter d which is assumed to change the sign of the relative velocity of the spheres at the point of their collision. In this case,

$$\sigma(kl \mid ij, g, \chi) = \frac{d^2}{4}, \quad \varepsilon_i = \frac{I\omega^2}{2kT}, \quad \varepsilon_{1j} = \frac{I_1\omega_1^2}{2kT}, \tag{10.6.1}$$

where I is the moment of inertia, ω is the angular velocity of rotation. The integral value $\langle F\rangle$ is defined as

$$\langle F\rangle = \left(\frac{kT}{\pi m}\right)^{1/2} Q^{-2}\frac{d^2}{4}\int_0^\infty F\zeta^3 \exp(-\zeta^2)\,d\zeta \int \exp\left(-\frac{I\omega^2}{2kT}\right)d\omega$$

$$\times \int \exp\left(-\frac{I_1\omega_1^2}{2kT}\right)d\omega_1 \int_0^\pi \sin\chi d\chi \int_0^{2\pi} d\varphi,$$

$$Q = \int \exp\left(-\frac{I\omega^2}{2kT}\right)d\omega = \left(\frac{2\pi kT}{I}\right)^{3/2}. \tag{10.6.2}$$

These expressions can be calculated if we pass from variables ω and ω_1 to the variables

$$\Omega = \frac{I_1\omega_1 + I\omega}{I_1 + I_2}, \qquad \omega_1 - \omega_2 \tag{10.6.3}$$

and express the value F in terms of the same variables using the relationships of the dynamics of collisions for the model of rough spheres. The successive procedure permits us to find the expressions of the kinetic coefficients in the function of the dimensionless parameter $K = 4I/md^2$ (the values of K range from zero, when the molecular mass is concentrated in the centre of the sphere, to the maximum value of 2/3, when the mass is evenly distributed on the spherical surface).

First of all consider such an important characteristic of the inelastic interaction as the relaxation time τ_E. Substituting $F = (\Delta\varepsilon)^2$ in terms of the variables of (10.6.3) into the general equation (10.6.2) and integrating, we have

$$\left\langle (\Delta\varepsilon)^2 \right\rangle = \left(\frac{kT}{\pi m}\right)^{1/2} \frac{4K}{(1+K)^2} \pi d^2$$

and for the relaxation time

$$n\tau_E = \frac{3}{16}\frac{(1+K)^2}{K}\left(\frac{m}{\pi kT}\right)^{1/2}\frac{1}{d^2}. \tag{10.6.4}$$

Then the volume viscosity ζ is defined by the expression

$$\zeta = \frac{kc^{rot}}{c_v^2} p\tau_E = \left(\frac{mkT}{\pi}\right)^{1/2} \frac{(1+K)^2}{32K^2}\frac{1}{d^2}, \tag{10.6.5}$$

which agrees with the result obtained in the framework of the strict classical theory for the model of rough spheres [18].

Similarly, calculations of the self-diffusion coefficient D_{11} and diffusion coefficient for the internal energy D^{int} result in

$$nD_{11} = \frac{3}{8}\left(\frac{kT}{\pi m}\right)^{1/2} \frac{1+K}{1+2K}\frac{1}{d^2},$$

$$nD^{int} = \frac{3}{8}\left(\frac{kT}{\pi m}\right)^{1/2} \frac{(1+K)^2}{1+K+2K^2}\frac{1}{d^2}. \tag{10.6.6}$$

For their ratio, we have

$$\frac{D^{int}}{D_{11}} = \frac{(1+K)(1+2K)}{1+K+2K^2}. \tag{10.6.7}$$

It is evident that at the value K growing from 0 to 2/3 this ratio monotonously rises from unity to 1.52.

The expressions (10.6.6) also agree with the results of the strict theory of rough spheres. However, the expressions calculated in this way to obtain the shear viscosity η and, hence, for λ somewhat differ from the results of the classical theory. In particular, the calculation of η proves that this coefficient does not depend on K, i.e. it has the same value as in the case of usual hard elastic spheres. The strict classical theory states [18]

$$\eta = \eta_{el} (1+K)^2 \frac{1}{1+\frac{13}{6}K}, \quad \eta_{el} = \frac{5}{16} \left(\frac{mkT}{\pi} \right)^{1/2} \frac{1}{d^2}. \qquad (10.6.8)$$

The maximum difference between η and η_{el} in the range of values K is 14%.

The difference is due to the fact that the theory based on the quasi-classical Wang Chang and Uhlenbeck equation suggests the symmetry relationships between direct and reverse collisions, whereas in the classical treatment of the rough spheres this requirement is not met[1].

Using (10.6.2) and (10.6.6) we can define Z by the relationship

$$Z = \frac{4}{10\pi} \frac{6+13K}{K}.$$

It is evident that the least value of Z in the model of rough spheres at $K = 2/3$ is equal to $44/5\pi = 2.8$ collisions.

The model of rough spheres is of particular interest since the special feature of an inelastic interaction, vis. the transfer of translational into the rotational energy, is most prominent as compared with other models. However, the models of "loaded spheres" and "spherocylinders" are more realistic. In contrast to the model of rough spheres which does not require any variables defining the molecule orientation in space, in the model of loaded spheres the dynamic state of the molecule is characterized, in addition to $\mathbf{v}$ and $\boldsymbol{\omega}$, by the polar and azimuthal angles defining the orientation of the vector $\boldsymbol{\xi} = \xi \mathbf{e}$ directed off the center of mass of the sphere, at distance ξ from the geometrical centers of the sphere, to the sphere's center. Because of this, calculation of the values $\langle F \rangle$ for the models mentioned becomes complicated. The advantage of these models is the existence of reverse collisions in them which permits one to ensure full adequacy of the results obtained both in the classical treatment and on the basis of expressions (10.4.13)–10.4.14). Sandler

[1]Mathematically it is expressed in the nonconjugated linearized operator of collision. If the stationary principle of entropy production is used as the variation principle in calculating the transport coefficients for rough spheres, then the derived approximations were shown [9] to approximate the solution of integral equations with symmetry collisional operator and the transport coefficients turn out to be slightly increased, which is confirmed by the results described.

[25] compared the results of calculating $\lambda^{tr}m/\eta$ and $\lambda^{int}m/\eta$ with those resulting from Eq. (10.4.14) with Z_{rot} obtained in the framework of the same models. It turned out that even at $Z_{rot} > 2.5$ the results agree satisfactorily when $D_{rot} = D_{11}$. For a better agrement at arbitrary Z_{rot}, an approximate formula

$$D_{rot} = D_{11}\left(1 + \frac{0.27}{Z_{rot}} - \frac{0.44}{Z_{rot}^2} - \frac{0.90}{Z_{rot}^3}\right) \tag{10.6.9}$$

is proposed [25].

10.7 Thermal conductivity of molecular gases

Estimates in the previous section prove that the contribution of inelastic collisions to the shear viscosity and self-diffusion are not substantial. This is also confirmed by calculations of more realistic interaction potentials performed for N_2 and CO with the help of the method of classical trajectories [26,27]. The situation is different in the case of the thermal conductivity of diatomic and polyatomic gases. Let us dwell on this problem at some length. Instead of the expression of thermal conductivity λ, it would be more convenient to use the Eucken factor

$$f = \frac{m\lambda}{c_v\eta} = \frac{c^{tr}}{c_v}f^{tr} + \frac{c^{int}}{c_v}f^{int}, \tag{10.7.1}$$

where $c^{tr} = (3/2)\,k$.

Using (10.4.13), we have

$$\begin{aligned} f^{tr} &= \frac{5}{2} - \frac{10}{3\pi}\frac{c^{int}}{kZ}\delta^{-1}\left(\frac{5}{2} - \beta^{int}\right), \\ f^{int} &= \beta^{int}\left[1 + \frac{2}{\pi Z}\delta^{-1}\left(\frac{5}{2} - \beta^{int}\right)\right], \end{aligned} \tag{10.7.2}$$

where

$$\delta = 1 + \frac{2}{\pi Z}\left(\beta^{int} + \frac{5}{3}\frac{c^{int}}{k}\right), \quad \beta^{int} = \frac{\rho D^{int}}{\eta}. \tag{10.7.3}$$

For the total Eucken factor we obtain

$$f = \frac{15}{4}\frac{k}{c_v} + \frac{c^{int}}{c_v}\beta^{int} - \frac{2}{\pi Z}\frac{c^{int}}{c_v}\delta^{-1}\left(\frac{5}{2} - \beta^{int}\right)^2. \tag{10.7.4}$$

In practice, at $Z > 5$, an expression for the Eucken factor linearized with respect to Z^{-1} is often used [13,25]

$$f = \frac{15}{4}\frac{k}{c_v} + \frac{c^{int}}{c_v}\beta^{int} - \frac{2}{\pi Z}\frac{c^{int}}{c_v}\left(\frac{5}{2} - \beta^{int}\right)^2 . \qquad (10.7.5)$$

At $Z \gg 1$ and $D^{int} = D_{11}$ the latter expression is often translated into a modified Eucken formula

$$f = \frac{15}{4}\frac{k}{c_v} + \frac{\rho D_{11}}{\eta}\frac{c^{int}}{c_v}, \qquad c_v = \frac{3}{2}k + c^{int}. \qquad (10.7.6)$$

An important feature of the expressions (10.7.2) and (10.7.4) is the dependence on the collision number Z and the diffusion coefficient for the internal energy inherent in them. As was already mentioned, the values of Z can be found either from independent theoretical calculations or from experiments on sound absorption. Other useful information can be gained from the so called thermo-molecular pressure difference in a gas flowing through a capillary or a porous medium, in which the value λ^{tr} is directly determined [28].

As to D^{int}/D_{11}, its value can be obtained from the formula (10.6.9) or, in the case of the known Z, the ratio is found from comparison of theoretical and experimental values of the Eucken factor in a given range of temperatures.

The expressions used above for the Eucken factor f refer to the case when only one kind of degrees of freedom, e.g. rotational, is considered. In the temperature range where vibrational degrees of freedom are essentially excited, some transformation of expressions used is required. Thus, if the common conditions

$$1 < Z_{rot} \ll Z_{vib} \qquad (10.7.7)$$

are met, and small terms inversely proportional to the collision number Z_{rv} which correspond to the processes in the simultaneous changes of rotational and vibrational states of the colliding molecules are omitted, the expression for the Eucken factor can be written as [20,29]

$$f = \frac{c^{tr}}{c_v}f^{tr} + \frac{c^{rot}}{c_v}f^{rot} + \frac{c^{vib}}{c_v}f^{vib}, \qquad (10.7.8)$$

where f^{tr} and f^{rot} are defined by expression (10.7.2) with replacement of Z for Z_{rot} and β^{int} for β^{rot} and f^{vib} is equal to

$$f^{vib} = \frac{\rho D^{vib}}{\eta}. \tag{10.7.9}$$

For the case of polyatomic molecules, when several vibrational modes are excited, formula (10.7.8) allows a simple generalization referring only to the last term, namely

$$f^{vib} = \sum_{K} \frac{\rho D^k}{\eta} \frac{c^k}{c_v}, \tag{10.7.10}$$

where c^k is the vibrational heat capacity and D^k is the diffusion coefficient of the vibrational energy corresponding to the kth mode.

10.8 Transport phenomena in a polyatomic gas mixture

Now we shall turn to the equations for $\rho_\alpha \mathbf{w}_\alpha$ (10.3.9) and $\pi_{\alpha rs}, \mathbf{h}_\alpha^{tr}, \mathbf{h}_\alpha^{int}$ (10.3.12)–(10.3.14). Calculation of moments with respect to the collision integrals (10.3.15) on the right-hand sides of these equations and expansion for $f_{\alpha i}$ (10.3.7) enable us to express them in terms of a linear set of the same moments of the distribution function which occur on the left-hand sides of the equations. The analysis shows that the order of coefficients at the moments on the right-hand side of the equations is mainly determined by the values of $\tau_{\alpha\beta}^{-1}$, $\left(c_\alpha^{int}/k\right) \tau_{\alpha\beta}^{-1}$ and the values of $Z_{\alpha\beta}\tau_{\alpha\beta}^{-1}$. It will be recalled that $\tau_{\alpha\beta}$ is the characteristic time between collisions for α and β molecules, defined by the relationship

$$\tau_{\alpha\beta}^{-1} = \frac{n_\beta kT}{\mu_{\alpha\beta} n \left[\mathfrak{D}_{\alpha\beta}\right]_1}$$

and the parameter $Z_{\alpha\beta}$ characterizes the ratio of the mean frequencies of elastic and inelastic collisions of molecules α and β. These considerations allow us to neglect the time derivatives of the values $\rho_\alpha \mathbf{w}_\alpha, \pi_{\alpha rs}, \mathbf{h}_\alpha^{tr}$ and $\mathbf{h}_\alpha^{int}$ on the left-hand sides of the equations of these values as well as nonlinear terms as we did in the case of monoatomic molecules and a pure polyatomic gas when discussing slightly inhomogeneous and slowly changing states of a gas if the condition of (4.3.1) and (10.3.16) is met. The resulting system of linear algebraic equations of the diffusion velocities, reduced heat fluxes and viscous-stress tensors of species is as follows

$$\sum_\beta \frac{n_\alpha n_\beta kT}{n[\mathfrak{D}_{\alpha\beta}]_1}(w_{\alpha r}-w_{\beta r}) = -p\frac{\partial y_\alpha}{\partial x_r} - \left(y_\alpha - \frac{\rho_\alpha}{\rho}\right)\frac{\partial p}{\partial x_r}$$
$$-\frac{2}{3}\left(\frac{\partial n_\alpha \Delta E_\alpha^{tr}}{\partial x_r} - \frac{\rho_\alpha}{\rho}\frac{\partial n\Delta E^{tr}}{\partial x_r}\right) - \left(\frac{\partial \pi_{\alpha rs}}{\partial x_s} - \frac{\rho_\alpha}{\rho}\frac{\partial \pi_{rs}}{\partial x_s}\right)$$
$$-\sum_\beta \frac{m_\beta}{m_\alpha+m_\beta}\frac{n_\alpha n_\beta kT}{n[\mathfrak{D}_{\alpha\beta}]_1}\left(\frac{6}{5}C_{\alpha\beta}^*-1\right)\left(\frac{h_{\alpha r}^{tr}}{p_\alpha} - \frac{m_\beta}{m_\beta}\frac{h_{\beta r}^{tr}}{p_\beta}\right) \tag{10.8.1}$$
$$-\sum_\beta \frac{n_\alpha n_\beta kT}{n[\mathfrak{D}_{\alpha\beta}]_1}\left(E_{\alpha\beta}^*\frac{h_{\alpha r}^{int}}{n_\alpha c_\alpha^{int}T} - E_{\beta\alpha}^*\frac{h_{\beta r}^{int}}{n_\beta c_\beta^{int}T}\right),$$

$$-2y_\alpha \varepsilon_{rs} - \frac{4}{5p}\left\{\frac{\partial q_{\alpha r}^{tr}}{\partial x_s}\right\} = \sum_\beta H_{\alpha\beta}\frac{\pi_{\beta rs}}{y_\beta}, \tag{10.8.2}$$

$$-y_\alpha\frac{\partial T}{\partial x_r} - \frac{T}{p}\left(\frac{2}{3}\frac{\partial \pi_{\alpha rs}}{\partial x_s} + \frac{2}{3}\frac{\partial n_\alpha \Delta E_\alpha^{tr}}{\partial x_r}\right)$$
$$-\frac{T}{p}\sum_\beta \frac{m_\beta}{m_\alpha+m_\beta}\frac{n_\alpha n_\beta kT}{n[\mathfrak{D}_{\alpha\beta}]_1}\left(\frac{6}{5}C_{\alpha\beta}^*-1\right)(w_{\alpha r}-w_{\beta r})$$
$$= \sum_\beta \Lambda_{\alpha\beta}\frac{h_{\beta r}^{tr}}{y_\beta} + \sum_\beta \frac{k}{c_\beta^{int}}\Lambda'_{\alpha\beta}\frac{h_{\beta r}^{int}}{y_\beta}, \tag{10.8.3}$$

$$-y_\alpha\frac{c_\alpha^{int}}{k}\frac{\partial T}{\partial x_r} - \frac{T}{p}\frac{\partial n_\alpha \Delta E_\alpha^{int}}{\partial x_r} - \frac{T}{p}\sum_\beta \frac{n_\alpha n_\beta kT}{n[\mathfrak{D}_{\alpha\beta}]_1}E_{\alpha\beta}^*(w_{\alpha r}-w_{\beta r})$$
$$= \sum_\beta G_{\alpha\beta}\frac{h_{\beta r}^{tr}}{y_\beta} + \sum_\beta \frac{k}{c_\beta^{int}}G'_{\alpha\beta}\frac{h_{\beta r}^{int}}{y_\beta} \tag{10.8.4}$$

and the coefficients $H_{\alpha\beta}, \Lambda_{\alpha\beta}, \Lambda'_{\alpha\beta}, G_{\alpha\beta}$ and $G'_{\alpha\beta}$ are defined by the expressions

$$H_{\alpha\alpha} = \frac{y_\alpha^2}{[\eta_{\alpha\alpha}]_1} + \sum_{\gamma\neq\alpha}\frac{2y_\alpha y_\gamma}{(m_\alpha+m_\gamma)n[\mathfrak{D}_{\alpha\gamma}]_1}\left(1+\frac{3}{5}\frac{m_\gamma}{m_\alpha}A_{\alpha\gamma}^*\right),$$

$$H_{\alpha\beta} = -\frac{2y_\alpha y_\beta}{(m_\alpha+m_\beta)n[\mathfrak{D}_{\alpha\beta}]_1}\left(1-\frac{3}{5}A_{\alpha\beta}^*\right), \qquad \beta\neq\alpha, \tag{10.8.5}$$

$$\Lambda_{\alpha\alpha} = \frac{4}{15}\frac{m_\alpha y_\alpha^2}{k[\eta_{\alpha\alpha}]_1}\left(1+\frac{5}{6}\frac{c_\alpha^{int}}{k}Z_{\alpha\alpha}^{-1}\right)$$
$$+\frac{4}{25}\sum_{\gamma\neq\alpha}\frac{2y_\alpha y_\gamma}{(m_\alpha+m_\gamma)n[\mathfrak{D}_{\alpha\gamma}]_1}\left[\frac{15}{2}m_\alpha^2+\left(\frac{25}{4}-3B_{\alpha\gamma}^*\right)m_\gamma^2,\right.$$
$$\left.+4A_{\alpha\gamma}^*\left(1+\frac{5}{6}\frac{c_\alpha^{int}}{k}Z_{\alpha\gamma}^{-1}\right)m_\alpha m_\gamma\right],$$

$$\Lambda'_{\alpha\alpha} = G_{\alpha\alpha} = -\sum_{\gamma\neq\alpha}\frac{m_\gamma y_\alpha y_\gamma}{(m_\alpha+m_\gamma)kn[\mathfrak{D}_{\alpha\gamma}]_1}$$
$$\times\left[\frac{4}{5}\frac{m_\alpha}{m_\gamma}\frac{c_\alpha^{int}}{k}A_{\alpha\gamma}^*\left(L_{\alpha\gamma}^*+M_{\alpha\gamma}^*\right)Z_{\alpha\gamma}^{-1}-F_{\alpha\gamma}^*\right]$$
$$-\frac{\pi}{3}\frac{m_\alpha y_\alpha^2}{3k[\eta_{\alpha\alpha}]_1}\frac{c_\alpha^{int}}{k}Z_{\alpha\alpha}^{-1}, \tag{10.8.6}$$

$$\Lambda'_{\alpha\beta} = G_{\beta\alpha} = -\frac{m_\beta y_\alpha y_\beta}{(m_\alpha+m_\gamma)kn[\mathfrak{D}_{\alpha\beta}]_1}$$
$$\times\left[\frac{4}{5}\frac{c_\alpha^{int}}{k}A_{\alpha\beta}^*\left(L_{\beta\alpha}^*+M_{\alpha\beta}^*\right)Z_{\alpha\beta}^{-1}-F_{\beta\alpha}^*\right],\quad \beta\neq\alpha,$$

$$\Lambda_{\alpha\beta} = -\frac{4}{25}\frac{m_\alpha m_\beta y_\alpha y_\beta}{(m_\alpha+m_\beta)^2kn[\mathfrak{D}_{\alpha\beta}]_1}$$
$$\times\left[\frac{55}{4}-3B_{\beta\alpha}^*-4A_{\alpha\beta}^*\left(1+\frac{5}{6}\frac{c_\alpha^{int}}{k}Z_{\alpha\beta}^{-1}\right)\right],\quad \beta\neq\alpha, \tag{10.8.7}$$

$$G'_{\alpha\alpha} = \frac{y_\alpha^2}{kn[\mathfrak{D}_{\alpha\alpha}]_1}\left(\frac{c_\alpha^{int}}{k}+K_{\alpha\alpha}^*-N_{\alpha\alpha}^*\right)$$
$$+\frac{\pi}{2}\frac{m_\alpha y_\alpha^2}{k[\eta_{\alpha\alpha}]_1}\frac{c_\alpha^{int}}{k}Z_{\alpha\alpha}^{-1}+\sum_{\gamma\neq\alpha}\frac{y_\alpha y_\gamma}{kn[\mathfrak{D}_{\alpha\beta}]_1}$$
$$\times\left(\frac{c_\alpha^{int}}{k}+\frac{6}{5}\frac{m_\alpha}{m_\gamma}\frac{c_\alpha^{int}}{k}A_{\alpha\gamma}^*L_{\alpha\gamma}Z_{\alpha\gamma}^{-1}+K_{\alpha\gamma}^*\right), \tag{10.8.8}$$

$$G'_{\alpha\beta} = \frac{y_\alpha y_\gamma}{kn[\mathfrak{D}_{\alpha\beta}]_1}\left(\frac{6}{5}\frac{c_\alpha^{int}}{k}A_{\alpha\beta}^*M_{\alpha\beta}^*Z_{\alpha\beta}^{-1}-N_{\alpha\beta}^*\right),\quad \beta\neq\alpha.$$

The coefficients $[\eta_{\alpha\alpha}]_1$, $[\mathfrak{D}_{\alpha\beta}]_1$, $A_{\alpha\beta}^*$, $B_{\alpha\beta}^*$ and $C_{\alpha\beta}^*$ as parts of the above equations are defined similarly to Eqs. (5.1.8)–(5.1.10) of Chapter 5, with the difference that $\Omega_{\alpha\beta}^{lr}$ is written as

$$\Omega^{11}_{\alpha\beta} = \left\langle \zeta^2 - \zeta\zeta' \cos\chi \right\rangle_{\alpha\beta}, \quad \Omega^{12}_{\alpha\beta} = \left\langle \zeta^2 \left(\zeta^2 - \zeta\zeta' \cos\chi \right) \right\rangle_{\alpha\beta},$$
$$\Omega^{13}_{\alpha\beta} = \left\langle \zeta^2 \left(\zeta^4 - \zeta\zeta'^3 \cos\chi \right) \right\rangle_{\alpha\beta},$$
$$\Omega^{22}_{\alpha\beta} = \left\langle \zeta^2 \left(\zeta^2 - \zeta\zeta'^2 \cos^2\chi \right) - \frac{1}{6}\Delta\varepsilon^2_{\alpha\beta} \right\rangle_{\alpha\beta}, \tag{10.8.9}$$

and they are a generalization of the Chapman–Cowling integrals.

The other values of (10.8.5) and (10.8.8) are essentially related to the contribution of inelastic collisions and the expressions for them look like

$$A^*_{\alpha\beta} Z^{-1}_{\alpha\beta} = \frac{5}{8}\frac{k}{c^{int}_\alpha}\frac{\left\langle \Delta\varepsilon^2_{\alpha\beta} \right\rangle_{\alpha\beta}}{\Omega^{11}_{\alpha\beta}},$$

$$E_{\alpha\beta} = \frac{1}{\Omega^{11}_{\alpha\beta}} \left\langle \left(\varepsilon_{\alpha i} - \langle \varepsilon_\alpha \rangle\right) \left(\zeta^2 - \zeta\zeta' \cos\chi \right) \right\rangle_{\alpha\beta},$$

$$F^*_{\alpha\beta} = \frac{1}{\Omega^{11}_{\alpha\beta}} \left\langle \left(\varepsilon_{\alpha i} - \langle \varepsilon_\alpha \rangle\right) \left[\frac{2}{5} \left(\zeta^4 - \zeta\zeta'^3 \cos\chi \right) - \left(\zeta^2 - \zeta\zeta' \cos\chi \right) \right] \right\rangle_{\alpha\beta},$$

$$K^*_{\alpha\beta} = \frac{1}{\Omega^{11}_{\alpha\beta}} \left\{ \left\langle \left(\varepsilon_{\alpha i} - \langle \varepsilon_\alpha \rangle\right) \left[\zeta^2 \left(\varepsilon_{\alpha i} - \langle \varepsilon_\alpha \rangle\right) - \zeta\zeta' \left(\varepsilon_{\alpha k} - \langle \varepsilon_\alpha \rangle\right) \cos\chi \right] \right\rangle_{\alpha\beta} - \frac{c^{int}_\alpha}{k} \left\langle \zeta^2 - \zeta\zeta' \cos\chi \right\rangle_{\alpha\beta} \right\},$$

$$N^*_{\alpha\beta} = \frac{1}{\Omega^{11}_{\alpha\beta}} \left\langle \left(\varepsilon_{1\beta j} - \langle \varepsilon_\beta \rangle\right) \left[\zeta^2 \left(\varepsilon_{\alpha i} - \langle \varepsilon_\alpha \rangle\right) - \zeta\zeta' \left(\varepsilon_{\alpha k} - \langle \varepsilon_\alpha \rangle\right) \cos\chi \right] \right\rangle_{\alpha\beta},$$

$$L^*_{\alpha\beta} = \frac{\langle \Delta\varepsilon^2_\alpha \rangle_{\alpha\beta}}{\left\langle \Delta\varepsilon^2_{\alpha\beta} \right\rangle_{\alpha\beta}}, \qquad M^*_{\alpha\beta} = \frac{\langle \Delta\varepsilon_\alpha \Delta\varepsilon_{1\beta} \rangle}{\left\langle \Delta\varepsilon^2_{\alpha\beta} \right\rangle_{\alpha\beta}}. \tag{10.8.10}$$

Note that because of $L^*_{\alpha\beta} + 2M^*_{\alpha\beta} + L^*_{\beta\alpha} = 1$, we have $0 \le L^*_{\alpha\beta} \le 1$, $0 \le M^*_{\alpha\beta} \le 1$. The coefficients $E^*_{\beta\alpha}$ and $F^*_{\beta\alpha}$ can be obtained from $E^*_{\alpha\beta}$ and $F^*_{\alpha\beta}$ with the substitution of indices α for β and αi for $1\beta j$.

Equations (10.8.1)–(10.8.4) form linear transport relationships describing diffusion, viscosity and heat transport in a polyatomic gas mixture and their interrelationship. They should be complemented by the equations for $n_\alpha \Delta E_\alpha^{tr}$ and $n_\alpha \Delta E_\alpha^{int}$ in which, following the condition (10.3.16), nonlinear terms can be eliminated. However, in the general case the derivatives $d\Delta E_\alpha^{tr}/dt$ and $n_\alpha \Delta E_\alpha^{int}/dt$ in these equations should be preserved since the order of the right-hand side of these equations is mainly defined by the values of $\tau_{E\alpha\beta}^{-1} \sim Z_{\alpha\beta}^{-1}\tau_{\alpha\beta}^{-1}$ and at $Z_{\alpha\beta} \gg 1$, i.e. for the slow energy exchange, it may happen that $\tau_{E\alpha\beta} \sim \tau_L$. The relaxation equations and expressions for the volume viscosity of the mixture resulting from them will be discussed in the next section.

When looking into the vector and tensorial transport properties, the contribution of the derivatives $\partial n_\gamma \Delta E_\gamma^{tr}/\partial x_r$ and $\partial n_\gamma \Delta E_\gamma^{int}/\partial x_r$ as well as $\partial q_{\gamma r}^{tr}/\partial x_r$ and $\partial q_{\gamma r}^{int}/\partial x_r$ can be neglected in the equations. The derivatives $\partial \pi_{\gamma r s}/\partial x_s$ can be kept if we are interested in the diffusion in the viscous flow of the mixture. Analysis shows however, that, if the corrections of the second order to the barodiffusion factor are neglected, then the results accounting for $\partial \pi_{\gamma r s}/\partial x_s$ are fully equivalent to the expressions obtained in the case of monoatomic molecules. Therefore, we do not intend to consider the effect of the viscous transport of momentum on diffusion.

Resolving Eqs. (10.8.3) and (10.8.4) with respect to $\mathbf{h}_\alpha^{tr}$ and $\mathbf{h}_\alpha^{int}$ and substituting the expressions obtained into (10.8.1) we are led to the diffusion equations in a polyatomic gas mixture of formally the same kind as Eqs. (5.4.4) for a mixture of monoatomic molecules. The difference arises only in the definition of the coefficients of $[\mathfrak{D}_{\alpha\beta}]_1$ and D_γ^T. The value $[\mathfrak{D}_{\alpha\beta}]_1 = 3kT/16n\mu_{\alpha\beta}\Omega_{\alpha\beta}^{11}$ differs only in the generalized form of writing $\Omega_{\alpha\beta}^{11}$ and for D_α^T we have

$$D_\alpha^T = -\frac{m_\alpha}{k}\left[\frac{m_\beta}{m_\alpha + m_\beta}\left(\frac{6}{5}C_{\alpha\beta}^* - 1\right)\lambda_\alpha^{tr} + \frac{k}{c_\alpha^{int}}E_{\alpha\beta}^*\lambda_\alpha^{int}\right] \quad (10.8.11)$$

and D_β^T is obtained from D_α^T by replacing the indices. The partial thermal conductivities λ_α^{tr} and λ_α^{int} are defined by the expressions

$$\lambda_\alpha^{tr} = -y_\alpha \Delta^{-1}\begin{vmatrix} \Lambda_{\beta\gamma} & \Lambda'_{\beta\gamma} & y_\beta \\ G_{\beta\gamma} & G'_{\beta\gamma} & \frac{y_\beta c_\beta^{int}}{k} \\ \delta_{\alpha\gamma} & 0 & 0 \end{vmatrix},$$

$$\lambda_\alpha^{int} = -y_\alpha \frac{c_\alpha^{int}}{k} \Delta^{-1} \begin{vmatrix} \Lambda_{\beta\gamma} & \Lambda'_{\beta\gamma} & y_\beta \\ G_{\beta\gamma} & G'_{\beta\gamma} & \frac{y_\beta c_\beta^{int}}{k} \\ 0 & \delta_{\alpha\gamma} & 0 \end{vmatrix}, \tag{10.8.12}$$

$$\Delta = \begin{vmatrix} \Lambda_{\beta\gamma} & \Lambda'_{\beta\gamma} \\ G_{\beta\gamma} & G'_{\beta\gamma} \end{vmatrix},$$

where for simplicity the notations under $\Lambda_{\beta\gamma}, \Lambda'_{\beta\gamma}, G_{\beta\gamma}$ and $G'_{\beta\gamma}$ imply blocks of the order of N of the relevant elements. The values of $y_\beta, y_\beta\left(c_\beta^{int}/k\right)$ mean the columns of N elements where N is the number of species.

The expression of the total heat flux in the mixture according to (10.2.20) and with the solutions of $\mathbf{h}_\alpha^{tr}$ and $\mathbf{h}_\alpha^{int}$ acquires the form

$$\mathbf{q} = -\lambda \nabla T + \sum_\alpha \left(\frac{5}{2} + \langle\varepsilon_\alpha\rangle\right) p_\alpha \mathbf{w}_\alpha + \sum_{\alpha,\beta} \frac{n_\beta k T D_\alpha^T}{m_\alpha n \left[\mathfrak{D}_{\alpha\beta}\right]_1} \left(\mathbf{w}_\alpha - \mathbf{w}_\beta\right), \tag{10.8.13}$$

where

$$\lambda = \sum_\alpha \left(\lambda_\alpha^{tr} + \lambda_\alpha^{int}\right).$$

As far as the expression of the viscous-stress tensor π_{rs} is concerned, its solution is of the same form as for a mixture of monoatomic molecules but the coefficients $H_{\alpha\alpha}$ and $H_{\alpha\beta}$, with the help of which the viscosity coefficient of the mixture η is expressed, are written in terms of $\Omega_{\alpha\beta}^{11}$ and $\Omega_{\alpha\beta}^{22}$ generalized following the definition of Eqs. (10.8.9).

Note that to calculate η requires the knowledge of the value of $A_{\alpha\beta}^*$ in addition to $[\eta_{\alpha\alpha}]_1$ and $[\mathfrak{D}_{\alpha\beta}]_1$. Analysis shows that the inelastic collisions slightly affect $A_{\alpha\beta}^*$ and, hence, it is possible to use its values calculated on the basis of conventional interaction potentials for monoatomic gases. Due to that fact the viscosity of the mixture, as well as the barodiffusion factor in the viscous flow $[\alpha_p]_1$, can be well approximated by the same expressions as in the case of a monoatomic mixture using the experimental values of the viscosity $\eta_{\alpha\alpha}$ and the binary diffusion coefficient $\mathfrak{D}_{\alpha\beta}$ for a realistic polyatomic gas.

It is more difficult to calculate the thermal conductivity and thermal diffusion coefficients. For small values of $Z_{\alpha\beta}^{-1}$ the λ_α^{tr} and λ_α^{int} expressions can be essentially simplified by linearizing over the small parameter similar to the passage from Eqs. (10.7.4) to (10.7.5) for a simple polyatomic gas. Yet, even in this case the expressions are too cumbersome. For the case of the slow energy exchange ($Z_{\alpha\beta} \gg 1$), the terms including $Z^{-1}_{\alpha\beta}$ can be neglected. In such a case the coefficients $E^*_{\alpha\beta}, F^*_{\alpha\beta}, K^*_{\alpha\beta}$ and $N^*_{\alpha\beta}$ are assumed to be small, at least their contribution progressively decreases with the decreasing significance of the inelastic collisions as some estimates for simple molecular modes show [13]. The coefficients are shown [13] to vanish when the relative velocity before and after collisions does not depend on the starting internal states of the colliding particles (e.g. with similar differential cross sections of the collisions for all scattering channels). Under these conditions, we can take

$$\Lambda'_{\alpha\alpha} = \Lambda'_{\alpha\beta} = G_{\alpha\alpha} = G'_{\alpha\beta} = 0,$$

$$G'_{\alpha\alpha} = \frac{c_\alpha^{int}}{k}\left[\frac{y_\alpha^2}{kn\,[\mathfrak{D}_{\alpha\alpha}]_1} + \sum_{\gamma\neq\alpha}\frac{y_\alpha y_\gamma}{kn\,[\mathfrak{D}_{\alpha\gamma}]_1}\right],$$

and $[\mathfrak{D}_{\alpha\beta}]_1$ and the values left in $\Lambda_{\alpha\alpha}$ and $\Lambda_{\alpha\beta}$ with the accuracy up to the negligible contribution of inelastic collisions are consistent with the values calculated on the basis of the theory of monoatomic gases and then we have

$$\lambda_\alpha^{tr} = y_\alpha \sum_{\beta=1}^{N}\frac{y_\beta\,|\Lambda|_{\beta\alpha}}{|\Lambda|}, \quad \lambda_\alpha^{int} = \frac{c_\alpha^{int}}{k}\frac{y_\alpha^2}{G'_{\alpha\alpha}} \tag{10.8.14}$$

and for the total thermal conductivity

$$\lambda_{ful} = \lambda + \sum_\alpha n\,[\mathfrak{D}_{\alpha\alpha}]_1\, c_\alpha^{int}\left(1 + \sum_{\beta\neq\alpha}\frac{y_\beta\,[\mathfrak{D}_{\alpha\alpha}]_1}{y_\alpha\,[\mathfrak{D}_{\alpha\beta}]_1}\right)^{-1}, \tag{10.8.15}$$

where λ coincides with the thermal conductivity of a mixture of monoatomic molecules (5.2.13) and an extra term justifies the expression introduced by Hirschfelder for the generalized Eucken correction in the case of a polyatomic mixture [30].

The diffusion equations for a polyatomic mixture (without the allowance for the corrections of the second approximation to the diffusion coefficients) can be represented as

$$\sum_\beta \frac{n_\alpha n_\beta kT}{n\,[\mathfrak{D}_{\alpha\beta}]_1}(\mathbf{w}_\alpha - \mathbf{w}_\beta) = -p\mathbf{d}_\alpha - \sum_\beta \frac{n_\alpha n_\beta}{n} kT\alpha^T_{\alpha\beta}\nabla \ln T, \tag{10.8.16}$$

where the generalized thermal diffusion factor $\alpha^T_{\alpha\beta}$ is defined by

$$\alpha^T_{\alpha\beta} = \frac{\mu_{\alpha\beta}}{kn\left[\mathfrak{D}_{\alpha\beta}\right]_1}\left(\frac{6}{5}C^*_{\alpha\beta}-1\right)\left(\frac{\lambda^{tr}_\beta}{y_\beta m_\beta}-\frac{\lambda^{tr}_\alpha}{y_\alpha m_\alpha}\right) + \frac{1}{n\left[\mathfrak{D}_{\alpha\beta}\right]_1}\left(E^*_{\beta\alpha}\frac{\lambda^{int}_\beta}{y_\beta c^{int}_\beta}-E^*_{\alpha\beta}\frac{\lambda^{int}_\alpha}{y_\alpha c^{int}_\alpha}\right). \quad (10.8.17)$$

An important consideration following from this expression is that the thermal diffusion factor along with the usual term proportional to $(6/5)C^*_{\alpha\beta}-1$ as in the case of a mixture of monoatomic molecules includes the terms proportional to $E^*_{\beta\alpha}$ and $E^*_{\alpha\beta}$ essentially related to the inelastic interaction of molecules. Besides, if the value of $C^*_{\alpha\beta}$ is defined as is often done on the basis of the experimental dependence of $\mathfrak{D}_{\alpha\beta}$ on temperature, then, following the relation [31]

$$\frac{6}{5}C^*_{\alpha\beta}-1=\frac{2}{5}\left[2-\left(\frac{\partial\ln\mathfrak{D}_{\alpha\beta}}{\partial\ln T}\right)-E^*_{\alpha\beta}-E^*_{\beta\alpha}\right]$$

the knowledge of $E^*_{\alpha\beta}$ and $E^*_{\beta\alpha}$ is essential when calculating that part of the thermal diffusion factor which is related to λ^{tr}_β and λ^{tr}_α. It is to be noted that at $Z_{\alpha\beta}\gg 1$ approximate expressions (10.8.14) for λ^{tr}_β and λ^{int}_β may be used. For mixtures with substantially differing masses and average cross section of the molecules of species, the translational part of the thermal diffusion factor proves to be greater than the one related to $E^*_{\alpha\beta}$ and $E^*_{\beta\alpha}$ (of course, if the value $(6/5)C^*_{\alpha\beta}-1$ does not cancel which happens in the case of Maxwellian molecules). It also explains the fact that for most mixtures calculations of the thermal diffusion factor based on the theory of monoatomic gases give, as a rule, fairly satisfactory results. However, for mixtures with close or practically the same mass of the molecules of the components (A–CO_2, D_2–HT etc.) the contribution of the inelastic terms may become quite significant. Thus, the calculations of the thermal diffusion of mixture D_2–HT, ^{4}He–HT, and ^{3}He–HD with the model of loaded spheres were considered [32,33]. For a purely elastic interaction of molecules, calculation for a mixtures D_2-HT gives the value $\alpha_T^{theor}=2.4\cdot 10^{-4}$ [32] which markedly differs from the experiment ($\alpha_T^{theor}=0.023$–0.028 [34,35]). Calculations for the model of the loaded spheres proves that the main contribution to α_T is related to the inelastic part of the thermal diffusion factor which is equivalent to the allowance for $E^*_{\beta\alpha}$ in the expression (10.8.17) in which the coefficient $E^*_{\alpha\beta}$ is identically equal to zero in the case under study. In this case, we have $\alpha_T^{theor}=0.052$. Similar calculations for mixtures ^{4}He–HT and ^{3}He–HD [33] allowed one to estimate the effect of the differences in the effective diameters of the molecules (dimensional effects) and purely inelastic contributions which also improved the agreement with the experiment.

To conclude, we make a few remarks concerning the effects of inelastic collisions of charged particles and molecules on transport properties. In a weakly ionized gas, the viscosity and thermal conductivity are mainly defined by collisions of neutral particles and, thus, the electrons and ions available in it do not in fact influence these transport coefficients. However, diffusion and thermal diffusion of charged particles in a molecular plasma will be generally determined by both elastic and inelastic collisions. The expressions obtained allow us to take into account the effect of inelastic collisions of electrons and ions with molecules on the appropriate kinetic coefficients. The Einstein relationship (3.5.7) being rigorously valid as before for the diffusion and mobility coefficients in a weakly ionized molecular plasma. To estimate real contributions to the diffusion and thermal diffusion coefficients due to excitation of rotational and vibrational degrees of molecular freedom in their collisions with electrons or molecular ions would require calculations based on the knowledge of specific potentials of particle interaction. Though such calculations have not been performed so far, some general estimations prove that the coefficients of ion mobility and ion diffusion in a molecular gas do not practically differ from the values calculated in terms of model potentials in which the effect of inelastic processes is neglected. Yet, this statement cannot be applied for thermal diffusion since in the polarization interaction of ion and atom ($U \sim r^{-4}$) the value $(6/5)\, C^*_{\alpha\beta} - 1 = 0$, i.e. thermal diffusion, because of purely elastic collisions, is not present at all.

Similar considerations seem to be valid in the case of electron transport properties. Quantum mechanical calculations of cross sections of rotational and vibrational excitation of molecules by the electron impact show that the inelastic diffusion cross sections are far smaller than the relevant cross sections of elastic collisions which results in a negligibly small contribution of inelastic processes to the electron diffusion coefficients and mobility. However, if the frequency of elastic electron–molecule collisions turns out to be approximated constant ($\nu_{e\beta}(v) =$ const), which corresponds to $\zeta_{e\beta} = (6/5)C^*_{\alpha\beta} - 1 = 0$, a possible effect of inelastic collisions on the electron thermal diffusion factor should not be excluded, either.

10.9 Energy relaxation and volume viscosity of a mixture

Now look into the analysis of the relaxation equations for scalar nonequilibrium parameters ΔE^{tr}_{α} and ΔE^{int}_{α}. As was pointed out, the coefficients in these parameters on the right-hand sides of the equations can be of the order of $Z_{\alpha\beta}\tau_{\alpha\beta}^{-1}$ and at $Z \gg 1$ the time derivatives of ΔE^{tr} and ΔE^{int} on the left-hand sides cannot be neglected, as we do in the analysis of vector and tensor transport phenomena. After eliminating the derivatives dn_α/dt and dT/dt from the left-hand side of the equations with the help of continuity and energy equations, neglecting the nonlinear terms (including the divergence of diffusion and heat fluxes) the relaxation equations can be written as

$$n_\alpha \frac{d\Delta E_\alpha^{tr}}{dt} + \frac{c^{int}}{c_v}\nabla\cdot\mathrm{u} = -\frac{2}{3}\sum_\beta c_{\alpha\beta}\frac{n_\beta \Delta E_\beta^{tr}}{y_\beta} + \sum_\beta \frac{k}{c_\beta^{int}} c'_{\alpha\beta}\frac{n_\beta \Delta E_\beta^{int}}{y_\beta},$$

$$n_\alpha \frac{d\Delta E_\alpha^{int}}{dt} - \frac{c_\alpha^{int}}{c_v}\nabla\cdot\mathrm{u} = \frac{2}{3}\sum_\beta d_{\alpha\beta}\frac{n_\beta \Delta E_\beta^{tr}}{y_\beta} + \sum_\beta \frac{k}{c_\beta^{int}} d'_{\alpha\beta}\frac{n_\beta \Delta E_\beta^{int}}{y_\beta}, \tag{10.9.1}$$

where

$$c_{\alpha\beta} = y_\alpha\left\{\delta_{\alpha\beta}\sum_\gamma \frac{2m_\gamma^2}{(m_\alpha+m_\gamma)^2}\left[\frac{c_\alpha^{int}}{k}\tau_{E\alpha\gamma}^{-1}(1+\delta_{\alpha\gamma}) + \frac{3}{2}\frac{m_\alpha}{m_\gamma}\right.\right.$$
$$\left.\times\,\tau_{\alpha\gamma}^{-1}(1-\delta_{\alpha\gamma})\right] + (1-\delta_{\alpha\beta})\frac{2m_\alpha m_\beta}{(m_\alpha+m_\gamma)^2}$$
$$\left.\times\left(\frac{c_\alpha^{int}}{k}\tau_{E\alpha\beta}^{-1} - \frac{3}{2}\tau_{\alpha\beta}^{-1}\right)\right\},$$

$$c'_{\alpha\beta} = y_\alpha\frac{c_\alpha^{int}}{k}\left[\delta_{\alpha\beta}\sum_\gamma \frac{2m_\gamma}{m_\alpha+m_\gamma}\left(L_{\alpha\gamma}^* + M_{\alpha\gamma}^*\right)\tau_{E\alpha\gamma}^{-1}\right.$$
$$\left.\times(1+\delta_{\alpha\gamma}) + (1-\delta_{\alpha\beta})\frac{2m_\beta}{m_\alpha+m_\beta}\left(L_{\beta\alpha}^* + M_{\alpha\beta}^*\right)\tau_{E\alpha\beta}^{-1}\right],$$

$$d_{\alpha\beta} = y_\alpha\frac{c_\alpha^{int}}{k}\left[\delta_{\alpha\beta}\sum_\gamma \frac{2m_\gamma}{m_\alpha+m_\gamma}\left(L_{\alpha\gamma}^* + M_{\alpha\gamma}^*\right)\tau_{E\alpha\gamma}^{-1}\right.$$
$$\left.\times(1+\delta_{\alpha\gamma}) + (1-\delta_{\alpha\beta})\frac{2m_\beta}{m_\alpha+m_\beta}\left(L_{\alpha\beta}^* + M_{\alpha\beta}^*\right)\tau_{E\alpha\beta}^{-1}\right],$$

$$d'_{\alpha\beta} = y_\alpha\frac{c_\alpha^{int}}{k}\left[\delta_{\alpha\beta}\left(\tau_{E\alpha\beta}^{-1} + \sum_\gamma 2L_{\alpha\gamma}^*\tau_{E\alpha\gamma}^{-1}\right) + (1-\delta_{\alpha\beta})\,2M_{\alpha\beta}^*\tau_{E\alpha\beta}^{-1}\right] \tag{10.9.2}$$

and

$$\tau_{E\alpha\beta}^{-1} = 4n_\beta \frac{k}{c_\alpha^{int}} \left\langle \Delta\varepsilon_{\alpha\beta}^2 \right\rangle_{\alpha\beta} = \frac{6}{5} A_{\alpha\beta}^* Z_{\alpha\beta}^{-1} \tau_{\alpha\beta}^{-1}.$$

If the energy exchange between translational and internal degrees of freedom is carried out quite easily ($Z_{\alpha\beta} \sim 1$), then, following the requirement of $\tau_E \ll \tau_L$ on the left-hand side of Eqs. (10.9.1), the derivatives $d\Delta E_\alpha^{tr}/dt$ and $d\Delta E_\alpha^{int}/dt$ are negligible. As a result, a system of linear algebraic equations can be obtained from which $n_\alpha \Delta E_\alpha^{tr}$ and $n_\alpha \Delta E_\alpha^{int}$ are derived.

By definition, the total pressure tensor of a mixture is equal to

$$P_{rs} = \left(p + \frac{2}{3} n\Delta E^{tr}\right) \delta_{rs} + \pi_{rs}$$

and in the case under investigation

$$\frac{2}{3} n\Delta E^{tr} = \frac{2}{3} \sum_\alpha n_\alpha \Delta E_\alpha^{tr} = -\zeta \nabla \cdot \mathbf{u}, \qquad (10.9.3)$$

where ζ is the volume viscosity of the mixture defined by

$$\zeta = -\frac{p}{c_V} \begin{vmatrix} c_{\alpha\beta}^{00} & c_{\alpha\beta}^{01} & y_\alpha c^{int} \\ c_{\alpha\beta}^{10} & c_{\alpha\beta}^{11} & y_\alpha c_\alpha^{int} \\ y_\beta & 0 & 0 \end{vmatrix} \times \begin{vmatrix} c_{\alpha\beta}^{00} & c_{\alpha\beta}^{01} \\ c_{\alpha\beta}^{10} & c_{\alpha\beta}^{11} \end{vmatrix}^{-1}, \qquad (10.9.4)$$

in which, as previously, $c_{\alpha\beta}^{mn}$ mean the square blocks of the Nth order of the elements, where, $y_\alpha c^{int}$, $y_\alpha c_\alpha^{int}$ and y_β indicate columns and lines of N elements.

The elements of the determinants $c_{\alpha\beta}^{mn}$ are expressed in terms of $c_{\alpha\beta}, c'_{\alpha\beta}, d_{\alpha\beta}$ and $d'_{\alpha\beta}$ with the accuracy up to arbitrary addends vanishing when the determinants are calculated

$$c_{\alpha\beta}^{00} = c_{\alpha\beta} - y_\beta k_\alpha^{(0)}, \qquad c_{\alpha\beta}^{01} = c'_{\alpha\beta} + \frac{2}{3} y_\beta \frac{c_\beta^{int}}{k} k_\alpha^{(0)},$$

$$c_{\alpha\beta}^{00} = d_{\alpha\beta} + y_\beta k_\alpha^{(1)}, \qquad c_{\alpha\beta}^{11} = d'_{\alpha\beta} - \frac{2}{3} y_\beta \frac{c_\beta^{int}}{k} k_\alpha^{(1)}. \qquad (10.9.5)$$

The form of the determinants follows the principle that the system of equations for $n_\alpha \Delta E_\alpha^{tr}$ and $n_\alpha \Delta E_\alpha^{int}$ must be additionally defined by condition (10.2.11).

It should be stressed that the linear relationship between $n_\alpha \Delta E_\alpha^{tr}$ and $\nabla \cdot \mathbf{u}$ occurs in the case of an easy energy exchange, when the condition $\tau_{\alpha\beta} \sim \tau_{E\alpha\beta} \ll \tau_L$ is met. At $Z_{\alpha\beta} \gg 1$, this condition is violated and in Eqs. (10.9.1) time derivatives must be kept. Notice that the coefficients $c'_{\alpha\beta}, d_{\alpha\beta}$ and $d'_{\alpha\beta}$ on the right-hand sides are of the order of $Z^{-1}_{\alpha\beta}\tau^{-1}_{\alpha\beta}$, whereas $c_{\alpha\beta} = c^D_{\alpha\beta} + c^E_{\alpha\beta}$, where $c^D_{\alpha\beta}$ does not comprise $Z^{-1}_{\alpha\beta}$ which corresponds to the intense energy exchange between the translational degrees of freedom of individual species. However, if $\sum_\alpha c^D_{\alpha\beta} = 0$, the right-hand sides of the equations for $\Delta E^{tr} = \sum_\alpha y_\alpha \Delta E^{tr}_\alpha$ retain only the terms $\sim Z^{-1}_{\alpha\beta}\tau^{-1}_{\alpha\beta}$ ensuring at $Z_{\alpha\beta} \gg 1$ a very slow energy exchange between the translational degrees of freedom of a mixture as a whole and the internal degrees of freedom of individual species. Therefore, the equation for ΔE^{tr}_α combined with the equation for ΔE^{tr} and in neglect of $\sim Z^{-1}_{\alpha\beta}$, can be conveniently written as

$$\frac{n_\alpha d\left(E^{tr}_\alpha - E^{tr}\right)}{dt} = -\frac{2}{3}\sum_\beta n c^D_{\alpha\beta}\left(E^{tr}_\beta - E^{tr}\right).$$

Hence, during the time of the order of the mean time between the collisions τ the translational energy of species E^{tr}_α relaxes to the common E^{tr}. Therefore, at states satisfying the condition $\tau_{\alpha\beta} \ll \tau_L$ we can assume that $E^{tr}_\alpha = E^{tr}$ for all α, whereas the relaxation E^{tr} to $E^{tr}_0 = (3/2)\,kT$ and relaxation of all the internal energy of species E^{int}_α to $E^{int}_{\alpha 0}$ are described by the system of equations

$$\frac{d\Delta E^{tr}}{dt} + kT\frac{c^{int}}{c_v}\nabla\cdot\mathbf{u} = \sum_\beta y_\beta \tau^{-1}_{E\alpha\beta}\left(\Delta E^{int}_\beta - \frac{2}{3}\frac{c^{int}_\beta}{k}\Delta E^{tr}\right),$$

$$\frac{d\Delta E^{int}_\alpha}{dt} - kT\frac{c^{int}}{c_v}\nabla\cdot\mathbf{u} = -\tau^{-1}_{E\alpha}\left(\Delta E^{int}_\alpha - \frac{2}{3}\frac{c^{int}_\alpha}{k}\Delta E^{tr}\right), \tag{10.9.6}$$

where

$$\tau^{-1}_{E\alpha} = \sum_\gamma 2L^*_{\alpha\gamma}\tau^{-1}_{\alpha\gamma}$$

and an additional assumption is taken according to which the complex collisions accompanied by the transition of both colliding particles occur less frequently than collisions with single particle transition ($M^*_{\alpha\beta} \ll 1$).

Equations (10.9.6) together with the equations of hydrodynamics and linear transport relations (10.8.2), (10.8.13) and (10.8.16) can be the basis for considering the transport and relaxation phenomena in a polyatomic gas mixture when $Z_{\alpha\beta} \gg 1$. A typical example is propagation of sound or relaxation behind the front of a shock wave for the gas of vibrational excited molecules. In more general cases, the relaxation equations (10.9.1) should be used.

Appendices

A1. General system of moment equations

The system of moment equations or the equations for the coefficients $n_\alpha b_{\alpha\beta}^{mn}$, used in Chapter 4, follows from the general transport equation (2.1.9) if we take $\psi_\alpha = G_\alpha^{mn}(\mathbf{c},\gamma_\alpha)$. Orthogonal tensorial polynomials $G_\alpha^{mn}(\mathbf{c},\gamma_\alpha)$, taking into account Eqs. (4.2.6) and (4.1.7), can be represented by

$$G_\alpha^{mn}(\mathbf{c},\gamma_\alpha) = (-1)^n n! m_\alpha \gamma_\alpha^{-(n+\frac{m}{2})} S_{m+\frac{1}{2}}^n\left(\frac{\gamma_\alpha c^2}{2}\right) P^{(m)}\left(\gamma_\alpha^{1/2}\mathbf{c}\right). \tag{A1.1}$$

Since the explicit dependence $G_\alpha^{mn}(\mathbf{c},\gamma_\alpha)$ on coordinates and time is expressed in terms of the variable $\gamma_\alpha = m_\alpha/kT_\alpha(\mathbf{r},t)$, it is convenient to separate the derivatives dT_α/dt and $\partial T_\alpha/\partial x_r$ in the explicit form in Eq. (2.1.9) and to put down original moment equations as

$$\begin{aligned} & \frac{d}{dt}\left(n_\alpha\left\langle G_\alpha^{mn}\right\rangle\right) + n_\alpha\left\langle G_\alpha^{mn}\right\rangle \nabla\cdot\mathbf{u} + \frac{\partial}{\partial x_l}\left(n_\alpha\left\langle c_l G_\alpha^{mn}\right\rangle\right) \\ & -n_\alpha\left[\left\langle\frac{F_{\alpha l}}{m_\alpha}\frac{\partial G_\alpha^{mn}}{\partial c_l}\right\rangle - \left\langle\frac{\partial G_\alpha^{mn}}{\partial c_l}\right\rangle\frac{du_l}{dt}\right] + n_\alpha\left\langle c_s\frac{\partial G_\alpha^{mn}}{\partial c_r}\right\rangle\frac{\partial u_r}{\partial x_s} \\ & +\frac{n_\alpha\gamma_\alpha}{T_\alpha}\left\langle\frac{\partial G_\alpha^{mn}}{\partial\gamma_\alpha}\right\rangle\frac{dT_\alpha}{dt} + \frac{n_\alpha\gamma_\alpha}{T_\alpha}\left\langle c_l\frac{\partial G_\alpha^{mn}}{\partial\gamma_\alpha}\right\rangle\frac{\partial T_\alpha}{\partial x_l} \\ = \; & R_\alpha^{mn} = \sum_\beta \int G_\alpha^{mn} J_{\alpha\beta} d\mathbf{c}_\alpha, \end{aligned} \tag{A1.2}$$

where

$$n_\alpha\left\langle\varphi_\alpha\right\rangle = \int \varphi_\alpha f_\alpha d\mathbf{c}_\alpha.$$

Using the expansion (4.1.9) and relationship (A1.1), let us represent the expansion of the distribution function over polynomials $G_\alpha^{mn}(\mathbf{c},\gamma_\alpha)$ as

$$f_\alpha = f_\alpha^{(0)} \sum_{m,n=0}^{\infty} 2^{2n} m_\alpha^{-2} \gamma_\alpha^{2n+m} \sigma_{mn} b_\alpha^{mn} G_\alpha^{mn} (\mathbf{c}, \gamma_\alpha). \qquad \text{(A1.3)}$$

In this expansion, $n_\alpha b_\alpha^{mn}$ are the moments of the distribution function of interest, which are defined in terms of the polynomials G_α^{mn} following

$$n_\alpha b_\alpha^{mn} = n_\alpha \langle G_\alpha^{mn} \rangle = \int G_\alpha^{mn} f_\alpha d\mathbf{c}_\alpha. \qquad \text{(A1.4)}$$

To pass from (A1.2) to the explicit form of the infinite hierarchy of the related equations for the coefficients $n_\alpha b_\alpha^{mn}$, the left-hand side of the equations should have the functions

$$c_l G_\alpha^{mn}, \qquad \frac{\partial G_\alpha^{mn}}{\partial c_l}, \qquad \frac{c_s \partial G_\alpha^{mn}}{\partial c_r}, \qquad \frac{\partial G_\alpha^{mn}}{\partial \gamma_\alpha}, \qquad \frac{c_l \partial G_\alpha^{mn}}{\partial \gamma_\alpha}$$

expressed in terms of linear combinations of the full set of orthogonal polynomials G_α^{mn}, whereas the right-hand side can be calculated by substituting expansions for f_α and f_β in the form of (A1.3) into the expression for R_α^{mn}.

First let us transform the left-hand side of Eq. (A1.2), which can be done using the properties of the Sonine polynomials $S_{m+1/2}^{n}(x)$ and tensorial polynomials $P^{(m)}(\boldsymbol{\xi})$ given below[1]. In particular, for the Sonine polynomials we have [1]

1) recurrent relationship and growth of the kth order

$$S_k^n(x) = S_{k+1}^n(x) - S_{k+1}^{n-1}(x),$$

2) the derivative

$$\frac{d}{dx} S_k^n(x) = -S_{k+1}^{n-1}(x),$$

3) multiplication by the argument and diminishing kth order

$$x S_k^n(x) = (n+k) S_{k-1}^n(x) - (n+1) S_{k-1}^{n+1}.$$

The symmetric irreducible polynomial $P^{(m)}(\boldsymbol{\xi})$ according to Grad [2] is defined as an irreducible projection of the tensorial monomial $\boldsymbol{\xi}^m = \xi_{r_1}\xi_{r_2}...\xi_{r_m}$ and is of the kind

$$P^{(m)} = \sum_{l=0}^{[m/2]} \frac{(-2)^l (2m-2l)! m!}{(2m)! (m-l)!} \boldsymbol{\xi}^{m-2l} \cdot \delta^l \xi^{2l},$$

[1]The transformation of the left-hand side of Eq. (A.1.2) described below was carried out by P.N. Yushmanov.

where the product of tensors means a symmetric summation over all the various permutations of the indices and, in particular

$$\xi \cdot \delta = \xi_r \delta_{sl} + \xi_s \delta_{rl} + \xi_l \delta_{rs},$$

$$\delta^2 = \delta_{rs}\delta_{tu} + \delta_{rt}\delta_{su} + \delta_{ru}\delta_{st},$$

$$\xi^2 \cdot \delta = \xi_r \xi_s \delta_{tu} + \xi_r \xi_t \delta_{su} + \xi_r \xi_u \delta_{st} + \xi_s \xi_u \delta_{rt} + \xi_t \xi_u \delta_{rs} + \xi_s \xi_t \delta_{ru}.$$

Using the irreducibility and symmetry of $P^{(m)}(\xi)$, we can obtain the following relationships which the polynomials meet:

1) recurrent relationship

$$P^{(m+1)} = \frac{1}{m+1}\xi P^{(m)} - \frac{2m\xi^2}{(m+1)(4m^2-1)}\delta P^{(m-1)},$$

2) contraction with ξ_r

$$\xi_r P_r^{(m)} = \frac{m\xi^2}{2m-1} P^{(m-1)},$$

3) derivative

$$\frac{\partial P^{(m)}}{\partial \xi_r} = \left\{P^{(m-1)}\delta\right\}_r \equiv \delta_r P^{(m-1)} - \frac{2}{2m-1}\delta P_r^{(m-1)},$$

4) multiplication by ξ_r

$$\xi_r P^{(m)} = P^{(m+1)} + \frac{\xi^2}{2m+1}\left\{P^{(m-1)}\delta\right\}_r, \tag{A1.5}$$

5) the second derivative

$$\frac{\partial^2 P^{(m)}}{\partial \xi_r \xi_s} = \left\{P^{(m)}\delta\delta\right\}_{rs} \equiv \frac{2m+5}{2m+3}\delta_r\delta_s P^{(m)} - \frac{2}{2m+3}\delta\delta P^{(m)} + \frac{2(2m+5)}{(2m+3)(2m+1)}\delta\delta P_{rs}^{(m)},$$

6) multiplication of ξ_s by the derivative

$$\xi_s \frac{\partial P^{(m)}}{\partial \xi_r} = \left\{P_s^{(m)}\delta\right\}_r + \frac{\xi^2}{2m-1}\left\{P^{(m-2)}\delta\delta\right\}_{rs},$$

7) multiplication by $\xi_r \xi_s$

$$\xi_r \xi_s P^{(m)} = P^{(m+2)} + \frac{\xi^2}{2m+3}\delta P^{(m)} - \frac{\xi^2}{2m-1}\delta P^{(m)}_{rs} + \frac{\xi^4}{4m^2-1}\left\{P^{(m-2)}\delta\delta\right\}_{rs}.$$

The product of tensors with indices implies that the indices do not participate in permutations and symmetric summations. In particular,

$$\xi\delta_r = \xi_s\delta_{rt} + \xi_t\delta_{rs}, \quad \delta_r\delta_s = \delta_{rt}\delta_{su} + \delta_{ru}\delta_{st} \quad \text{etc.}$$

Using the relationships (A1.5) given above and taking into account the definition of $G^{mn}_\alpha(\mathbf{c}, \gamma_\alpha)$ allows us to represent the expressions which occur on the left-hand sides of the moment equations (A1.2) in terms of linear combinations of the polynomials G^{mn}_α

$$\frac{\partial G^{mn}_\alpha}{\partial \gamma_\alpha} = \frac{1}{2\gamma^2_\alpha} n(2m+2n+1)\, G^{m,n-1}_\alpha,$$

$$c_s \frac{\partial G^{mn}_\alpha}{\partial \gamma_\alpha} = \frac{n(2m+2n+1)}{2\gamma^2_\alpha}\left[G^{m+1,n-1}_\alpha + \frac{n-1}{\gamma_\alpha} G^{m+1,n-2}_\alpha + \frac{2}{2m+1}\left(\frac{2m+2n-1}{2\gamma_\alpha}\left\{G^{m-1,n-1}_\alpha\delta\right\}_s + \left\{G^{m-1,n}_\alpha\delta\right\}_s\right)\right],$$

$$\frac{\partial G^{mn}_\alpha}{\partial c_s} = nG^{m+1,n-1}_\alpha + \frac{2m+2n+1}{2m+1}\left\{G^{m-1,n}_\alpha\delta\right\}_s,$$

$$c_s G^{mn}_\alpha = G^{m+1,n}_\alpha + \frac{n}{\gamma_\alpha} G^{m+1,n-1}_\alpha + \frac{2}{2m+1} \times \left[\frac{2m+2n+1}{2\gamma_\alpha}\left\{G^{m-1,n}_\alpha\delta\right\}_s + \left\{G^{m-1,n+1}_\alpha\delta\right\}_s\right],$$

$$c_s \frac{\partial G^{mn}_\alpha}{\partial c_r} = nG^{m+2,n-1}_\alpha + \frac{n(n-1)}{\gamma_\alpha} G^{m+2,n-2}_\alpha + \frac{2n}{2m+3}\delta G^{mn}_\alpha + \frac{n(2m+2n+1)}{(2m+3)\gamma_\alpha} G^{m,n-1}_\alpha\delta + \delta_r G^{mn}_{\alpha s} - \frac{2n+2}{2m-1}\delta G^{mn}_{\alpha rs} - \frac{n(2m+2n+1)}{(2m-1)\gamma_\alpha}\delta G^{m,n-1}_\alpha + \frac{2m+2n+1}{4m^2-1} \times \left[2\left\{G^{m-2,n+1}_\alpha\delta\delta\right\}_{rs} + \frac{2m+2n+1}{\gamma_\alpha}\left\{G^{m-2,n}_\alpha\delta\delta\right\}_{rs}\right]. \tag{A1.6}$$

Now, an infinite hierarchy of the moment equations can be obtained by expressing the average values of the kind $\langle c_r G_\alpha^{mn}\rangle$, $\langle \partial G_\alpha^{mn}/\partial\gamma_\alpha\rangle$ via the moments $n_\alpha b_\alpha^{mn}$ with different indices. Substituting the formulas (A1.6) into Eq. (A1.2) and after integration we have a full system of equations for $n_\alpha b_\alpha^{mn}$

$$\begin{aligned}
&\frac{d}{dt}\left(n_\alpha b_\alpha^{mn}\right) + n_\alpha b_\alpha^{mn}\nabla\cdot\mathbf{u} + \frac{\partial}{\partial x_r}\left[n_\alpha\left(b_{\alpha r}^{m+1,n} + \frac{n}{\gamma_\alpha}b_{\alpha r}^{m+1,n-1}\right)\right.\\
&\left.+\frac{2}{2m+1}\left(\frac{2m+2n+1}{2\gamma_\alpha}\left\{n_\alpha b_{\alpha r}^{m-1,n}\delta\right\}_r + \left\{n_\alpha b_\alpha^{m-1,n+1}\delta\right\}_r\right)\right]\\
&-n_\alpha\left(\frac{\langle F_{\alpha r}\rangle}{m_\alpha} - \frac{du_r}{dl}\right)\left(nb_{\alpha r}^{m+1,n-1} + \frac{2m+2n+1}{2m+1}\left\{b_\alpha^{m-1,n}\delta\right\}_r\right)\\
&-n_\alpha b_{\alpha s}^{mn}e_{rst}\omega_{\alpha t}\delta_r + n_\alpha\frac{\partial u_r}{\partial x_s}\left[nb_{\alpha rs}^{m+2,n-1} + \frac{n(n-1)}{\gamma_\alpha}b_{\alpha rs}^{m+2,n-2}\right.\\
&+\frac{2n}{2m+3}\delta b_\alpha^{mn} + \frac{n(2m+2n+1)}{(2m+3)\gamma_\alpha}\delta b_\alpha^{m,n-1} + b_{\alpha s}^{mn}\delta_r\\
&-\frac{2n+2}{2m-1}\delta b_{\alpha rs}^{mn} - \frac{n(2m+2n+1)}{(2m-1)\gamma_\alpha}\delta b_{\alpha rs}^{m,n-1} + \frac{2m+2n+1}{4m^2-1}\\
&\left.\times\left(2\left\{b_\alpha^{m-2,n+1}\delta\delta\right\}_{rs} + \frac{2m+2n-1}{\gamma_\alpha}\left\{b_\alpha^{m-2,n}\delta\delta\right\}_{rs}\right)\right]\\
&+\frac{n(2m+2n+1)}{2m_\alpha}n_\alpha\left\{b_\alpha^{m,n-1}\frac{dT_\alpha}{dt} + \left[b_{\alpha r}^{m+1,n-1}\right.\right.\\
&+\frac{n-1}{\gamma_\alpha}b_{\alpha r}^{m+1,n-2}\frac{2}{2m+1}\left(\frac{2m+2n+1}{2\gamma_\alpha}\left\{b_\alpha^{m-1,n-1}\delta\right\}_r\right.\\
&\left.\left.\left.+\left\{b_\alpha^{m-1,n}\delta\right\}_r\right)\right]\frac{\partial T_\alpha}{\partial x_r}\right\} = R_{\alpha r}^{mn}.
\end{aligned} \tag{A1.7}$$

Following the polynomial $P^{(m)}$ properties given above

$$\left\{b_\alpha^{mn}\delta\right\}_r = \delta_r b_\alpha^{mn} - \frac{2}{2m+1}\delta b_{\alpha r}^{mn},$$

$$\begin{aligned}
\left\{b_\alpha^{mn}\delta\delta\right\}_{rs} &= \frac{2m+5}{2m+3}\delta_r\delta_s b_\alpha^{mn} - \frac{2}{2m+3}\delta\delta b_\alpha^{mn}\\
&+\frac{2(2m+5)}{(2m+3)((2m+1)}\delta\delta b_{\alpha rs}^{mn}.
\end{aligned}$$

Apart from the calculation of the right-hand side of Eq. (A1.7), we should mention that in Section 4.3 we discussed the procedure of simplifying the nonlinear moment equations for the case of slight nonequilibrium plasma state when the plasma parameters change slowly in time and at lengths of the order of the characteristic mean time and mean free path of the particles between the collisions. In practice, it corresponds to the neglect of all the nonlinear terms on the left-hand side of Eq. (A1.7), including the products of the moments (or coefficients $n_\alpha b_\alpha^{mn}$) and small gradients of the relevant thermodynamic variables as well as small gradient potentials (weak external fields). After simplification, the system of moment equations looks like

$$\frac{d}{dt}\left(n_\alpha b_\alpha^{mn}\right) + \frac{\partial}{\partial x_r}\left[n_\alpha\left(b_{\alpha r}^{m+1,n} + \frac{n}{\gamma_\alpha} b_{\alpha r}^{m+1,n-1}\right.\right.$$
$$\left.\left. + \frac{2}{2m+1} n_\alpha \left(\frac{2m+2n+1}{2\gamma_\alpha}\left\{b_\alpha^{m-1,n}\delta\right\}_r + \left\{b_\alpha^{m-1,n+1}\delta\right\}_r\right)\right]$$
$$+ \rho_\alpha \nabla\cdot\mathbf{u}\,\delta_{m0}\delta_{n0} - \rho_\alpha\left(\frac{F_{\alpha r}}{m_\alpha} - \frac{du_r}{dt}\right)\delta_{m1}\delta_{n0} + 2p_\alpha \varepsilon_{rs}\delta_{m2}\delta_{n0}$$
$$+ p_\alpha \nabla\cdot\mathbf{u}\delta_{m0}\delta_{n1} - n_\alpha b_{\alpha s}^{mn} e_{rst}\omega_{\alpha t}\delta_r + \frac{3}{2} n_\alpha \frac{dT_\alpha}{dt}\delta_{m0}\delta_{n1}$$
$$+ \frac{5}{2}\frac{p_\alpha}{m_\alpha}\frac{\partial T_\alpha}{\partial x_r}\delta_{m1}\delta_{n1} = R_\alpha^{mn}. \tag{A1.8}$$

Under the same assumptions the right-hand sides of the moment equations are calculated preserving only the linear terms with respect to the moments which allows one to represent R_α^{mn} as

$$R_\alpha^{mn} = -\sum_\beta\sum_k \left(A_{\alpha\beta}^{mnk} b_\alpha^{mk} + B_{\alpha\beta}^{mnk} b_\beta^{mk}\right). \tag{A1.9}$$

The expressions for $A_{\alpha\beta}^{mnk}$ and $B_{\alpha\beta}^{mnk}$ will be given later.

Further simplification of the equations can be carried out by discarding the derivative $dn_\alpha b_\alpha^{mn}/dt$ on the left-hand side compared with the respective terms on the right-hand side (A1.8) as well as the terms with the space derivatives of $n_\alpha b_\alpha^{mn}$. The latter agrees with the first approximation over the small parameter $\varepsilon = \lambda/L$ used in the Chapman–Enskog method [1].

The resulting equations are

$$\left(\frac{d\rho_\alpha}{dt} + \rho_\alpha \nabla\cdot\mathbf{u}\right)\delta_{m0}\delta_{n0} + \left(\frac{3}{2} n_\alpha \frac{dT_\alpha}{dt} + p_\alpha \nabla\cdot\mathbf{u}\right)\delta_{m0}\delta_{n1}$$

$$\left(\frac{d\rho_\alpha}{dt}+\rho_\alpha\nabla\cdot\mathbf{u}\right)\delta_{m0}\delta_{n0}+\left(\frac{3}{2}n_\alpha\frac{dT_\alpha}{dt}+p_\alpha\nabla\cdot\mathbf{u}\right)\delta_{m0}\delta_{n1}$$
$$+\left(\frac{\partial p_\alpha}{\partial x_r}-n_\alpha\left\langle F^*_{\alpha r}\right\rangle\right)\delta_{m1}\delta_{n0}+2p_\alpha\varepsilon_{rs}\delta_{m2}\delta_{n0}+\frac{5}{2}\frac{p_\alpha}{m_\alpha}\frac{\partial T_\alpha}{\partial x_r}\delta_{m1}\delta_{n1}$$
$$-n_\alpha b^{mn}_{\alpha s}e_{rst}\omega_{\alpha t}\delta_r = -\sum_\beta\sum_k\left(A^{mnk}_{\alpha\beta}b^{mk}_\alpha+B^{mnk}_{\alpha\beta}b^{mk}_\beta\right). \tag{A1.10}$$

Equations (A1.10) with $m=0$ correspond to the well-known continuity equation ($n=0$) and energy equation ($n=1$) within the Euler approximation. At $m=1$, a related system of equations for diffusion velocities and reduced heat fluxes of the species and at $m=2$ the equations for partial viscous-stress tensors follow from Eqs. (A1.10). Similar equations correspondent to the linear approximation with respect to ε were obtained in Section 4.6 directly from the linearized Boltzmann equation.

Now turn to the right-hand side of Eq. (A1.2). Having represented the distribution function as $f_\alpha=f^{(0)}_\alpha(1+\varphi_\alpha)$, let us write the collision integral $J_{\alpha\beta}$ as

$$J_{\alpha\beta} = \int f^{(0)}_\alpha f^{(0)}_\beta\left(\varphi'_\alpha+\varphi'_\beta-\varphi_\alpha-\varphi_\beta+\varphi'_\alpha\varphi'_\beta-\varphi_\alpha\varphi_\alpha\right)$$
$$\times g\sigma_{\alpha\beta}(g,\chi)\,d\Omega d\mathbf{v}_\beta.$$

Assuming that the distribution function slightly deviates from the equilibrium, we can neglect φ_α and φ_β squared terms. Then

$$R^{mn}_\alpha = \sum_\beta\int G^{mn}_\alpha f^{(0)}_\alpha f^{(0)}_\beta\left(\varphi'_\alpha+\varphi'_\beta-\varphi_\alpha-\varphi_\beta\right)$$
$$\times g\sigma_{\alpha\beta}(g,\chi)\,d\Omega d\mathbf{v}_\alpha d\mathbf{v}_\beta. \tag{A1.11}$$

Substituting φ_α and φ_β, obtained from the expansion (A1.3), into Eq. (A1.11), we find

$$R^{mn}_\alpha = \sum_\beta\sum_{k,l}2^{2l}\sigma_{kl}\int f^{(0)}_\alpha f^{(0)}_\beta\left\{\gamma^{2l+k}_\alpha m^{-2}_\alpha\left[G^{kl}_\alpha\left(\mathbf{c}'_\alpha\right)\right.\right.$$
$$\left.-G^{kl}_\alpha(\mathbf{c}_\alpha)\right]b^{kl}_\alpha+\gamma^{2l+k}_\beta m^{-2}_\beta\left[G^{kl}_\beta\left(\mathbf{c}'_\beta\right)-G^{kl}_\beta(\mathbf{c}_\beta)\right]b^{kl}_\beta\Big\}$$
$$\times G^{mn}_\alpha(\mathbf{c}_\alpha)\,g\sigma_{\alpha\beta}(g,\chi)\,d\Omega d\mathbf{v}_\alpha d\mathbf{v}_\beta. \tag{A1.12}$$

Using the definitions for the partial bracket integrals [1,3]

$$n_\alpha n_\beta [F, G]'_{\alpha\beta} = \int f_\alpha^{(0)} f_\beta^{(0)} \left(F_\alpha - F'_\alpha\right) G_\alpha g \sigma_{\alpha\beta}(g, \chi)\, d\Omega d\mathbf{v}_\alpha d\mathbf{v}_\beta,$$

$$n_\alpha n_\beta [F, G]''_{\alpha\beta} = \int f_\alpha^{(0)} f_\beta^{(0)} \left(F_\beta - F'_\beta\right) G_\alpha g \sigma_{\alpha\beta}(g, \chi)\, d\Omega d\mathbf{v}_\alpha d\mathbf{v}_\beta,$$

we can write (A1.12) as

$$\begin{aligned} R_\alpha^{mn} &= -\sum_\beta \sum_{k,l} 2^{2l} \sigma_{kl} n_\alpha n_\beta \left\{ \left[G^{kl}, G^{mn}\right]'_{\alpha\beta} \frac{\gamma_\alpha^{2l+k}}{m_\alpha^2} b_\alpha^{kl} \right. \\ &\quad \left. + \left[G^{kl}, G^{mn}\right]''_{\alpha\beta} \frac{\gamma_\beta^{2l+k}}{m_\beta^2} b_\beta^{kl} \right\}. \end{aligned}$$

Allowing for (A1.1) and the evident condition

$$\left[P^{(k)}, P^{(m)}\right] = \frac{1}{2k+1} \left[P^{(k)}, P^{(k)}\right] \delta_{km},$$

the final expression for R_α^{mn} acquires the form of (A.1.9), where

$$\begin{aligned} A_{\alpha\beta}^{mnk} &= n_\alpha n_\beta Q_{mnk} \gamma_\alpha^{k-n} \\ &\quad \times \left[S_{m+\frac{1}{2}}^k \left(W^2\right) P^{(m)}(\mathbf{W}), S_{m+\frac{1}{2}}^n \left(W^2\right) P^{(m)}(\mathbf{W})\right]'_{ab}, \\ B_{\alpha\beta}^{mnk} &= n_\alpha n_\beta Q_{mnk} \left(\frac{m_\alpha \gamma_\beta^{k+\frac{m}{2}}}{m_\beta \gamma_\alpha^{k+\frac{m}{2}}} \right) \\ &\quad \times \left[S_{m+\frac{1}{2}}^k \left(W^2\right) P^{(m)}(\mathbf{W}), S_{m+\frac{1}{2}}^n \left(W^2\right) P^{(m)}(\mathbf{W})\right]''_{ab}, \end{aligned} \tag{A1.13}$$

where $\mathbf{W} = (m_\alpha/2kT)^{1/2}\, \mathbf{c}$ and the coefficients Q_{mnk} are of the form

$$Q_{mnk} = \frac{(2m)!\,(m+k)!\,n!}{(m!)^2\,(2m+2k+1)!} 2^{2k+m} (-1)^{n+k}. \tag{A1.14}$$

A2. Values of partial bracket integrals in terms of integrals $\Omega^{lr}_{\alpha\beta}$

Below are given some essential bracket integrals expressed in terms of integrals $\Omega^{lr}_{\alpha\beta}$ and $\mu_\alpha = m_\alpha/(m_\alpha + m_\beta)$, $\mu_\beta = m_\beta/(m_\alpha + m_\beta)$. The techniques of calculating them in any order of approximation with respect to Sonine polynomials are given in [1,3].

1. Partial bracket integrals required for calculating the vector transport properties (diffusion, thermal diffusion and thermal conductivity):

$$\left[S^n_{3/2}\left(^2\right)\mathbf{W}, S^k_{3/2}\left(W^2\right)\mathbf{W}\right]'_{\alpha\beta}, \quad n,k = 0,1,2.$$

$n = 0, k = 0$ $\quad 8\mu_\beta\Omega^{11}_{\alpha\beta}$,

$n = 0, k = 1$ $\quad 8\mu^2_\beta\left(\frac{5}{2}\Omega^{11}_{\alpha\beta} - \Omega^{12}_{\alpha\beta}\right)$,

$n = 1, k = 1,$
$$8\mu_\beta\left[\frac{5}{4}\left(6\mu^2_\alpha + 5\mu^2_\beta\right)\Omega^{11}_{\alpha\beta} - 5\mu^2_\beta\Omega^{12}_{\alpha\beta} + \mu^2_\beta\Omega^{13}_{\alpha\beta} + 2\mu_\alpha\mu_\beta\Omega^{22}_{\alpha\beta}\right],$$

$n = 0, k = 2,$
$$4\mu^3_\beta\left[\frac{35}{4}\Omega^{11}_{\alpha\beta} - 7\Omega^{12}_{\alpha\beta} + \Omega^{13}_{\alpha\beta}\right],$$

$n = 1, k = 2,$
$$8\mu^2_\beta\left[\frac{35}{16}\left(12\mu^2_\alpha + 5\mu^2_\beta\right)\Omega^{11}_{\alpha\beta} - \frac{21}{8}\left(4\mu^2_\alpha + 5\mu^2_\beta\right)\Omega^{12}_{\alpha\beta} + \frac{19}{4}\mu^2_\beta\Omega^{13}_{\alpha\beta} - \frac{1}{2}\mu^2_\beta\Omega^{14}_{\alpha\beta} + 7\mu_\alpha\mu_\beta\Omega^{22}_{\alpha\beta} - 2\mu_\alpha\mu_\beta\Omega^{23}_{\alpha\beta}\right],$$

$n = 2, k = 2,$
$$8\mu_\beta\left[\frac{35}{64}\left(40\mu^4_\alpha + 168\mu^2_\alpha\mu^2_\beta + 35\mu^4_\beta\right)\Omega^{11}_{\alpha\beta} - \frac{7}{8}\mu^2_\beta\left(84\mu^2_\alpha + 35\mu^2_\beta\right)\Omega^{12}_{\alpha\beta} + \frac{1}{8}\mu^2_\beta(108\mu^2_\alpha + 133\mu^2_\beta)\Omega^{13}_{\alpha\beta} - \frac{7}{2}\mu^4_\beta\Omega^{14}_{\alpha\beta} + \frac{1}{4}\mu^4_\beta\Omega^{15}_{\alpha\beta} + \frac{7}{2}\mu_\alpha\mu_\beta\left(4\mu^2_\alpha + 7\mu^2_\beta\right)\Omega^{22}_{\alpha\beta} - 14\mu_\alpha\mu^3_\beta\Omega^{23}_{\alpha\beta} + 2\mu_\alpha\mu^3_\beta\Omega^{24}_{\alpha\beta} + 2\mu^2_\alpha\mu^2_\beta\Omega^{33}_{\alpha\beta}\right].$$

$$\left[S_{3/2}^{n}\left(W^{2}\right)\mathbf{W},\, S_{3/2}^{k}\left(W^{2}\right)\mathbf{W}\right]_{\alpha\beta}'', \quad n,k=0,1,2.$$

$$n=0,k=0, \quad -8\mu_{\alpha}^{1/2}\mu_{\beta}^{1/2}\Omega_{\alpha\beta}^{11},$$

$$n=0,k=1, \quad -8\mu_{\alpha}^{3/2}\mu_{\beta}^{1/2}\left(\frac{5}{2}\Omega_{\alpha\beta}^{11}-\Omega_{\alpha\beta}^{12}\right),$$

$$n=1,k=1, \quad -8\mu_{\alpha}^{3/2}\mu_{\beta}^{3/2}\left(\frac{55}{4}\Omega_{\alpha\beta}^{11}-5\Omega_{\alpha\beta}^{12}+\Omega_{\alpha\beta}^{13}-2\Omega_{\alpha\beta}^{22}\right),$$

$$n=0,k=2, \quad -4\mu_{\alpha}^{5/2}\mu_{\beta}^{1/2}\left(\frac{35}{4}\Omega_{\alpha\beta}^{11}-7\Omega_{\alpha\beta}^{12}+\Omega_{\alpha\beta}^{13}\right),$$

$$n=1,k=2, \quad -8\mu_{\alpha}^{5/2}\mu_{\beta}^{3/2}\left(\frac{595}{16}\Omega_{\alpha\beta}^{11}-\frac{189}{8}\Omega_{\alpha\beta}^{12}+\frac{19}{4}\Omega_{\alpha\beta}^{13}\right.$$
$$\left.-\frac{1}{2}\Omega_{\alpha\beta}^{14}-7\Omega_{\alpha\beta}^{22}+2\Omega_{\alpha\beta}^{23}\right),$$

$$n=2,k=2, \quad -8\mu_{\alpha}^{5/2}\mu_{\beta}^{5/2}\left(\frac{8505}{64}\Omega_{\alpha\beta}^{11}-\frac{833}{8}\Omega_{\alpha\beta}^{12}+\frac{241}{8}\Omega_{\alpha\beta}^{13}\right.$$
$$\left.-\frac{7}{2}\Omega_{\alpha\beta}^{14}+\frac{1}{4}\Omega_{\alpha\beta}^{15}-\frac{77}{2}\Omega_{\alpha\beta}^{22}+14\Omega_{\alpha\beta}^{23}-2\Omega_{\alpha\beta}^{24}+2\Omega_{\alpha\beta}^{33}\right).$$

2. Partial bracket integrals required for calculation of the tensor transport properties (viscosity):

$$\left[S_{5/2}^{n}\left(W^{2}\right)\left(W_{r}W_{s}-\tfrac{1}{3}\delta_{rs}W^{2}\right),\, S_{5/2}^{k}\left(W^{2}\right)\left(W_{r}W_{s}-\tfrac{1}{3}\delta_{rs}W^{2}\right)\right]_{\alpha\beta}',$$
$$n,k=0,1.$$

$$n=0,k=0 \quad \frac{16}{3}\mu_{\beta}\left(5\mu_{\alpha}\Omega_{\alpha\beta}^{11}+\frac{3}{2}\mu_{\beta}\Omega_{\alpha\beta}^{22}\right),$$

$$n=0,k=1 \quad \frac{16}{3}\mu_{\beta}^{2}\left(\frac{35}{2}\mu_{\alpha}\Omega_{\alpha\beta}^{11}-7\mu_{\alpha}\Omega_{\alpha\beta}^{12}+\frac{21}{4}\mu_{\beta}\Omega_{\alpha\beta}^{22}-\frac{3}{2}\mu_{\beta}\Omega_{\alpha\beta}^{23}\right),$$

$$n=1,k=1 \quad \frac{16}{3}\mu_{\beta}\left[\frac{1}{4}\mu_{\alpha}\left(140\mu_{\alpha}^{2}+245\mu_{\beta}^{2}\right)\Omega_{\alpha\beta}^{11}-49\mu_{\alpha}\mu_{\beta}^{2}\Omega_{\alpha\beta}^{12}\right.$$
$$+8\mu_{\alpha}\mu_{\beta}^{2}\Omega_{\alpha\beta}^{13}+\frac{1}{8}\mu_{\beta}(154\mu_{\alpha}^{2}+147\mu_{\beta}^{2})\Omega_{\alpha\beta}^{22}$$
$$\left.-\frac{21}{2}\mu_{\beta}^{3}\Omega_{\alpha\beta}^{23}+\frac{3}{2}\mu_{\beta}^{3}\Omega_{\alpha\beta}^{24}+3\mu_{\alpha}\mu_{\beta}^{2}\Omega_{\alpha\beta}^{33}\right].$$

$$\left[S_{5/2}^{n}\left(W^2\right)\left(W_rW_s - \tfrac{1}{3}\delta_{rs}W^2\right), S_{5/2}^{k}\left(W^2\right)\left(W_rW_s - \tfrac{1}{3}\delta_{rs}W^2\right)\right]''_{\alpha\beta},$$

$$n, k = 0, 1.$$

$$n = 0, k = 0, \qquad -\frac{16}{3}\mu_\alpha\mu_\beta\left(5\Omega_{\alpha\beta}^{11} - \frac{3}{2}\Omega_{\alpha\beta}^{22}\right),$$

$$n = 0, k = 1, \qquad \frac{16}{3}\mu_\alpha^2\mu_\beta\left(-\frac{35}{2}\Omega_{\alpha\beta}^{11} + 7\Omega_{\alpha\beta}^{12} + \frac{21}{4}\Omega_{\alpha\beta}^{22} - \frac{3}{2}\Omega_{\alpha\beta}^{23}\right),$$

$$n = 1, k = 1, \qquad -\frac{16}{3}\mu_\alpha^2\mu_\beta^2\left(\frac{385}{4}\Omega_{\alpha\beta}^{11} - 49\Omega_{\alpha\beta}^{12} + 8\Omega_{\alpha\beta}^{13} - \frac{301}{8}\Omega_{\alpha\beta}^{22} + \frac{21}{2}\Omega_{\alpha\beta}^{23} - \frac{3}{2}\Omega_{\alpha\beta}^{24} + 3\Omega_{\alpha\beta}^{33}\right).$$

A3. Expressions for the elements q^{nk} and p^{nk}

At least a portion of the elements q^{nk} and p^{nk} (6.3.4) applied for the calculation of the electron transport properties can be obtained through the use of the expressions for the partial bracket integrals which were given in A2 ($n, k = 0, 1, 2$ for q^{nk} and $n, k = 0, 1$ for p^{nk}). It is convenient to express these coefficients together with elements of the higher order ($n, k = 3$) in terms of the average cross section $\overline{Q}_{\alpha\beta}^{(lr)}$ which are related to $\Omega_{\alpha\beta}^{lr}$ as

$$\Omega_{\alpha\beta}^{lr} = \left(\frac{kT}{2\pi\mu_{\alpha\beta}}\right)^{1/2}\frac{1}{2}(r+1)!\left[1 - \frac{1}{2}\frac{1 + (-1)^l}{l+1}\right]\overline{Q}_{\alpha\beta}^{(lr)}$$

so that $\overline{Q}_{\alpha\beta} = \pi d_{\alpha\beta}$ for the hard sphere model. For the case of electron–neutral and electron–ion interactions, $\mu_{\alpha\beta} = m_1$. In the expressions to be given below [4], it is understood that subscript 1 denotes the electrons and summation over β runs over all heavy species (i.e. from 2 to N). Recall that the elements q^{nk} and p^{nk} are symmetric in n and k.

$$q^{00} = 8\sum_\beta n_1 n_\beta \overline{Q}_{1\beta}^{(1,1)},$$

$$q^{01} = 8\sum_\beta n_1 n_\beta\left(\frac{5}{2}\overline{Q}_{1\beta}^{(1,1)} - 3\overline{Q}_{1\beta}^{(1,2)}\right),$$

$$q^{11} = 8\sqrt{2}n_1^2\overline{Q}_{11}^{(2,2)} + 8\sum_\beta n_1 n_\beta \times \left(\frac{25}{4}\overline{Q}_{1\beta}^{(1,1)} - 15\overline{Q}_{1\beta}^{(1,2)} + 12\overline{Q}_{1\beta}^{(1,3)}\right),$$

$$q^{02} = 8\sum_\beta n_1 n_\beta \left(\frac{35}{8}\overline{Q}_{1\beta}^{(1,1)} - \frac{21}{2}\overline{Q}_{1\beta}^{(1,2)} + 6\overline{Q}_{1\beta}^{(1,3)}\right),$$

$$q^{12} = 8\sqrt{2}n_1^2\left(\frac{7}{4}\overline{Q}_{11}^{(2,2)} - 2\overline{Q}_{11}^{(2,3)}\right) + 8\sum_\beta n_1 n_\beta \left(\frac{175}{16}\overline{Q}_{1\beta}^{(1,1)} - \frac{315}{8}\overline{Q}_{1\beta}^{(1,2)} + 57\overline{Q}_{1\beta}^{(1,3)} - 30\overline{Q}_{1\beta}^{(1,4)}\right),$$

$$q^{22} = 8\sqrt{2}n_1^2\left(\frac{77}{16}\overline{Q}_{11}^{(2,2)} - 7\overline{Q}_{11}^{(2,3)} + 5\overline{Q}_{11}^{(2,4)}\right) + 8\sum_\beta n_1 n_\beta \left(\frac{1225}{64}\overline{Q}_{1\beta}^{(1,1)} - \frac{735}{8}\overline{Q}_{1\beta}^{(1,2)} + \frac{399}{2}\overline{Q}_{1\beta}^{(1,3)} - 210\overline{Q}_{1\beta}^{(1,4)} + 90\overline{Q}_{1\beta}^{(1,5)}\right),$$

$$q^{03} = 8\sum_\beta n_1 n_\beta \left(\frac{105}{16}\overline{Q}_{1\beta}^{(1,1)} - \frac{189}{8}\overline{Q}_{1\beta}^{(1,2)} + 27\overline{Q}_{1\beta}^{(1,3)} - 10\overline{Q}_{1\beta}^{(1,4)}\right),$$

$$q^{13} = 8\sqrt{2}n_1^2\left(\frac{63}{32}\overline{Q}_{11}^{(2,2)} - \frac{9}{2}\overline{Q}_{11}^{(2,3)} + \frac{5}{2}\overline{Q}_{11}^{(2,4)}\right) - 8\sum_\beta n_1 n_\beta \left(\frac{525}{32}\overline{Q}_{1\beta}^{(1,1)} - \frac{315}{4}\overline{Q}_{1\beta}^{(1,2)} + 162\overline{Q}_{1\beta}^{(1,3)} - 160\overline{Q}_{1\beta}^{(1,4)} + 60\overline{Q}_{1\beta}^{(1,5)}\right),$$

$$q^{23} = 8\sqrt{2}n_1^2\left(\frac{945}{128}\overline{Q}_{11}^{(2,2)} - \frac{261}{16}\overline{Q}_{11}^{(2,3)} + \frac{125}{8}\overline{Q}_{11}^{(2,4)} - \frac{15}{2}\overline{Q}_{11}^{(2,5)}\right) + 8\sum_\beta n_1 n_\beta \left(\frac{3675}{128}\overline{Q}_{1\beta}^{(1,1)} - \frac{11025}{64}\overline{Q}_{1\beta}^{(1,2)} + \frac{1953}{4}\overline{Q}_{1\beta}^{(1,3)} - \frac{1505}{2}\overline{Q}_{1\beta}^{(1,4)} + 615\overline{Q}_{1\beta}^{(1,5)} - 210\overline{Q}_{1\beta}^{(1,6)}\right),$$

$$
\begin{aligned}
q^{33} = 8\sqrt{2}n_1^2 \Bigg(& \frac{14553}{1024}\overline{Q}_{11}^{(2,2)} - \frac{1215}{32}\overline{Q}_{11}^{(2,3)} + \frac{1565}{32}\overline{Q}_{11}^{(2,4)} \\
& - \frac{135}{4}\overline{Q}_{11}^{(2,5)} + \frac{105}{8}\overline{Q}_{11}^{(2,6)} + \overline{Q}_{11}^{(4,4)} \Bigg) \\
& + 8\sum_{\beta} n_1 n_\beta \Bigg(\frac{11025}{256}\overline{Q}_{1\beta}^{(1,1)} - \frac{19845}{64}\overline{Q}_{1\beta}^{(1,2)} + \frac{17577}{16}\overline{Q}_{1\beta}^{(1,3)} \\
& - \frac{4515}{2}\overline{Q}_{1\beta}^{(1,4)} + \frac{5535}{2}\overline{Q}_{1\beta}^{(1,5)} - 1890\overline{Q}_{1\beta}^{(1,6)} + 560\overline{Q}_{1\beta}^{(1,7)} \Bigg);
\end{aligned}
$$

$$
p^{00} = 8\sqrt{2}n_1^2\overline{Q}_{11}^{(2,2)} + 8\sum_{\beta} n_1 n_\beta \overline{Q}_{1\beta}^{(2,2)},
$$

$$
p^{01} = 8\sqrt{2}n_1^2\left(\frac{7}{4}\overline{Q}_{11}^{(2,2)} - 2\overline{Q}_{11}^{(2,3)}\right) + 8\sum_{\beta} n_1 n_\beta \left(7\overline{Q}_{1\beta}^{(1,1)} - 8\overline{Q}_{1\beta}^{(1,2)}\right),
$$

$$
\begin{aligned}
p^{11} \;=\; & 8\sqrt{2}n_1^2\left(\frac{301}{48}\overline{Q}_{11}^{(2,2)} - 7\overline{Q}_{11}^{(2,3)} + 5\overline{Q}_{11}^{(2,4)}\right) \\
& + 8\sum_{\beta} n_1 n_\beta \left(\frac{147}{16}\overline{Q}_{1\beta}^{(2,2)} - 56\overline{Q}_{1\beta}^{(2,3)} + 40\overline{Q}_{1\beta}^{(2,4)}\right).
\end{aligned}
$$

References

Introduction

1. S.Chapman and T.G.Cowling, *Mathematical Theory of Non-uniform Gases,* Cambridge Univ. Press, Cambridge, (1952,1970).
2. J.O.Hirschfelder, C.F.Curtiss and R.B.Bird, *Molecular Theory of Gases and Liquids,* John Wiley and Sons, New York, (1954).
3. J.H.Ferziger and H.G.Kaper, *Mathematical Theory of Transport Processes in Gases,* North-Holland, Amsterdam, (1972).
4. E.M.Lifshitz and L.P.Pitaevski, *Landau and Lifshitz, Cource of Theoretical Physics,* vol.10, *Physical Kinetics,* Pergamon Press, Oxford, (1965).
5. R.Balescu, *Statistical Mechanics of Charged Particles,* Interscience Publishers, John Wiley and Sons, London, (1963).
6. Yu.L.Klimontovich, *Kinetic Theory of Nonequilibrium Processes in Plasma,* Pergamon Press, Oxford, (1967).
7. V.P.Silin, *Vvedeniye v kineticheskuju teoriju gasov,* [transl. title: *Introduction to the Kinetic Theory of Gases*] (in Russian), Nauka, Moscow, (1971).
8. Yu.L.Klimontovich, *Kinetic Theory of Non-Ideal Gases and Non-Ideal Plasma,* Pergamon Press, Oxford, (1982).
9. I.P.Shkarofsky, T.W.Johnston and Bachynski, *The Particle Kinetics of Plasmas,* Addison-Wesley, London, (1966).
10. A.R.Hochstim and G.Massel, in *Kinetic Processes in Gases and Plasma,* ed. A.R.Hochstim, Academic Press, New York, (1969).
11. M.Mitchner and C.H.Kruger, Jr., *Partially Ionized Gases,* John Wiley & Sons, New York, (1973).
12. H.Grad, *Comm. Pure and Appl. Math.* **2,** 331 (1949).
13. H.Grad, in *Handbuch der Physik,* vol.XII, ed. S.Flügge, Springer, Berlin, (1958).
14. I.I.Kolodner, *Moment Description of Gas Mixture,* Report NYO-7980, New York University, (1957).

15. L.Herdan and B.S.Liley, *Rev. Mod. Phys.* **32,** 731 (1960).
16. V.Zhdanov, Yu.Kagan and A.Sazykin, *Zh. Exp. Teor. Phys.* (in Russian) **42,** 857 (1962), [Engl. transl.: *Sov. Phys. JETP.* **15,** 596 (1962)].
17. J.M.Burgers, in *Plasma Dynamics,* ed. F.Clauser, Addison-Wesley, Massachusetts, (1960).
18. V.M.Zhdanov, *Prikl. Mat. i Mekh.* (in Russian) **28,** 280 (1962), [Engl. transl.: *Appl. Math. Mech.* **26,** 401 (1962)].
19. M.Ya.Alievskii and V.M.Zhdanov, *Zh. Prikl. Mekh. i Tekhn. Fiz.* (in Russian) No.5, 11 (1963).
20. W.L.Everett, in *Rarefied Gas Dynamics,* ed. J.A.Laurman, Academic Press, New York, (1963).
21. K.Suchy, in *Rarefied Gas Dynamics,* ed. J.A.Laurman, Academic Press, New York, (1963).

Chapter 1

1. S.Chapman and T.G.Cowling, *Mathematical Theory of Non-uniform Gases,* Cambridge Univ. Press, Cambridge, (1970).
2. C.Cercignani, *Theory and Application of the Boltzmann Equation,* Scottish Acad. Press, Edinburgh, (1975).
3. C.S.Wang Chang and G.E.Uhlenbeck, *Transport Phenomena in Polyatomic Gases,* Eng. Res. Rept. No. CM-681, University of Michigan, (1951).
4. C.S.Wang Chang, G.E.Uhlenbeck and J.de Boer, in *Studies in Statistical Mechanics,* vol.2, p.24, ed. J.de Boer and G.E. Uhlenbeck, North-Holland, Amsterdam, (1964).
5. L.Waldmann, *Z. Naturforsch.* **12a,** 661 (1957); ibid. **13a,** 609 (1958); *Handbuch der Physik,* vol.XII, p.295, ed. S.Flügge, Springer-Verlag, Berlin, (1958).
6. R.F.Snider, *J. Chem. Phys.* **32,** 1051 (1960).
7. L.Waldmann, in *Statistical Mechanics of Equilibrium and Non-equilibrium,* vol.2, North-Holland, Amsterdam, (1964).
8. L.D.Landau and E.M.Lifshitz, *Mechanics,* Pergamon Press, London, (1977).
9. J.H.Ferziger and H.G.Kaper, *Mathematical Theory of Transport Processes in Gases,* North-Holland, Amsterdam, (1972).
10. R.H.Neynaber, *Phys. Rev.* **124,** 135 (1961).
11. R.B.Brode, *Rev. Mod. Phys.* **5,** 257 (1933).
12. C.Ramsauer, *Ann. Physik* **64,** 91 (1921).

13. D.E.Golden and H.W.Budel, *Phys. Rev.* **149**, 58 (1966).
14. S.C.Brown, *Basic Data of Plasma Physics*, M.I.T. Press, (1966).
15. M.Mitchner and C.H.Kruger, Jr., *Partially Ionized Gases*, John Wiley & Sons, New York, (1973).
16. A.Dalgarno, *Phil. Trans. Roy. Soc. (London)* **A250**, 426 (1958).
17. E.W.McDaniel and E.A.Mason, *The Mobility and Diffusion of Ions in Gases*, John Wiley & Sons, New York, (1973).
18. J.O.Hirschfelder, C.F.Curtiss and R.B.Bird, *Molecular Theory of Gases and Liquids*, John Wiley and Sons, New York, (1954).
19. E.A.Mason and W.E.Rice, *J. Chem. Phys.* **22**, 843 (1954).
20. A.Dalgarno, *Plan. Space Soc.* **3**, 217 (1961).
21. E.W.McDaniel, *Collision Phenomena in Ionized Gases*, John Wiley & Sons, New York, (1964).
22. N.A.Krall and A.W.Trivelpiece, *Principles of Plasma Physics*, McGraw-Hill Book Company, New York, (1973).
23. R.B.Liboff, *Phys. Fluids.* **2**, 40 (1959).
24. L.Spitzer, Jr., *Physics of Fully Ionized Gases*, Interscience, John Wiley & Sons, New York, (1962).
25. M.Rosenbluth, W.MacDonald and P.Judd, *Phys. Rev.* **107**, 1 (1957).
26. B.A.Trubnikov, in *Reviews of Plasma Physics*, vol.1, ed. M.A. Leontovich, Consultants Bureau, New York, (1965).
27. V.P.Silin, *Vvedeniye v kineticheskuju teoriju gasov*, [transl. title: *Introduction to the Kinetic Theory of Gases*] (in Russian), Nauka, Moscow, (1971).
28. L.D.Landau, *Phys. Z. der Sowjetunion.* **10**, 154 (1936).
29. Yu.L.Klimontovich, *Statistical Theory of Nonequilibrium Processes in Plasma*, Pergamon Press, Oxford, (1964).
30. Yu.L.Klimontovich, *Kinetic Theory of Non-Ideal Gases and Non-Ideal Plasmas*, Pergamon Press, Oxford, (1982).
31. P.Schram, *Kinetic Theory of Gases and Plasma*, Kluwer, Dordrecht, (1991).
32. R.Balescu, *Statistical Mechanics of Charged Particles*, Interscience Publishers, John Wiley and Sons, London, (1963).
33. V.P.Silin, *Nuclear Fusion* **2**, 125 (1962).
34. L.M.Gorbunov and V.P.Silin, *Sov. Phys., Tech. Phys.* **9**, 305 (1964).

Chapter 2

1. E.M.Lifshitz and L.P.Pitaevski, *Landau and Lifshitz, Course of Theoretical Physics,* vol.10, *Physical Kinetics,* Pergamon Press, Oxford, (1965).
2. S.R.de Groot and P.Mazur, *Nonequilibrium Thermodynamics,* North-Holland, Amsterdam, (1962).
3. J.E.Mayer and M.C.Mayer, *Statistical Mechanics,* Wiley-Intercience, John Wiley & Sons, New York, (1977).
4. S.Chapman and T.G.Cowling, *Mathematical Theory of Non-uniform Gases,* Cambridge Univ. Press, Cambridge, (1970).
5. H.Grad, *Comm. Pure and Appl. Math.* **2,** 331 (1949).
6. S.I.Braginskii, in *Reviews of Plasma Physics,* vol.1, ed. M.A. Leontovich, Consultants Bureau, New York, (1965).
7. R.Balescu, *Transport Processes in Plasmas,* vol.1, North-Holland, Amsterdam, (1989).
8. W.Ebeling, V.E.Fortov, Yu.L.Klimontovich *et al.*, *Transport Properties of Dense Plasmas,* Birkhäuser Basel, (1984).
9. J.O.Hirschfelder, C.F.Curtiss and R.B.Bird, *Molecular Theory of Gases and Liquids,* John Wiley and Sons, New York, (1954).

Chapter 3

1. I.P.Shkarofsky, T.W.Johnston and Bachynski, *The Particle Kinetics of Plasmas,* Addison-Wesley, London, (1966).
2. A.R.Hochstim and G.Massel, in *Kinetic Processes in Gases and Plasma,* ed. A.R.Hochstim, Academic Press, New York, (1969).
3. S.Chapman and T.G.Cowling, *Mathematical Theory of Non-uniform Gases,* Cambridge Univ. Press, Cambridge, (1970).
4. T.Kihara, O.Aono and Itikawa, *Journ. Phys. Soc. Japan* **18,** 1043 (1963).
5. T.Kihara and O.Aono, in *Kinetic Equation,* ed. R.L.Liboff and N.Rostoker, Gordon and Breach, (1971).
6. T.G.Cowling, *Magnetohydrodynamics,* Interscience, Inc., New York, (1957).
7. M.Mitchner and C.H.Kruger, Jr., *Partially Ionized Gases,* John Wiley & Sons, New York, (1964).
8. P.Polovin and K.P.Cherkasova, *Sov. Phys., Tech. Phys.* **32,** 649 (1962).
9. T.G.Cowling, *Proc. Roy. Soc.* **183A,** 453 (1945).

10. L.Spitzer, Jr., *Physics of Fully Ionized Gases*, Interscience, John Wiley and Sons, New York, (1962).
11. E.W.McDaniel and E.A.Mason, *The Mobility and Diffusion of Ions in Gases*, John Wiley and Sons, New York, (1973).
12. V.P.Allis, in *Handbuch der Physik*, vol.21, ed. S.Flügge, Springer, Berlin, (1956).
13. S.I.Braginskii, in *Reviews of Plasma Physics*, vol.1, ed. M.A. Leontovich, Consultants Bureau, New York, (1965).
14. J.B.Taylor, *Phys. Fluids* **4**, 1142 (1964).

Chapter 4

1. H.Grad, *Comm. Pure and Appl. Math.* **2**, 331 (1949).
2. H.Grad, in *Handbuch der Physik*, vol.XII, ed. S.Flügge, Springer, Berlin, (1958).
3. M.N.Kogan, *Dinamika Razrezhennogo Gasa*, (in Russian), Nauka, Moscow, (1967), [Engl. transl.: *Rarefied Gas Dynamics*, Plenum Press, New York, (1969)].
4. D.Mintzer, *Phys. Fluids* **18**, 1076 (1965).
5. S.Chapman and T.G.Cowling, *Mathematical Theory of Non-uniform Gases*, Cambridge Univ. Press, Cambridge, (1970).
6. J.H.Ferziger and H.G.Kaper, *Mathematical Theory of Transport Processes in Gases*, North-Holland, Amsterdam, (1972).
7. H.Grad, *Phys. Fluids* **6**, 147 (1963).
8. L.Waldmann, in *Handbuch der Physik*, vol.XII, ed. S.Flügge, Springer, Berlin, (1958).
9. I.I.Kolodner, *Moment Description of Gas Mixture*, Report NYO-7980, New York University, (1957).
10. L.Herdan and B.S.Liley, *Rev. Mod. Phys.* **32**, 731 (1960).
11. V.Zhdanov, Yu.Kagan and A.Sazykin, *Zh. Exp. Teor. Phys.* (in Russian) **42**, 857 (1962), [Engl. transl.: *Sov. Phys. JETP* **15**, 596 (1962)].
12. V.M.Zhdanov, *Prikl. Mat. i Mekh.* (in Russian) **28**, 280 (1962), [Engl. transl.: *Appl. Math. Mekh.* **26**, 401 (1962)].
13. M.Ya.Alievskii and V.M.Zhdanov, *Zh. Prikl. Mekh. i Tekhn. Fiz.* (in Russian) No.5, 11 (1963).
14. T.G.Cowling, *Magnetohydrodynamics*, Interscience, Inc., New York, (1957).
15. A.G.Kulikovskii and G.A.Lyubimov, *Magnitnaia Gidrodinamika*, (in Russian), Nauka, Moscow, (1962), [Engl. transl.: *Magnetohydrodynamics*, Addison-Wesley, London, (1965)].

16. G.W.Sutton and A.Sherman, *Engeenering Magnetohydrodynamics,* McGraw-Hill, (1965).
17. M.Mitchner and C.H.Kruger, Jr., *Partially Ionized Gases,* John Wiley & Sons, New York, (1964).
18. I.P.Shkarofsky, T.W.Johnston and Bachynski, *The Particle Kinetic of Plasma,* Addison-Wesley, London, (1966).
19. H.Grad, in *Proceedings of the I RHD Symposium,* ed. F.M. Devieene, Pergamon Press, Oxford, (1960).
20. S.I.Braginskii, in *Reviews of Plasma Physics,* vol.1, ed. M.A. Leontovich, Consultants Bureau, New York, (1965).
21. A.B.Mikhailovskii, *Teorija Plasmennykh Neustoichivstei,* tom 2, (in Russian), [Transl. title: *Theory of Plasma Instabilities*], Atomizdat, Moscow, (1977).

Chapter 5

1. V.Zhdanov, Yu.Kagan and A.Sazykin, *Zh.Exp. Teor. Phys.* (in Russian) **42,** 857 (1962), [Engl. transl.: *Sov. Phys. JETP* **15,** 596 (1962)].
2. S.Chapman and T.G.Cowling, *Mathematical Theory of Non-uniform Gases,* Cambridge Univ. Press, Cambridge, (1970).
3. J.O.Hirshfelder, C.F.Curtiss and R.B.Bird, *Molecular Theory of Gases and Liquids,* John Wiley & Sons, New York, (1954).
4. C.Muckenfuss and C.F.Curtiss, *J. Chem. Phys.* **29,** 1273 (1958).
5. J.H.Ferziger and H.G.Kaper, *Mathematical Theory of Transport Processes in Gases,* North-Holland, Amsterdam, (1972).
6. S.R. de Groot and P.Mazur, *Nonequilibrium Thermodynamics,* North-Holland, Amsterdam, (1962).
7. V.M.Zhdanov and V.I.Roldughin, *Uspekhi Fiz. Nauk* (in Russian) **168,** 407 (1998), [Engl. transl.: *Physics-Uspekhi.* **41,** 349 (1998).
8. L.Monchick, R.G.Munn and E.A.Mason, *J. Chem. Phys.* **45,** 3051 (1966).
9. C.Muckenfuss, *J. Chem. Phys.* **59,** 1747 (1973).
10. A.F.Kolesnikov and G.A.Tirskii, in *Molekuljarnaja Gazodinamika* (in Russian), [transl. title: *Molecular Gas Dynamics*], Nauka, Moscow, (1982).
11. L.Waldmann, in *Handbuch der Physik,* vol.XII, ed. S.Flügge, Springer, Berlin, (1958).
12. C.F.Curtiss, *J. Chem. Phys.* **49,** 2917 (1968).

13. V.M.Zhdanov and V.I.Roldughin, *Zh, Exp. Theor. Phys.*(in Russian) **113**, 2081 (1998); [Engl. transl.: *Sov. Phys. JETP*. **86**, 1141 (1998).

14. F.R.W.McCourt, J.J.J.Beenakker, W.E.Köhler and I.Kuščer, *Nonequilibrium Phenomena in Polyatomic Gases*, vol.1, p.61, Clarendon Press, Oxford, (1990).

15. I.Kuščer, *Physica* **A133**, 397 (1985).

16. V.M.Zhdanov and V.I.Roldughin, *Zh. Exp. Theor. Phys.* (in Russian) **109**, 1267 (1996); [Engl. transl.: *Sov. Phys. JETP* **82**, 683 (1996).

Chapter 6

1. M.Ya.Alievskii V.M.Zhdanov, *Zh. Prikl. Mekh. i Tekhn. Fiz.* (in Russian) No.5, 11 (1963).

2. M.Ya.Alievskii V.M.Zhdanov, and V.A.Poljanskii, *Zh. Prikl. Mekh. i Tekhn. Fiz.* (in Russian) No.3, 32 (1964).

3. S.I.Braginskii, in *Reviews of Plasma Physics*, vol.1, ed. M.A. Leontovich, Consultants Bureau, New York, (1965).

4. A.T.Demetriades and G.S.Argyropoulos, *Phys. Fluids* **9**, 2136 (1966).

5. V.A.Poljanskii, *Zh. Prikl. Mekh. i Tekhn. Fiz.* (in Russian) No.5, 11 (1964).

6. S.Chapman and T.G.Cowling, *Mathematical Theory of Nonuniform Gases*, Cambridge Univ. Press, Cambridge, (1970).

7. L.Spitzer, Jr., *Physics of Fully Ionized Gases*, Interscience, John Wiley & Sons, New York, (1962).

8. U.Daybelge, *J. Appl. Phys.* **41**, 2130 (1970).

9. J.H.Ferziger and H.G.Kaper, *Mathematical Theory of Transport Processes in Gases*, North-Holland, Amsterdam, (1972).

10. R.S.Devoto, *Phys. Fluids* **10**, 354 (1967).

11. R.S.Devoto, *Phys. Fluids* **10**, 2105 (1967).

12. M.Mitchner and C.H.Kruger, Jr., *Partially Ionized Gases*, John Wiley & Sons, New York, (1964).

Chapter 7

1. V.M.Zhdanov, *Prikl. Mat. i Mekh.* (in Russian) **28**, 280 (1962), [Engl. transl.: *Appl. Math. Mech.* **26**, 401 (1962)].
2. M.Ya.Alievskii, V.M.Zhdanov, and V.A.Poljanskii, *Zh. Prikl. Mekh. i Tekhn. Fiz.* (in Russian) No.3, 32 (1964).
3. S.I.Braginskii, in *Reviews of Plasma Physics,* vol.1, ed. M.A. Leontovich, Consultants Bureau, New York, (1965).
4. L.Herdan and B.S.Liley, *Rev. Mod. Phys.* **32**, 731 (1960).
5. J.H.Ferziger and H.G.Kaper, *Mathematical Theory of Transport Processes in Gases,* North-Holland, Amsterdam, (1972).
6. S.C.Brown, *Basic Data of Plasma Physics,* M.I.T. Press, (1966).
7. E.W.McDaniel, *Collision Phenomena in Ionized Gases,* John Willey & Sons, New York, (1964).

Chapter 8

1. V.M.Zhdanov and P.N.Yushmanov, *Zh. Prikl. Mekh. i Tekhn. Fiz.* (in Russian) No.4, 24 (1980).
2. S.I.Braginskii, in *Reviews of Plasma Physics,* vol.1, ed. M.A. Leontovich, Consultants Bureau, New York, (1965).
3. R.Balescu, *Transport Processes in Plasmas,* vol.2, North-Holland, Amsterdam, (1989).
4. S.P.Hirshman, *Phys. Fluids* **20**, 589 (1977).
5. L.A.Artsimovich *et al.*, in *Plasma Phys. and Contr. Nucl. Fusion Res.,* IAEA, Vienna, vol.1, 172 (1971).
6. M.Kruscal and M.Schwarzchild, *Proc. Roy. Soc.* **223A**, 3548 (1954).
7. V.D.Shafranov in *Reviews of Plasma Physics,* vol.2, ed. M.A. Leontovich, Consultants Bureau, New York, (1966).
8. D.Pfirsch and A.Schlüter, Max Planck Institute Rept., MPI/PA/7 /62 (1962).
9. A.A.Galeev and R.Z.Sagdeev, in *Reviews of Plasma Physics,* vol.7, ed. M.A.Leontovich, Consultants Bureau, New York, (1974).
10. V.D.Shafranov, *Plasma Phys.* **8**, 314 (1966).
11. F.L.Hinton and R.D.Hazeltin, *Rev. Mod. Phys.* **48**, 239 (1976).
12. S.P.Hirshman, D.T.Sigmar, *Nucl. Fusion* **21**, 1079 (1981).
13. R.D.Hazeltin and F.L.Hinton, *Phys. Fluids* **16**, 1883 (1973).
14. P.H.Rutherford, *Phys. Fluids.* **17**, 1732 (1974).
15. V.M.Zhdanov and P.N.Yushmanov, *Fizika Plasmy* (in Russian) **3**, 1193 (1977).
16. F.Engelman and A.Nocentini, *Nucl. Fusion* **16**, 694 (1976).

Chapter 9

1. R.J.Rosa, *Magnetohydrodynamic Energy Conversion,* McGraw-Hill, New York, (1968).
2. M.Mitchner and C.H.Kruger, Jr., *Partially Ionized Gases,* John Wiley & Sons, New York, (1973).
3. V.M.Zhdanov, *Zh. Prikl. Mekh. i Tekhn. Fiz.* (in Russian) No.4, 66 (1965).
4. S.Lin, E.L.Resler and A.Kantrowitz, *J. Appl. Phys.* **26**, 95 (1955).
5. L.S.Frost, *J. Appl. Phys.* **32**, 2029 (1961).
6. S.Schweitzer and M.Mitchner, *Phys. Fluids* **10**, 799 (1967).
7. B.Bonnevier, *Arkiv. Fys.* **33**, 255 (1966).
8. V.M.Zhdanov, A.I.Karchevskii, E.P.Potanin and A.L.Ustinov, *Zh. Tekh. Fiz.* (in Russian) **49**, 1879 (1979).
9. S.I.Braginskii, in *Reviews of Plasma Physics,* vol.1, ed. M.A. Leontovich, Consultants Bureau, New York, (1965).
10. V.M.Zhdanov and P.N.Yushmanov, *Zh. Prikl. Mekh. i Tekhn. Fiz.* (in Russian) No.4, 24 (1980).
11. V.M.Zhdanov and P.N.Yushmanov, *Fizika Plasmy* (in Russian) **3**, 1193 (1977).

Chapter 10

1. M.Mitchner and C.H.Kruger, Jr., *Partially Ionized Gases,* John Wiley & Sons, New York, (1973).
2. D.R.Bates, A.E.Kingston and R.V.P.Whirter, *Proc. Roy. Soc.* **A 267**, 297 (1962); **A 270**, 155 (1962).
3. J.F.Shaw, M.Mitchner and C.H.Kruger, *Phys. Fluids* **13**, 325, 339 (1970).
4. F.R.W.McCourt, J.J.J.Beenakker, W.E.Köhler and I.Kuščer, *Nonequilibrium Phenomena in Polyatomic Gases,* vol.1, p.61, Clarendon Press, Oxford, (1990).
5. C.S.Wang Chang and G.E.Uhlenbeck, *Transport Phenomena in Polyatomic Gases,* Eng. Res. Rept. No. C-681, University of Michigan, (1951).
6. C.S.Wang Chang, G.E.Uhlenbeck and J.de Boer, in *Studies in Statistical Mechanics,* vol.2, p.24, ed. J.de Boer and G.E. Uhlenbeck, North-Holland, Amsterdam, (1964).

7. Yu.Kagan and A.M.Afanas'ev, *Zh. Eksp. Teor. Fiz.* (in Russian) **41**, 1536 (1961), [Engl. transl.: *Sov. Phys. JETP* **14**, 1096 (1962)].
8. Yu.Kagan and L.A.Maksimov, *Zh. Eksp. Teor. Fiz.* (in Russian) **41**, 842 (1961), [Engl. transl.: *Sov. Phys. JETP* **14**, 604 (1962)].
9. D.W.Condiff, W.K.Lu and E.A.Mason, *J. Chem. Phys.* **42**, 3445 (1965).
10. J.S.Dahler and N.F.Sather, ibid. **38**, 2363 (1963).
11. N.F.Sather and J.S.Dahler, ibid. **43**, 1750 (1965).
12. L.Monchick, K.S.Yun and E.A.Mason, *J. Chem. Phys.* **39**, 654 (1963).
13. L.Monchick, A.N.G.Pereira and E.A.Mason, ibid. **42**, 3241 (1965).
14. V.M.Zhdanov, *Zh. Exp. Teor. Fiz.* (in Russian) **53**, 2099 (1967), [Engl. transl.: *Sov. Phys. JETP.* **26**, 1187 (1968)].
15. M.Ya.Alievskii and V.M.Zhdanov, ibid. **55**, 221 (1968), [Engl. transl.: *Sov. Phys. JETP* **28**, 116 (1969)].
16. F.J.McCormack, *Phys. Fluids* **11**, 2533 (1968).
17. L.Waldmann and E.Trübenbacher, *Z. Naturforsch.* **17a**, 363 (1962).
18. S.Chapman and T.G.Cowling, *Mathematical Theory of Non-uniform Gases,* Cambridge Univ. Press, Cambridge, (1970).
19. K.F.Herzfeld and T.A.Litovitz, *Absorption and Dispersion of Ultrasonic Waves,* Academic Press, New York, (1959).
20. V.M.Zhdanov and M.Ya.Alievskii, *Protsessy Perenosa i Relaksatsii v Molekuljarnikh Gazakh* (in Russian), [transl. title: *Transport and Relaxation Processes in Molecular Gases*], Nauka, Moscow, (1989).
21. W.E.Köhler, S.Hess and L.Waldmann, *Z. Naturforsch.* **25a**, 336 (1970).
22. W.E.Köhler and J.Schaefer, *J. Chem. Phys.* **78**, 4682 (1983).
23. F.R.W.McCourt, J.J.J.Beenakker, W.E.Köhler and I.Kuščer, *Nonequilibrium Phenomena in Polyatomic Gases,* vol.2, p.563, Clarendon Press, Oxford, (1991).
24. V.Vesovic and W.A.Wakeham, *Int. Rev. in Phys. Chem.* **11**, 161 (1992).
25. S.I.Sandler, *Phys. Fluids* **11**, 2546 (1968).
26. E.L.Heck and A.S.Dickinson, *Molec. Phys.* **81**, 1325 (1994).
27. E.L.Heck and A.S.Dickinson, *Physica* **A 217**, 107 (1995).
28. E.A.Mason and A.P.Malinauskas, *Gas Transport in Porous Media – the Dusty Gas Model,* Elsevier, Amsterdam, (1983).

29. W.F.Ahtye, *J. Chem. Phys.* **57**, 5542 (1972).
30. J.O.Hirschfelder, ibid. **26**, 282 (1957).
31. L.Monchick, R.J.Munn and E.A.Mason, ibid. **45**, 3051 (1966).
32. N.F.Sather and J.S.Dahler, ibid. **47**, 4653 (1967).
33. N.F.Sather and E.A.Mason, ibid. **47**, 2621 (1967).
34. J.Schirdewahn, A.Klemm and L.Waldmann, *Z. Naturforsch.* **16a**, 133 (1961).
35. C.J.G.Sliker, *Physica* **31**, 1388 (1965).

Appendices

1. S.Chapman and T.G.Cowling, *Mathematical Theory of Non-uniform Gases*, Cambridge Univ. Press, Cambridge, (1970).
2. H.Grad, *Phys. Fluids* **6**, 147 (1963).
3. J.H.Ferziger and H.G.Kaper, *Mathematical Theory of Transport Processes in Gases*, North-Holland, Amsterdam, (1972).
4. R.S.Devoto, *Phys. Fluids* **10**, 2105 (1967).

Index

For Product Safety Concerns and Information please contact our EU representative GPSR@taylorandfrancis.com
Taylor & Francis Verlag GmbH, Kaufingerstraße 24, 80331 München, Germany

www.ingramcontent.com/pod-product-compliance
Lightning Source LLC
Chambersburg PA
CBHW060741150926
51501CB00003B/67
9780415279208